Oregon Viticulture

Oregon Viticulture

edited by Edward W. Hellman

Oregon State University Press
Corvallis

Library of Congress Cataloging-in-Publication Data

Oregon viticulture / edited by Edward W. Hellman—1st ed.

p. cm.

ISBN 0-87071-554-2 (alk. paper)

1. Viticulture—Oregon. I. Hellman, Edward W. (Edward William)

SB387.76.O7 O74 2003

634.8'09795--dc21

2002153602

Oregon State University Press

500 Kerr Administration

Corvallis OR 97331

541-737-3166 • fax 541-737-3170

http://oregonstate.edu/dept/press

Contents

Contributors

Robert Burney, Proprietor and Irrigation Specialist, Kenworthy and Burney Vineyard Services, Santa Rosa, California; formerly in Newberg, Oregon

Alan Campbell, Ph.D., Instructor, Northwest Viticulture Center, Chemeketa Community College, Salem, Oregon

Steven Castagnoli, Area Extension Agent, Oregon State University, Hood River, Oregon

Ted Casteel, Owner and Vineyard Manager, Bethel Heights Vineyard, Salem, Oregon

Mark Chien, Wine Grape Agent, Penn State Cooperative Extension, Capital and Southeast Regions, Lancaster, Pennsylvania. Formerly Vineyard Manager, Temperance Hill Vineyard

Dai Crisp, Vineyard Manager, Temperance Hill Vineyard, Salem, Oregon

Robin Cross, Graduate Research Assistant, Agricultural and Resource Economics, Oregon State University, Corvallis, Oregon

Jack DeAngelis, Ph.D., Professor of Entomology, Oregon State University, Corvallis, Oregon

Catherine Durham, Ph.D., Assistant Professor of Agricultural and Resource Economics, Oregon State University, Food Innovation Center Experiment Station, Portland, Oregon

Bart Eleveld, Ph.D., Extension Economist, Farm Management, Associate Professor, Agricultural and Resource Economics, Oregon State University, Corvallis, Oregon

Daniel Fey, Vineyard Manager, Willakenzie Estate, Yamhill, Oregon

James R. Fisher, Ph.D., Research Entomologist, USDA Agricultural Research Service, Horticultural Crops Research Laboratory, Corvallis, Oregon

Alfonso Gardea, Ph.D., Horticulturist and Director, CIAD - Cuauhtemoc, Chihuahua, Mexico. Formerly Graduate Research Assistant, Oregon State University

Randy Gold, Owner and Vineyard Manager, Gold Vineyard, Talent, Oregon

Roberta Gruber, Agricultural Employer Consultant, Farm Employer Education & Legal Defense Service, Oregon Farm Bureau, Salem, Oregon

Edward W. Hellman, Associate Professor of Viticulture, Texas A&M University and Texas Tech University, Lubbock, Texas. Formerly Associate Professor of Viticulture and Extension Specialist, Oregon State University

Scott Hendricks, Vineyard Consultant, Owner Windrow Vineyards, Windrow Management, Milton Freewater, Oregon

Scott Henry III, Owner, Owner and Vineyard Manager, Henry Estate, Umpqua, Oregon

Richard J. Hilton, Faculty Research Associate, Entomology, Oregon State University, Medford, Oregon

Allen Holstein, Winegrower, Argyle Winery, Dundee, Oregon

Andy Humphrey, Viticulturist and Winemaker, Andy Humphrey Consulting, Inc., McMinnville, Oregon

H. Earl Jones, Owner and Winemaker, Abacela Vineyards and Winery, Roseburg, Oregon

Gregory Jones, Ph.D., Associate Professor of Geography, Southern Oregon University, Ashland, Oregon

Diane Kenworthy, Grower Relations and Vineyard Manager, Ravenswood Winery, Sonoma, California. Formerly General Manager, Oregon Grape Management, Inc.

W. Mark Kliewer, Ph.D., Professor Emeritus, Viticulture, University of California, Davis; Owner and Vineyard Manager, Kliewer's Weinberg Vineyard, Cheshire, Oregon

David Lett, Founder, owner and winemaker, The Eyrie Vineyards, Dundee, Oregon

Porter Lombard, Ph.D., Professor Emeritus, Horticulture, Oregon State University, Medford, Oregon

Jesse D. Lyon, Attorney, Davis Wright Tremaine LLP, Portland, Oregon

Al MacDonald, Instructor of Viticulture, Northwest Viticulture Center, Chemeketa Community College; Owner and Manager, Anden Vineyards, Salem, Oregon

Walter Mahaffee, Ph.D., Research Plant Pathologist, USDA Agricultural Research Service, Horticultural Crops Research Laboratory, Corvallis, Oregon

Robert Martin, Ph.D., Research Plant Pathologist, USDA Agricultural Research Service, Horticultural Crops Research Laboratory, Corvallis, Oregon

Mike McLain, McLain & Associates Vineyard Properties, Springhill Cellars Winery, Albany, Oregon

John Miller, President, Mahonia Vineyards & Nursery, Salem, Oregon

Don Moore, Co-owner, Quail Run and Griffin Creek Vineyards, Edenvale Winery, Talent, Oregon

Joel Myers, Viticulturist, Vinetenders, Dayton, Oregon

Dick O'Brien, Owner and Vineyard Manager, Elton Vineyards, Salem, Oregon

John Pinkerton, Ph.D., Research Plant Pathologist, USDA Agricultural Research Service, Horticultural Crops Research Laboratory, Corvallis, Oregon

Jay W. Pscheidt, Ph.D., Extension Plant Pathology Specialist, Oregon State University, Corvallis, Oregon

Scott Snyder, Illustrator, Amarillo, Texas
Susan Sokol Blosser, President, Sokol Blosser Winery, Dundee, Oregon
Rollin Soles, President, Argyle Winery, Dundee, Oregon
Bernadine Strik, Ph.D., Extension Berry Crops Professor, Berry Research Leader, North Willamette Research and Extension Center, Oregon State University, Corvallis, Oregon
David Sugar, Ph.D., Professor of Plant Pathology and Horticulture, Oregon State University, Medford, Oregon
Doug Tunnell, Proprietor, Winemaker, Brick House Vineyards, Newberg, Oregon
Philip VanBuskirk, Area Extension Agent, Oregon State University, Medford, Oregon
M. Carmo Vasconcelos, Ph.D., Associate Professor of Viticulture, Oregon State University, Corvallis, Oregon
Esteban Vega-H, M.S., M.B.A., Oregon State University, Food Innovation Center Experiment Station, Portland, Oregon
Barney Watson, Extension Specialist and Instructor of Enology, Oregon State University; Owner and Winemaker, Tyee Wine Cellars, Corvallis, Oregon
Ray William, Ph.D., Extension Specialist and Professor of Horticulture, Oregon State University, Corvallis, Oregon

Acknowledgments

This book represents the collective effort of many individuals. The editor and publishers gratefully acknowledge the contributions of the chapter authors. Their common desire to support the continued success of the Oregon wine industry made this book possible. Thanks to the many grape growers and winemakers who shared their knowledge and experience, and to the faculty of Oregon State University and scientists at the USDA Horticulture Crops Research Lab in Corvallis who contributed their considerable expertise.

Thanks are extended to the following for reviewing and providing helpful comments on various chapters of the book: John Baham, Nick Dokoozlian, Andy Gallagher, Randy Gold, Scott Hendricks, Allen Holstein, Earl Jones, Porter Lombard, Kurt Lotspeich, Laura Lotspeich, Paul Schreiner, Philip VanBuskirk, Robert Wample, and Larry Williams. Thanks also to Scott Snyder for many of the illustrations and to Dai Crisp for the cover photograph and the color photographs inside the book.

The editor wishes to thank the officers and staff of the Oregon Winegrowers' Association and the Oregon State University Press for their invaluable assistance in publishing this book.

Finally, a special thanks to the editor's family, Pam, Will, and Michael, for their support and understanding during this project.

History and Character of the Oregon Wine Industry

Susan Sokol Blosser

In the last few months of 1970, Bill Blosser and I were living in Portland and spending every weekend driving the countryside, looking for potential vineyard land. For a reason I still don't completely understand, we were driven by the belief that we needed to grow winegrapes and start a winery. Whether it was our fate or our folly, we found an 18-acre dilapidated prune orchard that looked perfect and we bought it in December, two weeks before we had our first child. The stars must have been perfectly aligned for us to have given birth in one short month to two enterprises that would completely direct our lives from then on—our family and our vineyard. What follows is a history of the Oregon wine industry from the inside, including my analysis of the milestones, from the perspective of time.

In 1970 there was one vinifera winery in Oregon and there were less than 100 acres of vinifera vineyards. By 2000 there were 168 wineries and 10,500 acres of vinifera vineyards. Hillsides in the northern Willamette Valley that used to be planted with walnut, filbert, prune, and cherry orchards, and dotted with prune and nut dryers, are now covered with vineyards and architecturally imposing wineries as well. In the last decades of the twentieth century, canneries left the area or cut back, and fruit and nut dryers closed and the transformation began. The wine industry in Oregon has changed the face of the land. The change has not been too hard for those who were able to take advantage of the new wave. Neighbors who initially ferociously opposed Sokol Blosser Winery in the mid-1970s found they could make good money selling their land as vineyard property when the orchard economy went sour.

The wine industry has brought good things to Oregon on many fronts. The vineyards are scenic as well as being one of the most environmentally friendly agricultural sectors. Much of the acreage is farmed sustainably with minimal use of synthetic chemicals and fertilizers. Oregon has a worldwide reputation for its excellent wines, and its wineries have become a tourist destination, bringing visitors from all over the world to appreciate the beauty and quality of Oregon wine country.

History

Oregon's first post-Prohibition winegrowers came north from California during the 1960s, seeking a cooler climate for Pinot noir. David Lett, Dick Erath, and Charles Coury settled in the northern Willamette Valley, and Richard Sommer in the Umpqua Valley. These men, trained in viticulture and enology, brought their families north to what they saw as a promised land. Developing the vineyard was a family project, and, though the children were too young to contribute, wives worked side by side with their husbands.

These pioneers were followed in the early 1970s by a wave of young, urban professionals with a love of wine but little or no agricultural background. Plenty of intelligence, resourcefulness, and energy compensated for their lack of experience. The Bjellands and Wisnovskys settled in the Rogue Valley in southern Oregon; the Girardets in the Umpqua Valley; the Ponzis, Blossers, Adelsheims, Campbells, and Fullers in the northern Willamette Valley. They started on very low budgets, bought used equipment, and worked other jobs to make ends meet.

These early Oregon growers met often to share information as they adapted California and French techniques to Oregon conditions. In those early days the whole industry could fit into one small living room, and they met in their homes. They experimented in their vineyards with clones, spacing, and trellising, participated in county land use planning to preserve prime hillside land for agriculture, wrote and promoted strict wine labeling regulations for Oregon, and obtained state funding (based on self-imposed grape taxes) for winegrape research. The growers took their role with great seriousness, realizing that they were building the foundation of Oregon's modern wine industry.

Growers in northern and southern Oregon first organized regionally and then communicated and tried to work together. They overcame the territoriality of their two grower organizations and agreed to merge into a unified statewide trade organization, the Oregon Winegrowers' Association. Although the earliest vineyards had been in southern Oregon, development in the northern Willamette Valley grew faster in both size and reputation. Friction between the two regions became an inevitable undercurrent in all statewide programs, especially where funding was involved.

In the 1980s and 90s, what had started as a trickle became a flood. New vineyards and wineries multiplied and out-of-state investment flowed into Oregon. The wine world at large sat up and took notice in 1988, when Robert Drouhin, the prominent Burgundian négociant, started vineyards and a winery in the Red Hills of Dundee, declaring there were only two places in the world he would grow Pinot noir—Burgundy and Oregon. Domaine Drouhin of Oregon triggered a series of increasingly expensive and sophisticated winery facilities. Many of these, like WillaKenzie Estate, King Estate, Domaine Serene, and Lemelson, were built by individuals who had personal fortunes. Others, like Archery Summit and Willamette Valley Vineyards, were built by groups of investors.

The days of used equipment and individual owner sweat-equity were long gone, and the original pioneers found they needed to keep up or be left in the dust. The state of the art, both in grape growing and winemaking, had evolved considerably, and the new players had all the bells and whistles—grape sorting conveyors, gravity flow processing, barrel caves, commercial kitchens, entertainment facilities, and more. In 1978, Sokol Blosser, for example, set a standard that went beyond the converted garage practice of the day by having noted Oregon architect John Storrs (Salishan Lodge, Western Forestry Center) design and build their new tasting room. It had a very small kitchen, a very small office, two small bathrooms, and a room for visitors. It was a bold new statement for the industry. Rex Hill moved the bar higher in the early 1980s with a commercial kitchen, a small amphitheater, and landscaped gardens. Each new facility outdid the last.

But despite its rapid growth and increasing professionalism, the character of Oregon's wine industry has changed surprisingly little. The small "mom and pop" operations continue. And regardless of personal fortune, Oregon wine producers are driven by passion rather than dollars and are in jeans more often than silk. It is the passion and commitment to quality that have enabled Oregon wines, with virtually no traditional advertising, to command high prices and win a coveted niche on wine lists in the best restaurants in the country. Although almost every state in this country has a wine industry, Oregon is one of only a very few whose wines have achieved national and international distribution.

By the turn of the twenty-first century, Oregon boasted 168 bonded wineries and 10,500 acres of vinifera grapes. It was second in the United States in number of wineries and fourth in gallons produced.

Milestones

There have been many important happenings over the course of establishing Oregon's wine industry, but some stand out as milestones. Here are four events from the past several decades that have most significantly shaped the Oregon wine industry.

The first was the passage of Oregon's land use planning bill (Senate Bill 100, passed in 1972), which fortuitously coincided with the early winegrowers' arrival in the state. The bill mandated that each county work with citizen groups to create a land use plan. The winegrowers of the 1970s became actively involved in this process and were able to persuade planners to set aside hillside land, previously zoned for development as "view property," as agricultural land for vineyards. The effort was most successful in Yamhill County, which consequently has the largest vineyard acreage today. The hillside soils are not as deep and fertile as the soils on the valley floor, so the county's assumption was that they were not "prime" agricultural land. But, in fact, hillside vineyard land in the northern Willamette Valley is now worth many times more than land on the valley floor. This would not have happened without the synchronicity of Senate Bill 100 and the early winegrowers seizing the opportunity it presented.

The second was the development and adoption of Oregon's wine labeling regulations, which were designed by the state's winegrowers in the mid-1970s to protect the integrity of their small, super-premium industry. Early winegrower activists set the nation's strictest labeling standards, which remain a unique and distinctive feature of the Oregon wine industry today. For example, Oregon law specified that in order to label a wine by the name of the grape (called a "varietal" designation), 90% of the grapes had to be of that variety. (The only exception is Cabernet Sauvignon.) During the same period, federal law permitted varietal labeling if a wine contained only 51% of the variety named on the label. A decade after Oregon raised the standard, the federal government followed suit, raising the required percentage from 51% to 75% for a varietal designation.

Another then-unique aspect of Oregon's labeling regulations was aimed at the way California, New York, and other winegrowing areas misrepresented European geographical areas such as Burgundy, Chablis, and Rhine on their labels. The use of such terms was prohibited on Oregon labels unless used as an appellation, meaning that all the grapes came from the named area. These labeling regulations were significant not only because they protected the integrity of a small, then politically powerless industry, but also because they represented a level of cooperation and compromise that is rare in the cutthroat world of business.

The third history-making event, in 1985, shook up the world wine press and focused attention on Pinot noirs from Oregon. The scene was a tasting at the International Wine Center in New York City to compare approximately fifteen of the top 1983 Oregon Pinot noirs with a similar number of high-quality French Burgundies of the same vintage. Members of the press and wine buyers from prominent restaurants and wine shops tasted wines blind. Tasters were asked to do two things: label each wine as either French or Oregon, and rank their five favorites. The surprise came when the wines were unveiled and the results tallied. This august group of wine tasters was not able to distinguish Oregon Pinot noir from French Burgundy. And, not only was the number-one ranked wine from Oregon, all five of the top ranked wines were from Oregon.

The tasting had an immediate impact on sales of Oregon Pinot noir, as wine writers spread what they considered hot news. The average retail price of the top Oregon Pinot noirs at that time was less than $8 a bottle, and many wineries were sitting on large inventories. The New York tasting cleaned out those inventories in a few short months and a price climb began. Intrigued by the New York tasting, many wine appreciation groups around the country created similar tastings on their own and came up with substantially the same results. Oregon Pinot noir gained prestige every time. Previous tastings (such as the Gault Millau tasting in Paris in 1979, in which the Eyrie Vineyards 1975 Pinot noir stood the wine press on its ear by scoring in the top ten) helped set the stage. But it took the development of a critical mass of wineries, all with top quality, to really get the wine world's attention.

The final milestone has been the unprecedented success of the International Pinot Noir Celebration (IPNC), an annual event created in 1987 by a small group of Oregon wineries and business people from the city of McMinnville. The IPNC brings top Pinot noir producers in the world together, featuring their wines with outstanding meals prepared by prominent chefs. Owners and winemakers get the opportunity to talk about their wines and mingle with consumers and members of the wine trade. This three-day event has become so popular and prestigious that tickets are now distributed by lottery. International in scope, the IPNC has enhanced Oregon's position in the wine world.

In addition to the big milestones, there are some especially synchronous trends that favored the growth of the wine industry. In the early 1970s, for example, the pole bean farmers in the Willamette Valley were switching to bush beans and put all their used poles and wire up for sale. This gave the early growers an easy solution to vineyard trellising. We sorted through huge piles of old wire and stakes to find the best ones and considered it a great boon to be able to save money by buying used poles and wire. Of course, it all had to be replaced in time, but we got many good years out of it.

Another synchronicity was the availability of labor, primarily Hispanic, for harvest. It turned out that grape harvest occupied a vacant niche between broccoli and Christmas trees. Virtually all of the vineyard labor has been Hispanic. In recognition of how critical this labor is to a growing industry, eighteen Oregon wineries started a Pinot noir barrel auction in 1992 to raise money for health care services for vineyard workers. This auction, called ¡Salud!, has shown benefits on many fronts. It has enabled Tuality Healthcare Foundation to provide extensive health care services for vineyard workers who would otherwise not get them; it has become a premier auction with forty wineries that sells out every year; it has showcased Oregon Pinot noir; and it has made the ¡Salud! wines collector's items.

Not all trends have favored growth; the wine industry has faced its share of challenges. Until the mid 1990s, commercial banks were decidedly uninterested in loaning money to vineyards and wineries. Loans that were made had high interest rates, usually 3–4% over prime. The recession in the early 1980s hit the early wineries hard as interest rates skyrocketed. At Sokol Blosser we felt good when, after mountains of paperwork, we were able to get a bank loan with an SBA (Small Business Administration) guarantee for 2% over prime. Good, that is, until prime hit 19% and we were paying 21% interest on our loan. We survived to live through subsequent downturns in the economy, but nothing surpasses the memory of having to pay all of our income to the bank just to cover interest.

In the vineyard, phylloxera, the dreaded root louse, loomed over us from the beginning, although when we started it had not been found in any commercial vineyard in Oregon. Rumor was that it existed in an abandoned vineyard in southern Oregon, but not in the Willamette Valley. Even though growers knew that phylloxera existed, Sokol Blosser consciously made the decision to plant self-rooted vines. It was a calculated risk and I'm convinced it was the right decision. Any rootstock available in 1971 would have been the wrong choice for our growing conditions. We had over twenty years to find the right rootstocks. Phylloxera was discovered in several spots in the northern Willamette Valley in the early 1990s, and I think it now can be considered to be virtually everywhere. Sokol Blosser was one of the earliest vineyards where it was identified, and, after the initial fear and frustration, we embraced the opportunity to reconfigure and replant, making use of what we had learned over the past years. By the late 1990s virtually all planting was done on the small collection of rootstock that had been identified as appropriate for Oregon.

The Future

The Oregon industry is a story by itself, but it is also a significant part of an international wine industry rebirth taking place simultaneously throughout the United States, and in Australia, New Zealand, South Africa, Italy, France, and Germany. It began in the early 1970s with individual countries doing their own thing. Now the wine industry has become genuinely global. This is demonstrated not only through international ownership but also through cooperation and friendship, international symposiums, and joint research that has been helpful to all. In addition, wine-country tourism has become a popular international pastime, stimulating the development in every country of wine country inns, bed-and-breakfasts, related tourist attractions, and fine restaurants serving wine-friendly cuisine.

As the Oregon wine industry enters the twenty-first century, most of the early winegrowers who developed the industry are still around. The years have taken their toll. Fortunes and marriages have faltered, but there is little talk of regrets. What a success story they have to tell. Out of nothing they built an industry that has given Oregon a reputation and visitor destination more romantic and glamorous than its stereotypical rugged outdoorsiness; an industry that contributes millions of dollars to Oregon's economy through its many facets—agriculture, winemaking, tourism, and sales. The proximity of metropolitan Portland to the wine country of the northern Willamette Valley has mutually enhanced wine country tourism, the bed-and-breakfast industry, top-quality restaurants, and markets committed to locally grown fruits and vegetables. It is no accident, for example, that the short main strip of the town of Dundee (which calls itself the epicenter of the Oregon wine industry) boasts a surprising number of successful white tablecloth restaurants (Tina's, Dundee Bistro, Red Hills Provincial). Visitors come to wine country to tour and taste, go out for a gourmet dinner, and stay at a bed-and-breakfast.

For a lucky few of the old timers, the second generation is stepping up to take its place at the helm, working with their parents in some cases and taking over in others. They have a rich inheritance; their parents created a whole new industry for Oregon. When their parents started, no one noticed or cared what they were doing in the northwest corner of the United States. Today, Oregon wine, especially Oregon Pinot noir, is well known. The next generation's challenge is to keep the momentum moving forward, to build on that success without losing the passion and focus on quality that fueled their parents, and to integrate newcomers into the Oregon wine community. And the world will be watching.

Grapevine Structure and Function

Edward W. Hellman

This chapter presents an overview of grapevine structure and function to provide a basic understanding of how grapevines grow. Such understanding is the foundation of good vineyard management, and the practical application of this knowledge is emphasized throughout this book. The reader should consult the references cited in this chapter for more complete coverage of these topics. Much of the common viticultural terminology is introduced in this chapter.

Grapevine Structure

Cells and Tissues

The basic unit of plant structure and function is the *cell.* All cells have the same general organization, consisting of a cell wall, protoplasm (liquid-filled region containing living organelles), and the vacuole (region containing the cell sap). This basic cell structure is modified to create different cell types that are capable of specialized functions. Organized groups of specialized cells that perform specific functions are called *tissues.* For example, the outside protective "skin" of grape leaves, the *epidermis,* consists of one to several layers of specialized cells. A thorough discussion of the cell and tissue anatomy of grapevines has been prepared by Pratt (1974), or a general discussion of plant anatomy can be found in any introductory botany textbook.

Meristems

Certain plant cells, termed *meristematic cells,* perform the specialized function of growth by the creation of new cells through cell division. Groups of these cells are organized into *meristems* (or *growing points*), positioned at various locations on the vine. The *apical meristem* is a tiny growing point, hidden from view within the unfolding leaves at the tip of an expanding shoot. In addition to the apical meristem, the shoot produces many additional growing points at the base of each leaf, called buds; these are described in more detail below. Each root tip also contains a growing point. Two specialized meristems, the vascular cambium and the cork cambium, are responsible for the radial growth (diameter increase) of woody parts of the vine.

New xylem and phloem tissues (described below under Vascular System) are produced every year from a specialized meristem called the *vascular cambium* (or simply *cambium).* The location and arrangement of cambium, xylem, and phloem vary between plant parts (e.g., shoots and roots) and with the developmental stage of the part. The cambium consists of a single layer of meristematic cells, which produce xylem cells to the inside and phloem cells outside (Figure 1). Thus, the annual increase in girth of woody tissues such as the trunk is a result of the addition of new xylem and phloem cells from the cambium. Xylem cells are larger and produced in much greater abundance than phloem cells, which form tissue only a few cell layers thick. This causes the cambium always to be positioned close to the outer surface of a woody stem. Older xylem can remain functional for up to seven years, but is mostly inactivated after two or three years. Some phloem cells continue to function for three to four years.

The exterior of woody parts of the vine is protected by *periderm,* which comprises *cork* cells and is covered by an outer *bark* consisting of dead tissues. Once a year, some of the cells within the outer, nonfunctional phloem become meristematic, creating the *cork cambium.* The cork cambium produces a layer of new cells that soon become impregnated with an impermeable substance, cutting off the water supply to the cork cells and older phloem that are external to the layer. These cells die and add to the layers of bark. Older bark cracks from the expansion growth of new bark beneath it, creating the peeling bark that is characteristic of older wood on grapevines.

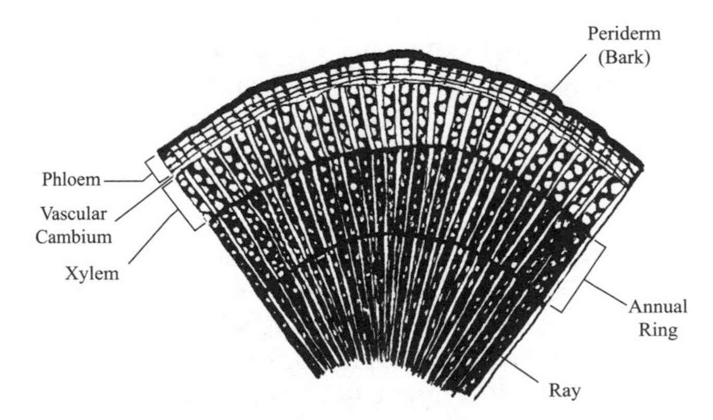

Figure 1. Cross section of 3-year-old grapevine arm. Redrawn from Esau (1948) by Scott Snyder.

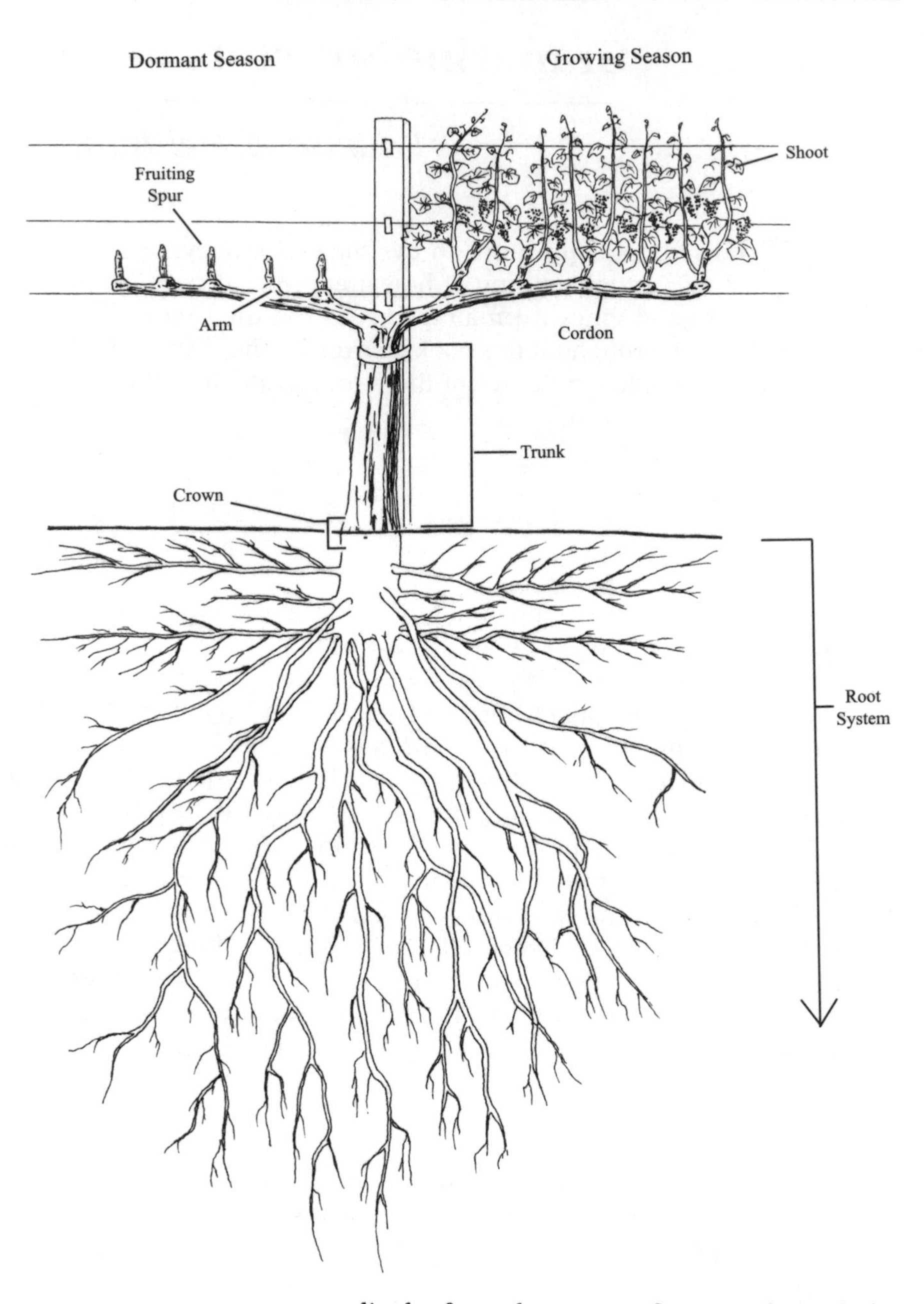

Figure 2. Grapevine structures and features: self-rooted vine. Drawing by Scott Snyder.

Vascular System

The interior of all the plant parts described below contains groups of specialized cells organized into a *vascular system* that conducts water and dissolved solids throughout the vine. There are two principal parts to the vascular system: *xylem* is the conducting system that transports water and dissolved nutrients absorbed by the roots to the rest of the vine, and *phloem* is the food-conducting system that transports the products of photosynthesis from leaves to other parts of the vine. The xylem and phloem tissues each consist of several different types of cells, some of which create a continuous conduit throughout the plant, and others provide support functions to the conducting cells, such as the storage of food products in xylem cells. A group of specialized cells are arranged in narrow bands of tissue called *rays*, which extend out perpendicular from the center of a stem, through the xylem and phloem. Ray cells facilitate the radial transfer of water and dissolved substances between and among xylem and phloem cells and are a site for storage of food reserves. The vascular system constitutes the *wood* of older stems, and the thick cell walls of the xylem provide the principal structural support for all plant parts.

Parts of the Vine

The shape of a cultivated grapevine is created by pruning and training the vine into a specific arrangement of parts according to one of many *training systems*. Over the centuries, innumerable training systems have been developed and modified in efforts to facilitate vine management and provide a favorable growing environment for the production of grapes.

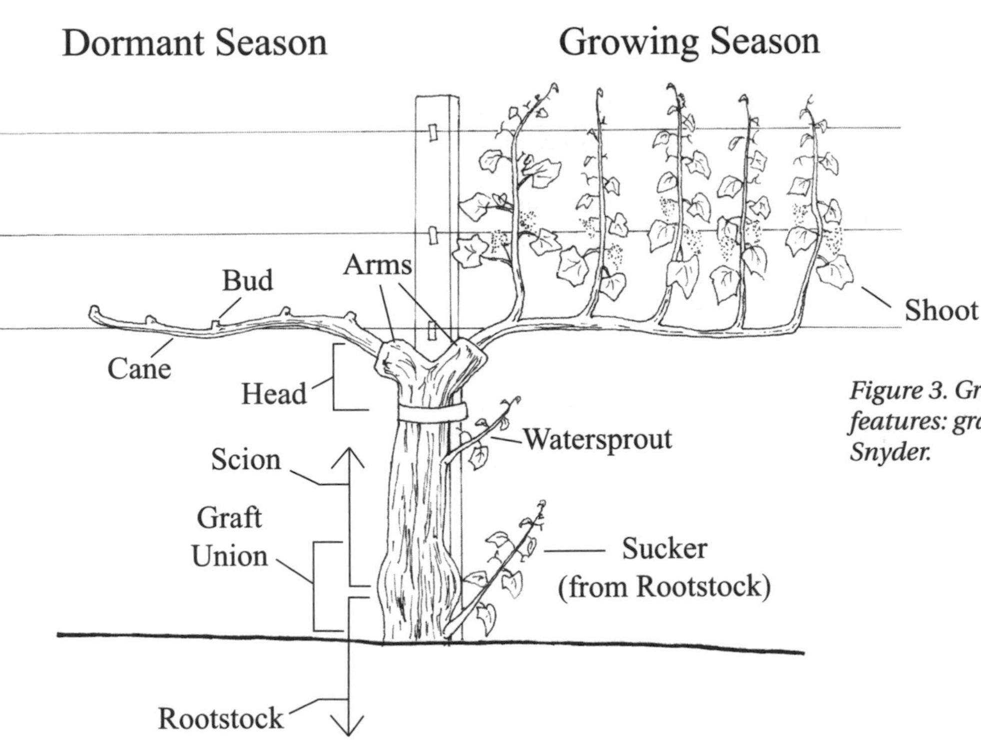

Figure 3. Grapevine structures and features: grafted vine. Drawing by Scott Snyder.

Figures 2 and 3 illustrate a mature grapevine as it might appear at two representative time periods, trained to two different systems. The parts of the vine are labeled with commonly used viticultural terms that often reflect how we manage the vine rather than describing distinct morphological structures as defined by botanists.

The Root System

In addition to anchoring the vine, roots absorb water and nutrients, store carbohydrates, other foods, and nutrients for the vine's future use, and produce hormones that control plant functions. The root system of a mature grapevine consists of a woody framework of older roots (Richards, 1983) from which permanent roots arise and grow either horizontally or vertically. These roots are typically multi-branching, producing lateral roots that can further branch into smaller lateral roots. Lateral roots produce many short, fine roots, which has the effect of increasing the area of soil exploited. Certain soil fungi, *mycorrhizae,* live in a natural, mutually beneficial association with grape roots. Mycorrhizae influence grapevine nutrition and growth and have been shown to increase the uptake of phosphorus.

The majority of the grapevine root system is usually reported to be within the top 3 feet of the soil, although individual roots can grow much deeper under favorable soil conditions. Distribution of roots is influenced by soil characteristics, the presence of hardpans or other impermeable layers, the rootstock variety (see below), and cultural practices such as the type of irrigation system.

Grapevines can be grown "naturally" on their own root system (*own-rooted or self-rooted vines*) or they may be grafted onto a rootstock. A *grafted vine* (Figure 3) consists of two general parts, the *scion* variety (e.g., Pinot noir), which produces the fruit, and the *rootstock* variety (often denoted by numbers, e.g., 101-14), which provides the root system and lower part of the trunk. The position on the trunk where the two varieties were joined by grafting and subsequently grew together is called the *graft union.* Successful healing of the graft union requires that the vascular cambiums of the stock and scion be in contact with each other, since these are the only tissues having the meristematic activity necessary for the production of new cells to complete the graft union. Healing of the graft union often results in the production of abundant *callus* (a wound healing tissue composed of large thin-walled cells that develop in response to injury) tissue, often making the area somewhat larger than adjacent parts of the trunk. Because rootstock and scion varieties may grow at different rates, trunk diameter can vary above and below the graft union.

Rootstock varieties were developed primarily to provide a root system for *Vitis vinifera* L. ("European" winegrape) varieties that is resistant or tolerant to *phylloxera,* a North American insect to which *V. vinifera* roots have no natural resistance. Most phylloxera-resistant rootstocks are either native North American species or hybrids of two or more of these species, including *V. riparia, V. berlandieri,* and *V. rupestris.* The rooting pattern and depth, as well as other root system characteristics, vary among the species and hybrid rootstocks, so the rootstock can influence aspects of vine growth, including vigor, drought tolerance, nutrient uptake efficiency, and pest resistance. Rootstock variety selection is, therefore, an important factor in vineyard development.

The Trunk

The *trunk*, formerly an individual shoot, is permanent and supports the aboveground vegetative and reproductive structures of the vine. The height of the trunk varies among training systems, and the top of the trunk is referred to as the *head*. The height of the head is determined by pruning during the initial stages of training a young grapevine. The trunk of a mature vine has *arms*, short branches from which canes or spurs (defined below) originate; arms are located in different positions depending on the system. Some training systems utilize *cordons* (Figure 2), semi-permanent branches of the trunk, usually trained horizontally along a trellis wire, with arms spaced at regular intervals along their length. Other systems utilize *canes* (Figure 3), one-year-old wood arising from arms usually located near the head of the vine. The *crown* refers to the region of the trunk near the ground, from slightly below to slightly above ground level.

Shoots and Canes

The *shoot* is the primary unit of vine growth and the principal focus of many viticultural practices. Shoots are the stemlike green growth arising from a bud. *Primary shoots* arise from primary buds (described below) and are normally the fruit-producing shoots on the vine. The components of the shoot are illustrated in Figure 4, and the stages of grapevine growth and flower and fruit development are shown in Figure 5. The main axis of the shoot consists of structural support tissues and conducting tissues to transport water, nutrients, and the products of photosynthesis. Arranged along the shoot in regular patterns are leaves, tendrils, flower or fruit clusters, and buds. General areas of the shoot are described as *basal* (closest to its point of origin), *mid-shoot*, and *apex* (tip). The term *canopy* is used to denote the collective arrangement of the vine's shoots, leaves and fruit; some viticulturists also consider the trunk, cordons, and canes to be parts of the canopy.

Shoot Tip. The shoot has many points of growth, but the extension growth of the shoot occurs from the *shoot tip* (growing tip). New leaves and tendrils unfold from the tip as the shoot grows. Growth rate of the shoot varies during the season. Grapevine shoots do not stop expanding by forming a terminal bud as some plants do; they may continue to grow if there is sufficient heat, soil moisture, and nutrients.

Leaves. Leaves are produced at the apical meristem. The shoot produces two or more closely spaced *bracts* (small scalelike leaves) at its base before it produces the first true foliage leaf. Leaves are attached at the slightly enlarged area on the shoot referred to as a *node*. The area between nodes is called the *internode*. The distance between nodes is an indicator of the rate of shoot growth, so internode length varies along the cane corresponding to varying growth rates during the season.

Leaves consist of the *blade*, the broad, flat part of the leaf designed to absorb sunlight and CO_2 in the food manufacturing process of photosynthesis (see below), and the *petiole*, the stemlike structure that connects the leaf to the shoot. The lower surface of leaf blades contains thousands of microscopic pores called *stomata* (s., *stomate*), through which diffusion of CO_2, O_2, and water vapor occurs. Stomata are open in the light and closed in the dark. The petiole conducts water and food material to and from the leaf blade and maintains the orientation of the leaf blade to perform its functions.

Flowers and Fruit. A *fruitful shoot* usually produces from one to three *flower clusters* (inflorescences) depending on variety, but typically two under Oregon conditions. Flower clusters develop opposite the leaves, typically at the third to sixth nodes from the base of the shoot, depending on the variety. If three

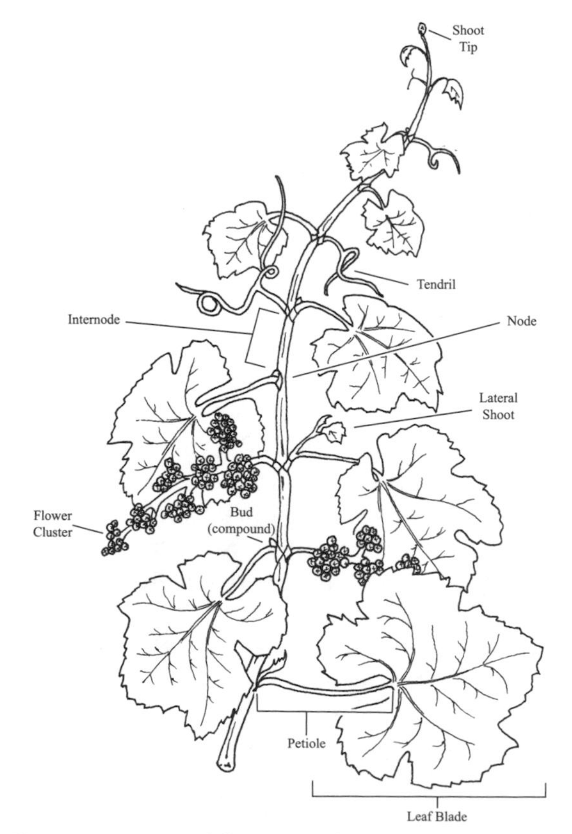

Figure 4. Principal features of a grapevine shoot prior to bloom. Drawing by Scott Snyder.

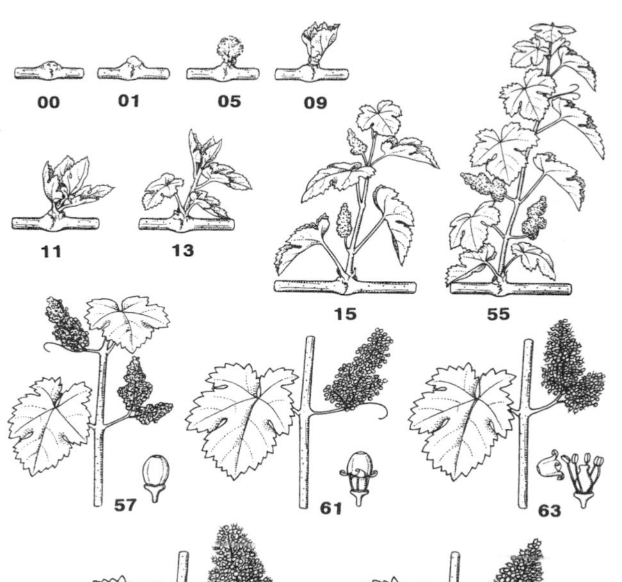

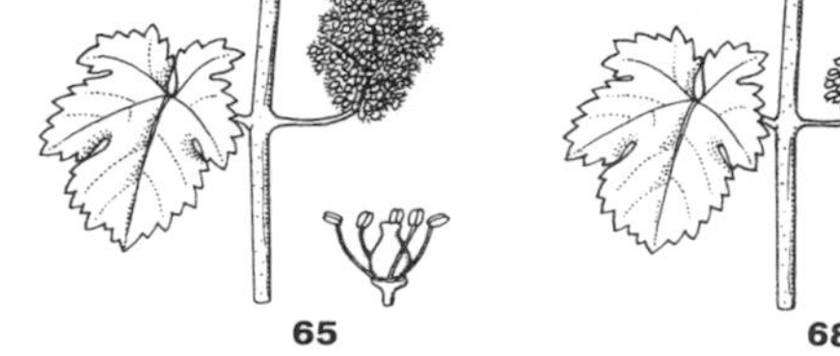

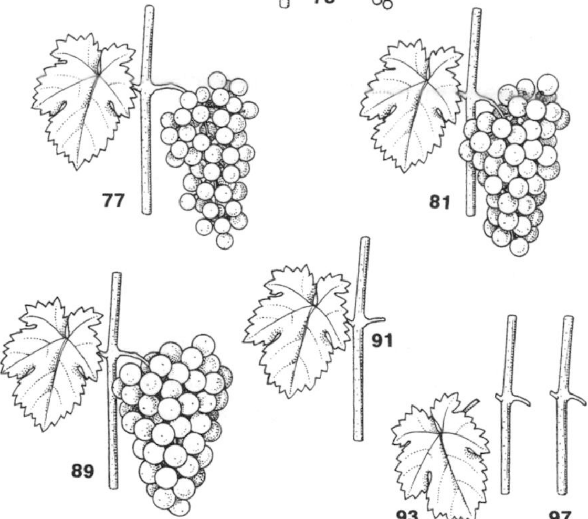

Figure 5. Stages of grapevine growth. Adapted with permission from Meier (2001).

Principal growth stage 0: Sprouting/Bud development

00	Dormancy: buds pointed to round, light or dark brown according to variety; bud scales more or less closed according to variety
01	Beginning of bud swelling: buds begin to expand inside the bud scales
03	End of bud swelling: buds swollen, but not green
05	"Wool Stage": brown wool clearly visible
07	Beginning of bud burst: green shoot tips just visible
08	Bud burst: green shoot tips clearly visible

Principal growth stage 1: Leaf development

11	First leaf unfolded and spread away from shoot
12	2nd leaves unfolded
13	3rd leaves unfolded
14	Stages continuous until...
19	9 or more leaves unfolded

Principal growth stage 5: Inflorescence emerge

53	Inflorescences clearly visible
55	Inflorescences swelling, flowers closely pressed together
57	Inflorescences fully developed; flowers separating

Principal growth stage 6: Flowering

60	First caps detached from the receptacle
61	Beginning of flowering: 10% of caps fallen
62	20% of caps fallen
63	Early flowering: 30% of caps fallen
64	40% of caps fallen
65	Full flowering: 50% of caps fallen
66	60% of caps fallen
67	70% of caps fallen
68	80% of caps fallen
69	End of flowering

Principal growth stage 7: Development of fruits

71	Fruit set: young fruits begin to swell, remains of flowers lost
73	Berries swelling, clusters begin to hang
75	Berries pea-sized, clusters hang
77	Berries beginning to touch
79	Majority of berries touching

Principal growth stage 8: Ripening of berries

81	Beginning of ripening: berries begin to develop variety-specific color
83	Berries developing color
85	Softening of berries
89	Berries ripe for harvest

Principal growth stage 9: Senescence

91	After harvest; end of wood maturation
92	Beginning of leaf discoloration
93	Beginning of leaf-fall
95	50% of leaves fallen
97	End of leaf-fall
99	Harvested product

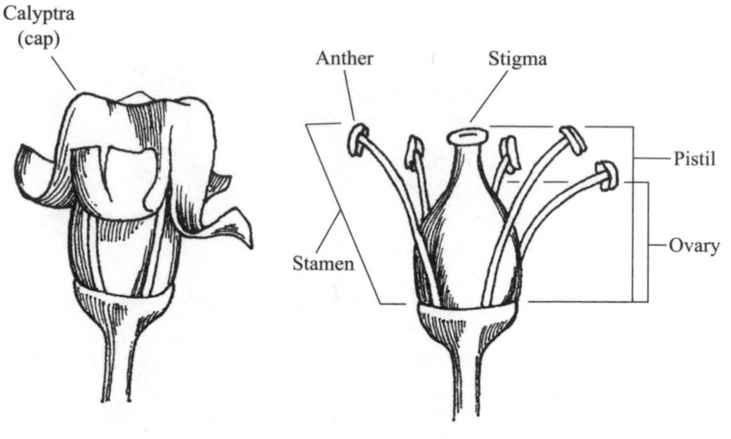

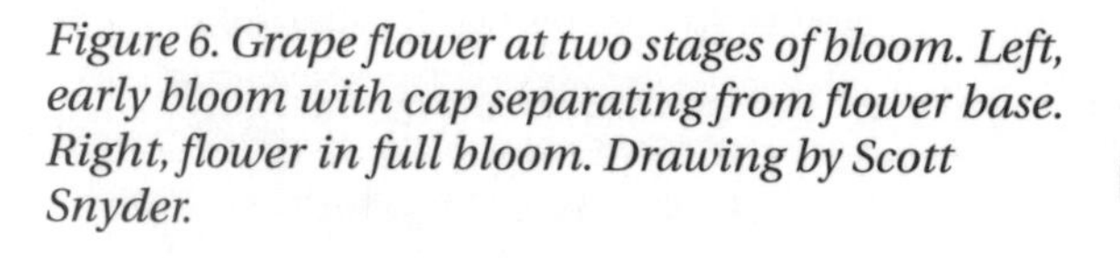
Figure 6. Grape flower at two stages of bloom. Left, early bloom with cap separating from flower base. Right, flower in full bloom. Drawing by Scott Snyder.

flower clusters develop, two develop on adjacent nodes, the next node has none, and the following node has the third flower cluster. The number of flower clusters on a shoot is dependent upon the grape variety and the conditions of the previous season under which the dormant bud (that produced the primary shoot) developed. A cluster may contain several to many hundreds of individual flowers, depending on variety.

The grape flower does not have conspicuous petals (Figure 6); instead, the petals are fused into a green structure termed the *calyptra* but commonly referred to as the *cap*. The cap encloses the reproductive organs and other tissues within the flower. A flower consists of a single *pistil* (female organ) and five *stamens*, each tipped with an *anther* (male organ). The pistil is roughly conical in shape, with the base disproportionately larger than the top and the tip (the *stigma*) slightly flared. The broad base of the pistil is the *ovary*, which consists of two internal compartments, each having two ovules containing an embryo sac with a single egg. The anthers produce many yellow *pollen* grains, which contain the sperm.

The time during which flowers are open (the calyptra has fallen) is called *bloom* (also flowering or anthesis) and can last from one to three weeks depending on weather. Viticulturists variously refer to *full bloom* as the stage at which either roughly one-half or two-thirds of the caps have loosened or fallen from the flowers. Bloom typically occurs between 50 and 80 days after budburst in Oregon.

The stages of bloom (60-69) are illustrated in Figure 5. When the flower opens, the cap separates from the base of the flower, becomes dislodged, and usually falls off, exposing the pistil and anthers. The anthers may release their pollen either before or after capfall. Pollen grains randomly land upon the stigma of the pistil. This event is termed *pollination*. Multiple pollen grains can germinate, each growing a pollen tube down the pistil to the ovary and entering an ovule, where a sperm unites with an egg to form an embryo. The successful union is termed *fertilization*, and the subsequent growth of berries is called *fruit set*. The berry develops from the tissues of the pistil, primarily the ovary. The ovule together with its enclosed embryo develops into the seed.

Because there are four ovules per flower, there is a maximum potential of four seeds per berry. Unfavorable environmental conditions during bloom, such as cool, rainy weather, can reduce both fruit set (number of berries) and berry size. Berry size is related to the number of seeds within the berry but can also be influenced by growing conditions and practices, particularly water management. Some immature berries may be retained by a cluster without completing their normal growth and development, a phenomenon known as *millerandage* or *"hens and chicks."* See Pratt (1971) for a more complete botanical description of grapevine reproductive anatomy and process.

Tendrils. The shoot also produces *tendrils*—slender structures that coil around smaller objects (e.g., trellis wires, small stakes, and other shoots) to provide support for growing shoots. Tendrils grow opposite a leaf in the absence of a flower cluster, except the first two or three leaves and thereafter skipping every third leaf. Flower clusters and tendrils have a common developmental origin (Mullins et al., 1992), so occasionally a few flowers develop on the end of a tendril.

Buds. A *bud* is a growing point that develops in the leaf axil, the area just above the point of connection between the petiole and shoot. The single bud that develops in this area is described in botanical terms as an *axillary bud*. It is important to understand that on grapevines a bud develops in every leaf axil, including the inconspicuous basal bracts (scalelike leaves). In viticultural terminology, we describe two buds associated with a leaf—the *lateral bud*, and the *dormant bud* (or latent bud). The lateral bud is the true axillary bud of the foliage leaf, and the dormant bud forms in the bract axil of the lateral bud. Because of

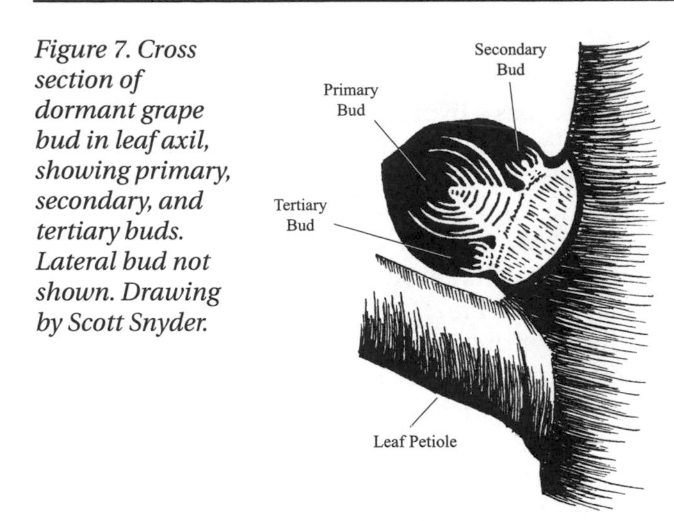

Figure 7. Cross section of dormant grape bud in leaf axil, showing primary, secondary, and tertiary buds. Lateral bud not shown. Drawing by Scott Snyder.

their developmental association, the two buds are situated side-by-side in the main leaf axil.

Although the dormant bud (sometimes called an *eye*) looks like a simple structure, it is actually a compound bud consisting of three growing points, sometimes referred to as the *primary, secondary,* and *tertiary buds.* The distinction between secondary and tertiary buds is sometimes difficult to make and often of little importance, so it is common to refer to both of the smaller buds as secondary buds. The collection of buds is packaged together within a group of external protective bud scales (Figure 7). Continuing the bud development pattern, the primary growing point is the axillary bud of the lateral bud; the secondary and tertiary growing points are the axillary buds of the first two bracts of the primary growing point.

The dormant bud is of major concern at pruning, since it contains cluster primordia (the fruit-producing potential for the next season). It is referred to as dormant to reflect the fact that it does not normally grow out in the same season in which it develops.

The dormant bud undergoes considerable development during the growing season. The three growing points each produce a rudimentary shoot that ultimately will contain *primordia* (organs in their earliest stages of development) of the same basic components of the current season's fully grown shoot: leaves, tendrils, and in some cases flower clusters. The primary bud develops first, so it is the largest and most fully developed. If it is produced under favorable environmental and growing conditions, it will contain flower cluster primordia before the end of the growing season. The flower cluster primordia thus represent the fruiting potential of the bud in the following season. Reflecting the sequence of development, the secondary and tertiary buds are progressively smaller and less developed. They are generally less fruitful (have fewer and smaller clusters) than the primary bud. *Bud fruitfulness* (potential to produce fruit) is a function of the variety, environmental conditions, and growing practices. Dormant buds that develop under unfavorable conditions produce fewer flower cluster primordia.

In most cases, only the primary bud grows, producing the primary shoot. The secondary bud can be thought of as a "backup system" for the vine; normally, it grows only when the primary bud or young shoot has been damaged, often from freeze or frost. However, under some conditions such as severe pruning, destruction of part of the vine, or boron deficiency, it is possible for two or all three of the buds to produce shoots (Winkler et al., 1974). Tertiary buds provide additional backup if both the primary and secondary buds are damaged, but they usually have no flower clusters. If only the primary shoot grows, the secondary and tertiary buds remain alive, but dormant, at the base of the shoot.

The lateral bud grows in the current season, but growth may either cease soon after formation of the basal bract or continue, producing a *lateral shoot* (summer lateral) of highly variable length. Regardless of the extent of lateral bud development, a compound bud develops in the basal bract, forming the dormant bud. Long lateral shoots sometimes produce flower clusters and fruit, which is known as *second crop.* But because they develop later in the season than fruit on the primary shoot, second crop fruit does not mature fully in Oregon. If a lateral bud does not grow in the current season, it will die.

Suckers and Watersprouts. Shoots may also arise from bud locations on older wood such as cordons and trunks. *Suckers* are shoots that grow from the crown area of the trunk. *Watersprout* is a term sometimes used to refer to a shoot arising from the upper regions of the trunk or from cordons. Buds growing from older wood are not newly initiated buds; rather, they developed on green shoots as axillary buds that never grew out. These buds are known as *latent buds,* because they can remain dormant indefinitely until an extreme event such as injury to the vine or severe pruning stimulates renewed development and shoot growth.

Suckers often arise from latent buds at underground node positions on the trunk. In routine vine management, suckers are removed early in the season before axillary buds can mature in basal bracts of the sucker shoots. Similarly, aboveground suckers are typically stripped off the trunk manually so that a pruning stub does not remain to harbor additional latent buds that could produce more suckers in the following year.

Latent buds come into use when trunk, cordon, or spur renewal is necessary. Generally, numerous latent buds exist at the "renewal positions" (a pruning term) on the trunk or cordons. Dormant secondary and tertiary buds exist in the stubs that remain after canes or spurs have been removed by pruning.

Canes. The shoot begins a transitional phase about midseason, when it begins to mature, or *ripen*. Shoot maturation begins as periderm develops, starting at the shoot base, appearing initially as a yellow, smooth "skin." Periderm continues to extend development toward the shoot tip through summer and fall. As periderm develops, it changes from yellow to brown and becomes a dry, hard, smooth layer of bark. During shoot maturation, the cell walls of ray tissues thicken and there is an accumulation of starch (storage carbohydrates) in all living cells of the wood and bark. Once the leaves fall off at the beginning of the dormant season, the mature shoot is considered a *cane*.

The cane is the principal structure of concern in the dormant season, when the practice of pruning is employed to manage vine size and shape and to control the quantity of potential crop in the coming season. Because a cane is simply a mature shoot, the same terms are used to describe its parts. Pruning severity is often described in terms of the number of buds retained per vine, or *bud count*. This refers to the dormant buds, containing three growing points, described above. The "crown" of buds observed at the base of a cane includes the secondary and tertiary growing points of the compound bud that gave rise to the primary shoot, as well as the axillary buds of the shoot's basal bracts (Pratt, 1974). These basal buds are generally not fruitful and do not grow out, so they are not included in bud counts and may be referred to as *noncount buds*.

Canes can be pruned to varying lengths, and when they consist of only one to four buds they are referred to as *spurs*, or often as *fruiting spurs* since fruitful shoots arise from spur buds. Grapevine spurs should not be confused with true spurs produced by apple, cherry, and other fruit trees, which are the natural fruit-bearing structures of these trees. On grapevines, spurs are created by short-pruning of canes. Figure 2 illustrates a vine that is cordon-trained, spur-pruned. Training systems that use cane-pruning (Figure 3) sometimes also use spurs for the purpose of growing shoots to be trained for *fruiting canes* in the following season. These spurs are known as *renewal spurs*, indicating their role in replacing the old fruiting cane.

Major Physiological Processes

Photosynthesis

Grapevines, like other green plants, have the capacity to manufacture their own food by capturing the energy within sunlight and converting it to chemical energy (food). This multi-stage process is called *photosynthesis*. In simple terms, sunlight energy is used to split water molecules (H_2O), releasing molecular oxygen (O_2) as a byproduct. The hydrogen (H) atoms donate electrons to a series of chemical reactions that ultimately provide the energy to convert carbon dioxide (CO_2) into carbohydrates (CH_2O). Details of this complex process are beyond the scope of this chapter and have been summarized elsewhere (Mullins et al., 1992).

Photosynthesis occurs in chloroplasts, highly specialized organelles containing molecules called *chlorophyll*, which are abundant in leaf cells. The structure of a leaf is well adapted to carry out its function as the primary site of photosynthesis. Leaves provide a large sunlight receptor surface, an abundance of specialized cells containing many chloroplasts, numerous stomata to enable uptake of atmospheric carbon dioxide, and a vascular system to transport water and nutrients into the leaf and export food out.

The products of photosynthesis are generally referred to as *photosynthates* (or assimilates), which include sugar (mostly sucrose) and other carbohydrates. Sucrose is easily transported throughout the plant and can be used directly as an energy source or converted into other carbohydrates, proteins, fats, and other compounds. The synthesis of other compounds often requires the combination of carbon (C) based products with mineral nutrients such as nitrogen, phosphorus, sulfur, iron, and others that are taken up by the roots. Starch, a carbohydrate, is the principal form of food energy that the vine stores in reserve for later use. The carbohydrates cellulose and hemicellulose are the principal structural materials used to build plant cells. Organic acids (malic, tartaric, citric) are another early product of photosynthesis and are used directly or converted into amino acids by the addition of nitrogen. Amino acids can be stored or combined to form proteins.

Photosynthates are the food energy used to fuel plant growth and maintain plant function. The allocation of photosynthates to different parts of the vine is described in terms of "sources" and "sinks." Leaves are the *source* of photosynthates, and any plant part—such as shoots, fruit, or roots—or metabolic process that utilizes photosynthates is considered a *sink*. The amounts of food materials moved to different points of need (sinks) varies through the season, depending upon photosynthate production and demand from the various plant parts (Williams, 1996). Thus, the majority of foods and food materials are first sent to actively growing areas such as shoot tips, developing fruit, and root tips. Later, when growth has slowed and a full canopy is producing more photosynthates than are demanded by growing points, increasing quantities of food are directed to the roots, trunk, and other woody tissues for storage as reserves. However, during the ripening phase of fruit development, the fruit cluster is the main sink for photosynthates, and only surpluses go to reserves. After harvest, all woody tissues, especially roots, are the principal sinks. Food reserves in the roots and woody parts of the vine provide the energy for initial shoot growth in the spring, before new leaves are capable of producing more food than they consume.

Sunlight. The process of photosynthesis is obviously dependent upon sunlight, and it is generally assumed that between one-third and two-thirds full sunlight is needed to maximize the rate of photosynthesis. The optimization of sunlight captured by the vine is an important component of canopy management that not only affects the rate of photosynthesis but also directly influences fruit quality. Sunlight exposure on a vine is highly dependent upon the training system and the shoot density and can be influenced by the orientation of the rows and row spacing. The term *canopy management* encompasses many vineyard practices designed to optimize the sunlight exposure of the grapevine.

Other Environmental Influences. The rate of photosynthesis in grapevines is also influenced by leaf temperature; the apparently broad optimum range of 25–35°C (77–95°F) may be attributable to differences in grape variety, growing conditions, or seasonal variation (Williams et al., 1994). Leaf temperature can be highly dependent upon vine water status but otherwise cannot be influenced to the same extent as sunlight exposure in the canopy, so it is of less concern to vineyard management.

Water status of grapevines can have a strong impact on photosynthetic rate through its control over the closing of leaf stomata, the sites of gas exchange critical for photosynthesis. A *water deficit* exists when the plant loses more water (via transpiration, described below) than it takes up from the soil. One consequence of water deficits is the closure of stomata, which reduces water loss but also reduces the uptake of CO_2 necessary for photosynthesis. The extent of stomatal closure, and therefore the impact on photosynthetic rate, is related to the severity of water deficit. Vines are considered to be under *water stress* when the deficit is extreme enough to reduce plant functions significantly. The major impact of water deficits on vine photosynthesis is the reduction of leaf area (Williams, 1996).

Inadequate supply of certain nutrients (nitrogen and phosphorus) may also limit photosynthesis directly, or indirectly by reduced availability of elements (iron and magnesium) for the synthesis of chlorophyll.

Respiration

The process by which the stored energy within food is released for the plant's use is called *respiration.* In simple terms, the end result can be considered to be the reverse chemical reaction of the photosynthetic process, although the multiple reactions and sites of activity are completely different. Respiration involves the reaction of oxygen with the carbon and hydrogen of organic compounds, such as carbohydrates, to form water and CO_2 and release energy. Many forms of carbohydrates, including sugars and starch, can be oxidized (broken down) by respiration, as can fats, amino acids, organic acids, and other substances. The decrease of malic acid, and to a lesser extent tartaric acid, in ripening fruit is largely attributed to respiration.

The respiration process has been reviewed by Mullins et al. (1992), and this summary is primarily based upon their review. Respiration can be considered to perform two functions: supplying energy for growth, and supplying energy for organ maintenance. It is probable that a large portion of the daily photosynthate produced by a grapevine is consumed in maintenance respiration. The food energy demands of maintenance respiration are considerable even during times of little vine growth, and it is significant that respiration, unlike photosynthesis, occurs continuously. The energy derived from maintenance respiration is used to meet the demands of many physiological processes, including carbohydrate translocation, protein turnover, nitrogen assimilation, and nutrient uptake in the roots. The synthesis of substances integral to vine maintenance and growth, including proteins, enzymes, colors, aromas, flavors, acids, and tannins, is fueled by respiration.

The rate of maintenance respiration is dependant upon grapevine size, whereas growth respiration rates vary with the level of growth activity. Temperature is the most influential environmental factor affecting the rate of respiration. Increasing temperatures cause a progressive increase in respiration rate up to a point where tissue damage occurs. At 50°F (10°C), respiration of a mature grape leaf is close to zero, but respiration rate approximately doubles with every 18°F (10°C) increase in temperature.

Translocation

The long-distance movement of water, mineral nutrients, food, and other materials through the vascular system is called *translocation.* Water and dissolved mineral nutrients absorbed by the roots are moved upward in the xylem to all parts of the grapevine. The phloem is the conduit primarily for food materials and their derivatives to be moved throughout the plant.

Movement of photosynthates in the phloem throughout the growing season has been described by Kliewer (1981) and Williams (1996) and is summarized here. Beginning at budburst and continuing for about two to three weeks, carbohydrates and nitrogenous compounds are moved upward from their storage locations in roots and woody parts of the vine to support the new shoot growth. When the shoot and leaves develop to the point that some leaves (those greater than 50% of their final size) produce more photosynthates then they consume, food materials begin to move in both directions in the phloem. Mature leaves from the apical portion of the shoot supply the growing shoot tip, and the remaining leaves export photosynthates out of the shoot to the parent vine: canes, arms, trunk, and roots. This pattern continues until about bloom, when growth from the shoot tip generally begins to slow down. From fruit set until the beginning of fruit ripening, photosynthates move primarily to three sinks: shoot tip, fruit cluster, and the parent vine. The fruit cluster is the primary sink from the start of ripening until harvest; the parent vine and growing tips of primary and lateral shoots are weaker sinks. After harvest most of the photosynthates moves out of the shoot into the storage reserve parts of the vine: roots and woody tissues. Generally there is a period of root growth after harvest, so the growing root tips would further favor carbohydrate movement to the roots.

In grapevines, sucrose is the main carbohydrate translocated, so starch and other carbohydrates must first be broken down to release sucrose for transport. Plant hormones, which have a role in controlling plant functions, are also moved through the xylem and phloem. Some cross-movement (radial translocation) of water and materials between the xylem and phloem occurs through vascular rays, which also function as storage sites for food reserves.

Mineral nutrients absorbed by the roots (see discussion below) are moved into the xylem of the root, and from there they are translocated upward to the shoot and distributed in the plant to the areas of use. Nutrient reserves are stored in the roots and woody parts of the vine and are remobilized and translocated in the phloem when uptake from soil is inadequate to meet the current need. Remobilization from storage reserves is an important source of nutrients, especially nitrogen, during the early stages of shoot growth in the spring, before roots have begun active growth.

Transpiration

Transpiration is the loss of water, in the form of vapor, through open stomata. Stomatal pores open into the empty spaces between mesophyll (interior cells) cells of the leaf. This creates an uninterrupted path between the outside environment and the inner environment of the leaf. The outside environment almost always has a lower relative humidity than the protected interior of the leaf, which is assumed to be 100%. Thus a vapor pressure gradient exists, causing water vapor to move out of the leaf from the area of high vapor pressure (high water content) to the area of lower vapor pressure. When the thin-walled mesophyll cells lose water from transpiration, their absorptive power is increased due to concentration of the dissolved solids in the cell sap and partial drying of solid and semisolid materials of the cell. The partially dried cells then have a greater potential to absorb water, which they obtain from the xylem. Thus, the absorptive force, called *transpirational pull*, is applied to the continuous column of water *(transpiration stream)* in the xylem that extends from the leaves to the roots.

The rate of transpiration is dependent upon the extent to which stomata are open, which is primarily related to light levels and secondarily influenced by external environmental conditions: humidity, temperature, and wind. Stomata can, however, be partially or completed closed in response to varying degrees of water deficit, overriding the influence of light and other environmental conditions. Transpiration also has an evaporative cooling effect on the leaf because water molecules absorb heat energy during the conversion of water from the liquid phase to the gas phase within the leaf.

Absorption of Water and Nutrients

Water. The suction force of transpirational water loss is transmitted throughout the unbroken column of water in the xylem all the way to the roots, providing the major mechanism by which water is taken up from the soil and moved throughout the vine. Water is pulled into the root from the soil. Young roots absorb the majority of water, primarily through root hairs and other epidermal (outer layer) cells. But older suberized ("woody") roots uptake water at a lower, but constant, rate. Water then moves through the cells of the inner tissues of the root and into the xylem ducts, where it continues its movement upward, reaching all parts of the vine, and is eventually lost via the stomata.

The effect of transpiration on the rate and quantity of water uptake is obvious, but new root growth is also necessary because roots eventually deplete the available water in their immediate area and soil water movement is slow at best. Therefore, conditions that influence root growth affect the rate of water uptake.

Nutrients. Mineral nutrients must be dissolved in water for uptake by roots. Nutrient uptake often occurs against a concentration gradient; that is, the concentration of a mineral nutrient in the soil solution is usually much lower than its concentration in root cells. Thus an active process, consuming energy, is required to move nutrients against the concentration gradient. Active transport is a selective method of nutrient uptake, and some nutrients can be taken up in much greater quantity than others. Nitrates and potassium are absorbed several times as rapidly as calcium, magnesium, or sulfate. There are also interactions between nutrient ions that influence their absorption. For example, potassium uptake is affected by the presence of calcium and magnesium. In rapidly transpiring vines, nutrient uptake also occurs by mass flow (a passive process) with water from the soil solution (Mullins et al., 1992).

Major Developmental Processes

Shoot Growth

Shoot growth begins with *budburst* (or *budbreak),* when previously dormant buds begin to grow after they have received adequate heat in the spring. This usually occurs when average daily temperature reaches about 50ºF. Representative stages in the growth and fruiting of a grapevine are illustrated in Figure 5. At budburst, the primary growing point usually contains 10–12 leaf primordia and one or two cluster primordia, located opposite leaf primordia at node positions three to six. Development of these structures continues as the shoot grows out from the bud. Early shoot growth is relatively slow, but soon it enters a phase of rapid growth called the *grand period of growth,* which typically continues until just after fruit set. Even when the shoot is only a few inches long, developing flower clusters can be seen opposite the young leaves.

As each new leaf unfolds, the lateral bud and dormant bud begin to develop in its axil. Some lateral buds in the leaf axils grow into lateral shoots, but many produce only one or a few small leaves, then stop growing. Other laterals grow out to varying lengths. Under some circumstances, such as excessive vine vigor, or in response to summer pruning (tipping or hedging) of primary shoots, the lateral shoot grows out with substantial vigor.

After fruit set, shoot growth generally continues to slow, to a halt or nearly so, by about the time the fruit begins to ripen. Under circumstances of high vigor, however, shoot growth may continue at a steady rate throughout the season. This situation can arise from one or more of the following causes: abundant water, excessive nitrogen fertilization, severe pruning, or extreme undercropping. Smart and Robinson (1991) describe the "ideal" shoot to be 2–3 feet long with 10–15 full-sized leaves.

Flower Cluster Initiation

As the shoot grows, considerable development takes place within the dormant buds in the leaf axils. Of greatest interest is the development of flower cluster primordia, since they represent the fruiting potential of the vine for the following season. The period at which flower cluster primordia begin to form on the rudimentary shoot is called *flower cluster initiation.* The process occurs first in the midsection of the primary shoot at node positions four through eight, beginning soon after bloom of the current season's flower clusters (initiated in the previous season) and continuing for up to six weeks. The buds at basal nodes one to three undergo cluster initiation a little later, and initiation continues progressively in buds toward the growing tip. Usually, by the end of the season, fruitful buds exist along the cane to the extent to where it is fully ripened.

Grape flower initiation is described and illustrated in Mullins et al. (1992) and summarized below. Flower development in *V. vinifera* is described as a three-step process, occurring within the developing dormant buds. The first step is the formation of uncommitted primordia by the growing points of developing dormant buds (which are not dormant at this early developmental stage) in leaf axils of the current season's shoots. The primordia are described as uncommitted at this point because they can develop into either flower clusters or tendrils, depending on environmental and growing conditions experienced by the specific bud and the shoot in general. In the second stage the primordia become committed to becoming a flower cluster or a tendril. Mullins et al. report flower cluster initiation coinciding with the beginning of periderm development on the shoot, but others have found it to begin before bloom with some varieties (L. E. Williams, personal communication) or at about the time of bloom (Winkler et al., 1974). Cluster primordia develop during the current season, and the final step, formation of flowers from the cluster primordia, begins after budburst in the following spring. The later stages of flower development are completed as bloom time is approached.

Sunlight and temperature are the most influential environmental factors on grapevine flower cluster initiation, although opinions vary on which is the dominant factor. According to Williams et al. (1994), the development of uncommitted primordia into either flower clusters or tendrils is dependent upon the amount of sunlight striking the bud during development. The number and size of cluster primordia increase with increasing sunlight levels. Mullins et al. (1992) conclude that it is probable that a combination of exposure to high temperature and high light intensity is necessary for maximum fruitfulness of dormant buds. They also report that sunlight and temperature requirements for initiation of flower cluster primordia are known to vary among varieties. From a vineyard management perspective, it appears that, for a grape variety with demonstrated adaptation to a region's temperatures, sunlight exposure of the developing buds is the most critical concern. Thus training systems and canopy management practices that facilitate good sunlight exposure promote better fruitfulness than those that create conditions of shade.

Dormancy, Acclimation, and Cold Hardiness

In autumn, the vine enters *dormancy*—the stage with no leaves or growth activity, which extends until budburst the following spring. Despite the apparent inactivity of this stage, it can be a critical time for grapevines when they may be exposed to potentially damaging low temperatures. The ability of a dormant grapevine to tolerate cold temperatures is referred to as its *cold hardiness*. Grapevine cold hardiness is a highly dynamic condition, influenced by environmental and growing conditions, and varying among grapevine varieties and tissues and over time. Therefore, cold hardiness *cannot* be viewed or described in absolute terms such as "Variety X is cold hardy to -8°F."

There are three stages of the dormant season: *acclimation*, the period of transition from the non-hardy to the fully hardy condition; midwinter, the period of most severe cold and greatest cold hardiness; and *deacclimation*, the period of transition from fully hardy to the non-hardy condition and active growth.

Acclimation is a gradual process, beginning after shoot growth ceases and continuing through autumn and early winter. The combination of declining day length and decreasing temperatures in autumn are important factors influencing acclimation and cold hardiness. The process of acclimation in grapevines is not well understood, but it involves many simultaneous activities that collectively increase cold hardiness. Water content of some tissues decreases, while increases occur in cells' solute (dissolved solids) concentration, membrane permeability, and the thermal stability of several enzymes.

Howell (2000) has reviewed the mechanisms by which grapevines survive cold temperatures. The primordial tissues of dormant buds survive by avoiding the formation of ice crystals in the tissue by *supercooling*—a process by which a liquid remains fluid below its normal freezing temperature. Other tissues survive by increasing their capacity to tolerate both ice in the tissue and increased concentration of solutes in the cell. Increased solute concentration in the cell lowers its freezing point.

Because of the different mechanisms involved, tissues vary in tolerance to freezing temperatures. Woody tissues of the trunk, cordon, and canes generally have greater cold hardiness than dormant buds and roots. In comparisons of grapevine woody tissues, the vascular cambium is thought to be the last tissue to be damaged by cold temperatures, followed in sequence by younger xylem, older xylem, and phloem (Wample et al., 2000). Within dormant buds, primary buds are typically less cold hardy than secondary buds, and tertiary buds are the most hardy.

Species and varieties of grapes exhibit a broad range of potential cold hardiness based on their inherent genetic characteristics. This fundamental genetic potential for cold hardiness is influenced by both environmental conditions and the circumstances under which the vine grew in the previous season. Poor management practices or growing conditions can inhibit the acclimation process, resulting in reduced cold hardiness. Acclimation is promoted by exposure of shoots and leaves to sunlight and is associated with periderm development and low relative water content. Cold hardiness can vary considerably between and within vines. Reduced hardiness has been associated with large, dense (shaded) canopies, canes with either long internodes or large internode diameter, and canes with large persistent lateral canes. Additionally, heavy fruit loads or defoliation (early leaf fall due to stress, disease, or pest activity) inhibit acclimation, probably through reduced availability of photosynthates. Contrary to popular belief, neither nitrogen fertilization nor irrigation practices reduce grapevine cold hardiness, unless nonstandard practices are used that encourage continued late-season growth, which inhibits acclimation (Wample et al., 2000).

Cold hardiness of buds is fairly stable through the winter months, but sharp increases in temperature can cause buds to deacclimate and lose hardiness, and the extent of deacclimation can vary by variety or species. Bud hardiness has been correlated with air temperature of the preceding five-day period. Cold hardiness decreases as the grapevine rapidly deacclimates in response to warm temperatures in the spring. Deacclimation is much less gradual than cold acclimation in the fall, and the rate of deacclimation accelerates through the dormant season.

Fruit Growth

Berry development commences after successful pollination and fertilization of ovules within a flower. Flowers with unfertilized ovules soon shrivel and die, while those remaining begin growth into berries. Many of these tiny berries, *abscise* (drop off) within the first two to three weeks. Following this drop period (called *shatter*), the retained berries generally continue to develop to maturity. Commonly, only 20–30% of flowers on a cluster develop into mature berries, but this is adequate to produce a full cluster of fruit.

The berry develops from the tissues of the pistil, primarily the ovary. Although pollination and fertilization initiate fruit growth, seed development seems to provide the greatest growth stimulus, as evidenced by the relationship of fruit size to the number of seeds within the berry. The maximum number of seeds is four, but lack of ovule fertilization or ovule abortion reduces the number of developing seeds, generally resulting in smaller berry size.

Berry growth occurs in three general stages—rapid initial growth, followed by a shorter period of slow growth, and finishing with another period of rapid growth. A graph of grape berry growth thus appears as a double sigmoid pattern. Berry growth during the first stage is due to a rapid increase in cell numbers during the first three to four weeks, followed by two to three weeks of rapid cell enlargement. During this stage the berries are firm, dark green in color, and rapidly accumulating acid. Seeds have attained their full size by the end of the first growth stage.

The middle stage, called the *lag phase*, is a time of slow growth. The embryo is rapidly developing within each seed, and the seed coat becomes hardened. Berries reach their highest level of acid content and begin to accumulate sugar slowly. Toward the end of lag phase, berries undergo a reduction in chlorophyll content, causing their color to change to a lighter green.

The final stage of berry growth coincides with the beginning of fruit maturation *(ripening)*. The beginning of ripening, referred to by the French term *veraison*, is discernable by the start of color development and softening of the berry. The color change is most easily visible on dark-colored varieties, but "white" varieties continue to become lighter green, and some varieties turn a yellowish or whitish-green color by harvest. Softening of the berry and rapid sugar accumulation occur abruptly and simultaneously. Berry growth, occurring by cell enlargement, becomes rapid again in this final stage of ripening. It is thought that most of the water entering the berry after veraison comes from phloem sap, since xylem at the junction of the berry and its *pedicel* (stem) appears to become blocked at this time (Coombe, 1992).

During ripening, acid content declines and sugar content increases. It is widely believed that flavors develop in the later stages of ripening. Berries begin to accumulate sugar rapidly at the start of the ripening period, and the rate tends to remain steady until accumulation slows as the end of the maturation period is approached. Sugar is translocated as sucrose to the fruit, where it is quickly converted into glucose and fructose. Both sugars and acids primarily accumulate in cells constituting the *pulp* (flesh) of the berry, although a small amount of sugar accumulates in the skin.

The skin (epidermis) and the thin tissue layer immediately below it contain most of the color, aroma and flavor constituents, and tannins contained in the berry. Thus, all things being equal, small berries have greater color, tannins, and flavor constituents than large berries because the skin constitutes a larger percentage of the total mass of small berries. Seeds also contain tannins that can contribute to the overall astringency of wine.

The chemical composition of grape berries is complex, consisting of hundreds of compounds, many in tiny quantities, which may contribute to fruit quality attributes. The single largest component is water, followed by the sugars fructose and glucose, then the acids tartaric and malic. Other important classes of chemical compounds within grape berries include amino acids, proteins, phenolics, anthocyanins, and flavonols. The reader is referred to a review of the biochemistry of grape ripening by Kanellis and Roubelakis-Angelakis (1993) for a thorough discussion of this topic.

Berries are considered to be fully *ripe* when they achieve the desired degree of development for their intended purpose, and they are generally harvested at this time. Ripeness factors of the fruit that are typically considered when scheduling harvest are sugar content, acid content, pH, color, and flavor. The combination of these factors determines the *fruit quality* of the harvest. Ripening processes in the fruit cease upon harvest, but while fruit is on the vine ripening is a continuous process. So there is usually a short time, influenced by weather, during which the fruit remains within the desired ripeness parameters. Berries can become *overripe* if harvest is delayed until the fruit has developed beyond the desired range of ripeness. Consider also that ripeness parameters can vary considerably depending on the intended use. For example, Pinot noir grapes for sparkling wine production are harvested much earlier, at lower sugar and higher acid content, than Pinot noir for non-sparkling red wine. Thus, the terms "fruit ripeness" and "fruit quality" do not have absolute values but are defined subjectively.

Fruit ripening can be delayed, and the attainment of desired ripeness parameters inhibited, by an excessive *crop load* (amount of fruit per vine). A vine that is allowed to produce more fruit than it can develop to the desired level of ripeness is considered to be *overcropped*. Severe overcropping can negatively impact vine health as well as fruit quality by precluding

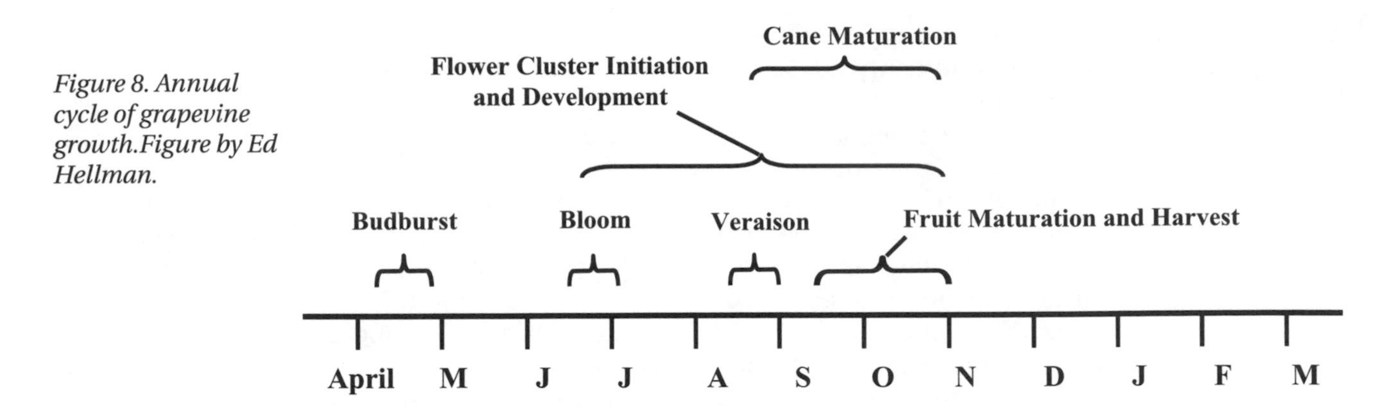

Figure 8. Annual cycle of grapevine growth.Figure by Ed Hellman.

the vine from allocating adequate photosynthates to weaker sinks: shoots, roots, and storage reserves. Viticulturists generally seek to attain *vine balance,* the condition of having a canopy of adequate, but not excessive, leaf area to support the intended crop load to the desired level of fruit ripeness.

Climatic factors, particularly temperature, have long been recognized to have a major influence on the fruit quality of grapes and subsequent wine quality. The principal effect is on the rates of change in the constituents of the fruit during development and the composition at maturity. Hot climates favor higher sugar content and lower acidity; cool climates tend to slow sugar accumulation and retain more acidity. Grape varieties tend to ripen their fruit with a desirable combination of quality components most consistently in specific climates. Thus, some varieties, such as Pinot noir and Gewürztraminer, are considered to be "cool climate varieties," whereas others such as Carignane and Souzão are considered to be "warm" or "hot climate varieties." A few varieties, most notably Chardonnay, are capable of producing high-quality wines in different climates by adjusting the wine style for the varying expression of fruit characteristics in each climate. The relationship of climate, and in particular temperature, to fruit ripening and wine quality has been incorporated into methods of matching grape varieties to climate; Winkler's heat summation (degree days) system for California (Winkler et al., 1974) is one such system, and there are other more elaborate methods (see, e.g., Jackson and Cherry, 1988; Gladstones, 1992). *Phenology* is the study of the relationship between climatic factors and the progression of plant growth stages and developmental events that recur seasonally.

Thus, the first step in the production of high-quality winegrapes is the selection of a site with appropriate climatic characteristics for fruit ripening of the varieties to be grown. Vineyard practices, including training systems and canopy management, are utilized to optimize the sunlight and temperature characteristics of the canopy for fruit ripening. In cool climates, canopy management practices that provide good exposure of leaves and fruit to sunlight have generally improved grape and wine composition. Vines in which the canopy interiors are well exposed to sunlight usually produce fruit with higher rates of sugar accumulation, greater concentrations of anthocyanins and total phenols, lower pH, and decreased levels of malic acid and potassium compared to vines with little interior canopy exposure (Williams et al., 1994). Improved fruit quality under such circumstances may be due to higher temperatures in addition to better sunlight exposure, but it is extremely difficult to separate these factors.

The Annual Cycle of Growth

The annual growth cycle of the grapevine involves many processes and events that have been briefly introduced above. Figure 8 illustrates the sequence of major processes and events in a timeline. It should be recognized that the timing and duration of developmental events are subject to variations due to the grape variety, local climate, and seasonal weather, but the sequence of events remains constant. It is significant that many of these events overlap others for a period of time, requiring the vine to allocate its resources among competing activities. For example, during the time that the vine is developing and ripening the current season's fruit, flower cluster initiation and development is underway in dormant buds and carbohydrates are being moved into storage reserves. Therefore, it is of critical importance for the long-term growth and productivity of grapevines that adequate photosynthates be produced to supply the complete needs of the vine. This goal can be achieved by supplying adequate water and nutrients to the vine, maintaining a healthy canopy, providing good sunlight exposure, and developing an appropriate balance between crop load and canopy size.

Acknowledgment

The author gratefully acknowledges Nick Dokoozlian, Larry Williams, and Robert Wample for their critical review of this chapter.

References

Coombe, B. G. 1992. Research on development and ripening of the grape berry. *American Journal of Enology and Viticulture* 43:101-110.

Esau, K. 1948. Phloem structure in the grapevine and its seasonal changes. *Hilgardia* 18:217-296.

Gladstones, J. 1992. *Viticulture and Environment.* Winetitles, Adelaide, Australia.

Howell, G. S. 2000. Grapevine cold hardiness: mechanisms of cold acclimation, mid-winter hardiness maintenance, and spring deacclimation. Proceedings of the ASEV Fiftieth Anniversary Meeting, Seattle, Washington. American Society of Enology and Viticulture, Davis, Calif.

Jackson, D. I., and N. J. Cherry. 1988. Prediction of a district's grape-ripening capacity using a latitude-temperature index (LTI). *American Journal of Enology and Viticulture* 39:19-26.

Kanellis, A. K., and K. A. Roubelakis-Angelakis. 1993. Grape. In G. B. Seymour, J. E. Taylor, and G. A. Tucker (eds.), *Biochemistry of Fruit Ripening.* Chapman and Hall, London.

Kliewer, W. M. 1981. *Grapevine Physiology: How Does a Grapevine Make Sugar?* Leaflet 21231, Division of Agricultural Sciences, University of California, Davis.

Meier, U. 2001. Grapevine. In Growth stages of mono- and dicotyledonous plants. BBCH Monograph. Federal Biological Research Centre for Agriculture and Forestry, Berlin, Germany.

Mullins, M. G., A. Bouquet, and L. E. Williams. 1992. *Biology of the Grapevine.* Cambridge University Press, Cambridge.

Pratt, C. 1971. Reproductive anatomy in cultivated grapes: a review. *American Journal of Enology and Viticulture* 22:92-109.

Pratt, C. 1974. Vegetative anatomy in cultivated grapes: a review. *American Journal of Enology and Viticulture* 25:131-150.

Richards, D. 1983. The grape root system. *Horticultural Reviews* 5:127-168.

Smart, R., and M. Robinson. 1991. *Sunlight into Wine.* Winetitles, Adelaide, Australia.

Wample, R. L., S. Hartley, and L. Mills. 2000. Dynamics of grapevine cold hardiness. Proceedings of the ASEV Fiftieth Anniversary Meeting, Seattle, Washington. American Society of Enology and Viticulture, Davis, Calif.

Williams, L. E. 1996. Grape. In E. Zamski and A. A. Schaffer (eds.), *Photoassimilate Distribution in Plants and Crops: Source-Sink Relationships.* Marcel Dekker, New York.

Williams, L. E., N. K. Dokoozlian, and R. Wample. 1994. Grape. In B. Schaffer and P. C. Andersen (eds.), *Handbook of Environmental Physiology of Fruit Crops,* Vol. 1: *Temperate Crops.* CRC Press, Boca Raton, Fla.

Winkler, A. J., J. A. Cook, W. M. Kliewer, and L. L. Lider. 1974. *General Viticulture.* University of California Press, Berkeley.

I

Vineyard Planning

Development and management of a vineyard has been compared to raising children. The analogy certainly has merit, because it is not uncommon to hear grape growers lovingly refer to their young vines as their babies. We nourish young vines and train them, and in the early years their performance can be somewhat erratic. Later, as the vines gain maturity, their growth and production habits tend to stabilize and be more predictable. Vineyards, like children, are a long-term commitment—physically, emotionally, and financially.

There are numerous viticulture texts that offer methods and guidelines for growing grapes, but none, including this book, can offer a recipe to be followed for success. Grape growing requires an integration of knowledge of climate, site characteristics, grape varieties, production practices, personnel management, economics, and marketing. Those are just the main categories of expertise required; numerous skills and abilities are also necessary. This book addresses most of these topics, and a recurring theme is that every vineyard is different and requires thoughtful consideration of the significant influential factors. Vineyard management involves a continuous series of decision-making events, and this book offers guidelines on the factors that must be considered when making decisions, as well as descriptions of growing practices and methods to achieve the desired objectives.

Before planting a vineyard, or even purchasing land for a potential site, there are some major factors to consider, the first of which is economics. As a business investment, vineyards generally are not highly profitable, except perhaps when they are sold. The first chapter of this section provides an analysis of the costs involved with establishing and maintaining a vineyard in the Willamette Valley. Many assumptions must be made when conducting such an economic analysis, and the particular set of assumptions used here will not match any existing or future vineyard. Each vineyard will have a unique set of circumstances and numerous decision-making opportunities along the way that affect the economics of the operation. The chapter should be used as an example of the costs and returns associated with vineyards. Prospective growers are encouraged to conduct their own economic analysis prior to investing in vineyard development. Winery operations are not the focus of this book, but many vineyards have plans to eventually establish their own winery. Therefore, an analysis of production costs and returns for an actual winery (a case study, if you can pardon the pun) is provided in Chapter 2.

Once the economic issues have been sufficiently examined and plans for the vineyard continue, the next critical item is to identify a suitable site for the vineyard. The climate and characteristics of a site are major determinants of the grape varieties that can be grown and the potential fruit quality that can be expected. Thus, site selection should always be done with the intended grape varieties in mind. There is much involved in assessing the suitability of a vineyard site, and Chapter 3 provides a thorough discussion of this topic. This is followed by extensive descriptions of the grape-growing regions of Oregon, and descriptions of areas within each region that have potential for vineyard development. The concept of matching grape varieties to the climate and site is the overall theme of Chapter 5. Then, Chapter 6 discusses rootstock varieties and the role that they play in influencing certain vine characteristics, particularly pest resistance. The vineyard planning section concludes with discussions of the important considerations for developing the vineyard design and selection of appropriate training systems. Again, the unique circumstances of each vineyard influence these important decisions.

1

Vineyard Economics

Bart Eleveld, M.Carmo Vasconcelos, and Edward W. Hellman

Oregon's winegrape acreage has expanded dramatically during the past decade or so. As shown in Figure 1 and Table 1, in 1987 there were 3,016 acres of winegrapes harvested in Oregon. By 2000, acres of harvested winegrapes had increased to 8,100 acres (Rowley et al., 2001). This represents a 169% increase in acreage in only fourteen years. The value of farm sales for Oregon winegrapes grew by 471%, which is a reflection of a steady increase in prices received as well as the acreage expansion. Again, the dramatic growth of the industry is apparent.

Vineyard numbers have increased with winegrape acreage and sales. In 1987, 263 vineyards were reported (Rowley et al., 2001). By 2000, Oregon's vineyards had increased to 480. The establishment of new vineyards indicates an increase in growers new to the Oregon winegrape industry.

The rapid expansion of the winegrape industry has created a need to update the vineyard economics report released in 1993, which estimated the costs of establishing and producing winegrapes (Satouf et al., 1993). This study meets that need by estimating costs for a typical, newly established vineyard. Since much of the vineyard expansion is occurring in the Willamette Valley, we assume that is where this typical vineyard is located.

Many individuals were involved in this study, including growers, consultants, researchers, and Extension staff. This study was prepared primarily to benefit new and potential vineyard owners. Existing vineyard owners who are considering expanding their acreage will, however, also find these analyses useful for budgeting and comparison purposes. We address three objectives: (1) an estimation of the economic costs and returns associated with winegrape production; (2) an estimation of the cash flows required to establish a new vineyard; and (3) an estimation of returns per acre for alternative prices and yields in a mature 20-acre vineyard.

Although the first two objectives sound similar, they require two different approaches. For the first, we develop enterprise budgets for winegrape establishment and production to estimate economic costs and returns. Enterprise budgets include variable costs for such items as chemicals, labor, and fuel as well as such fixed costs as depreciation, interest, and land payments. These budgets are useful for comparing the economic profitability of winegrapes with alternative farm and non-farm investments.

For the second objective, we examine the cash requirements of establishing a new vineyard. In this section, cash expenses include operating expenditures for such items as chemicals, labor, and fuel. In addition, purchases of capital assets are included to illustrate the total cash flows required to start a new vineyard. This cash flow analysis strictly examines the flow of funds in the vineyard and provides no information as to vineyard profitability. Cash flow budgets enable a manager to plan for financing and to gain control over the cash position of the vineyard.

In our final section, a sensitivity analysis evaluates the effects of changes in prices and yields on projected returns. This illustrates the impacts of production and marketing risk on profitability.

Assumptions

Our analyses necessarily create an example vineyard scenario that does not correspond to any specific real situation. Considerable variations typically exist in vineyard site conditions, personal choices, costs, and returns. This study can be used as a general guide to

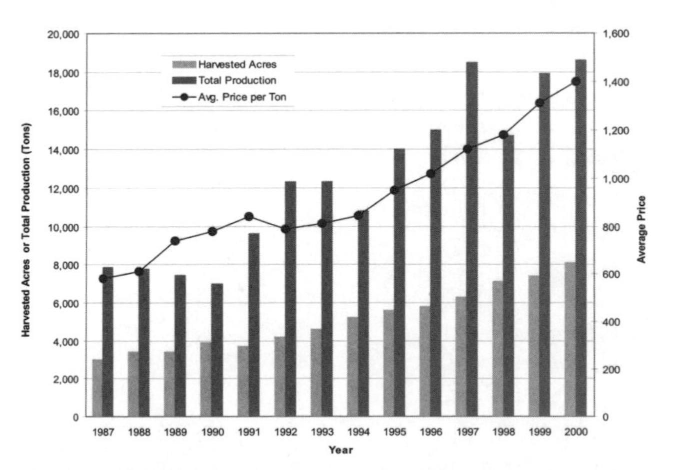

Figure 1. Oregon winegrape acreage, production, and price, 1987–2000.

the economics of vineyard establishment and production in the Willamette Valley, but both prospective and actual vineyard operators should conduct their own economic analysis to use as a tool for business management.

We make several broad assumptions to provide a common basis for analysis. First, we assume we are establishing a new vineyard. All prices for capital assets and resources are in year 2000 dollar values. All machinery, equipment, and improvements are solely for winegrape production. A 30-acre site is chosen, of which 20 acres will be planted. The remaining 10 acres are for roads, buildings, and access areas. The land costs $8,000 per acre, for a total of $240,000. The land is purchased bare and requires a minimal amount of clearing.

Next, we assume the vineyard will produce Pinot noir, Pinot gris, and Chardonnay grapes grown on a rootstock. It should be recognized that the three varieties have different average yields and prices. Therefore, the distribution of varieties among the 20 acres will influence the total potential return for the vineyard. Furthermore, price received varies with supply and demand and is particularly sensitive to fruit quality. Fruit quality is a function of site characteristics, vineyard management, and weather. The yield and price assumptions used in this analysis are based on recent average yields and prices for these varieties. It is assumed that the vineyard consists of 12 acres of Pinot noir and 4 acres each of Pinot gris and Chardonnay.

The vineyard is expected to be in production for 25 years after a 5-year establishment period. Vines are trained to a standard vertical shoot-positioned upright trellis. The operations and resources are typical of similar vineyards in the Willamette Valley.

All labor for winegrape establishment and production is valued at $10.50 per hour. This wage rate is the net cost to growers for hired labor, which is paid at $8.00 per hour, with an additional $2.50 per hour for payroll expenses (withholding taxes, record keeping, preparing W-2 forms, etc.). Alternatively, this wage rate can represent the opportunity cost of labor performed by the vineyard grower. Individuals engaged in various vineyard operations could work elsewhere for $10.50 per hour. In either case, labor is a resource used and must be paid. The cash flow analysis assumes all labor is hired as a cash expense.

Machinery and equipment operation costs are based on agricultural engineering estimates (McGrann, 1986). Purchase prices, salvage values, useful lives, average annual hours of use, and field capacities were obtained from growers and machinery dealers.

Opportunity costs of capital are charged at a rate of 10% for operating capital as well as intermediate and long-term capital provided by the owner. Operating capital is treated as a variable, cash expense; intermediate and long-term capital interest is treated as a fixed, non-cash expense.

Managers or consultants are commonly hired to establish and operate vineyards. We include a cost of $300 per acre, which is on the low end of going rates. Some growers hire managers only during the first few years of the establishment process.

Using these general assumptions, we develop enterprise budgets for each of the vineyard establishment years plus a typical full production year.

Table 1. Oregon Winegrapes Harvested Acreage, Yield, Production, Price, and Value of Farm Sales, 1987–2000

Year	*Harvested Acres*	*Yield per Harv. Acre*	*Total Production*	*Ave. Price per Ton*	*Value of Production*
1987	3,016	2.61	7,861	580	4,559
1988	3,400	2.28	7,750	610	4,728
1989	3,400	2.19	7,450	740	5,513
1990	3,900	1.79	7,000	780	5,460
1991	3,700	2.59	9,600	840	8,064
1992	4,200	2.93	12,300	790	9,717
1993	4,600	2.67	12,300	810	9,840
1994	5,200	2.08	10,800	845	9,126
1995	5,600	2.50	14,000	950	13,300
1996	5,800	2.59	15,000	1,020	15,300
1997	6,300	2.94	18,500	1,120	20,720
1998	7,100	2.07	14,700	1,180	17,346
1999	7,400	2.42	17,900	1,310	23,449
2000	8,100	2.30	18,600	1,400	26,040

Source: Rowley et al. (2001), and earlier editions of the "Oregon Vineyard Report."

Enterprise Budgets

An enterprise budget includes *all costs and returns* associated with producing one enterprise in some manner (Cross and Eleveld, 1988). In this study, winegrapes are the enterprise, and separate budgets are developed for each establishment year and a typical production year. The budgets are presented in an operations format, in which each operation is listed in the order it is performed. Costs are broken down by labor, machinery, and materials expenses for each operation. Each enterprise budget is for a calendar year, and all budgets are prepared as of the end of their respective years.

Table 2 is a summary of all the enterprise budgets. The detailed budgets are presented in Tables 4–9. Please note that the total cost of establishing the vineyard is amortized over its productive life. This is explained in greater detail under "Production Years." Production cycle profit is simply the annual profit during a typical full production year multiplied times the number of full production years.

Field Preparation: Year 0

Site acquisition and preparation take place during the year preceding the planting of vines. We refer to this establishment year as year 0 because the vines are not planted at this time. The budget for year 1 includes the first growing year for the vines, year 2 is the second growing year or second leaf, and so on.

Several improvements are made before any land preparation begins. First, a road is established by hiring a road grader and spreading rock or gravel. We assume the road is 1/4-mile long, and road establishment costs are $2.50 per foot. Total road grading and rock costs are $3,300. A well is also required, at a cost of $10,000 including pump and wiring. A 30- by 80-foot machine shed is constructed for $10,000. Electrical and telephone utilities are run 1/4-mile to the site and cost $5 per foot, or $6,600 for installation. The electrical cost assumes a single-phase 220-volt power source is sufficient for the vineyard. Keep in mind that a winery will likely require a three-phase, 440-volt power source, which costs substantially more than the smaller power source we used.

Table 3 highlights the annual costs associated with the machine shed and other site improvements. These costs are included on a per acre basis in Tables 4–9, which list the enterprise budget for years 0 through full production. Site improvement costs are spread over a 27–30-year period (depending on when they enter service) to show they are "used up" over 5 years of establishment and 25 years of production. To estimate economic costs and returns, we must charge to the winegrapes the portion of site improvements that are "used up" during each year of establishment and production.

Land preparation can begin while the site is being improved. Starting with previously farmed land can reduce development expenses substantially. Clearing

Table 2. Summary of Winegrapes Establishment and Production Budgets "Break-even Calculator" in spread-sheet form.

Crop and location	***Wine Grapes***
Number of full production years	***25.0***
Interest rate for ammortization	***10.00***

Item	Establishment Year 0	Year 1	Year 2	Year 3	Year 4	Full Production Year(s)
Price	$1,550.00	$0.00	0.00	1,550.00	1,550.00	1,550.00
Yield (Tons)	0.00	0.00	0.00	1.50	2.00	2.50
Gross revenue	**0.00**	**0.00**	**0.00**	**2,325.00**	**3,100.00**	**3,875.00**
Non-yield-related variable costs	727.98	5,802.91	1,636.73	2,133.97	2,384.46	2,514.48
Yield-related var. costs per ton	0.00	0.00	0.00	279.27	223.20	219.56
Total variable costs, current year	**727.98**	**5,802.91**	**1,636.73**	**2,552.87**	**2,830.86**	**3,063.38**
Returns above variable cost	-727.98	-5,802.91	-1,636.73	-227.87	269.14	811.62
Fixed costs, current year	**1,650.40**	**1,957.98**	**2,335.32**	**2,416.22**	**2,443.00**	**2,464.94**
Net return, current year	**-2,378.38**	**-7,760.89**	**-3,972.05**	**-2,644.09**	**-2,173.86**	**-1,653.32**
Interest carryover from prev. years		-237.84	-1,037.71	-1,538.69	-1,956.96	
Gain (loss) + interest carryover		-7,998.73	-5,009.76	-4,182.78	-4,130.82	
Cumulative establishment cost	-2,378.38	-10,377.11	-15,386.87	-19,569.65	-23,700.47	
Amortized establishment payment					-2,611.04	-2,611.04
Annual Profit						-4,264.36
Production Cycle Profit						-106,608.88

Yearly interest carryover

Amortized establishment costs are deducted as a fixed cost during full production years

Table 3. Annual Costs of Site Improvements and Miscellaneous Equipment

	Machine Shed	*Irrigation System*	*Trellis System*	*Small Tools*	*Bird Repeller*
Year entering service	0	1	2	0	3
Variable Cost					
Repairs & maintenance	68	52	62		
Utilities	250				
Fixed Cost					
Depreciation	300	1,288	1,786	429	65
Interest	985	3,671	4,911	279	167
Taxes & insurance	300	374	500	30	17
TOTAL	1,903	5,385	7,259	738	249

Table 4. Winegrape, Establishment Year 0, $/Acre Economic Costs and Returns

VARIABLE COST Description	*Labor*	*Machinery*	*Materials*	*Total*
LAND PREPARATION				
Plow, Custom	0.00	0.00	15.00	15.00
Disc, Custom	0.00	0.00	12.00	12.00
Soil Analysis	0.00	0.00	9.00	9.00
Nematodes (0.2 smpl x 20.00 = 4.00)				
Nutrients (0.1 smpl x 50.00 = 5.00)				
Lime, Custom	0.00	0.00	80.00	80.00
Disc, Custom	0.00	0.00	12.00	12.00
Tile, Custom	0.00	0.00	150.00	150.00
Total LAND PREPARATION				278.00
General Labor	73.50	0.00	0.00	73.50
Vineyard Management	300.00	0.00	0.00	300.00
Machine Shed	3.38	0.00	12.50	15.88
Pickup	0.00	21.15	0.00	21.15
Miscellaneous	0.00	0.00	25.00	25.00
Operating Capital Interest	0.00	0.00	14.45	14.45
Total VARIABLE COST				727.98

FIXED COST Description	*Unit*	*Total*
CASH Cost		
Insurance	acre	44.75
Property Taxes & License	acre	30.00
Total CASH Cost		74.75
NONCASH Cost		
Land Interest Charge	acre	1200.00
Depreciation & Interest	acre	375.65
Total NONCASH Cost		1575.65
Total FIXED Cost		1650.40
Total of ALL Cost		**2378.38**

rocks and trees may cost an additional $700–$1,000 per acre. Such costs are not included in our establishment costs.

The site we purchased was clear, and we started our land preparation by hiring a custom operator to plow and disc the land. Next, soil analyses were obtained for essential nutrients, pH, and nematodes. A soil test for nutrients and pH analysis costs $50.00 per sample, and two samples are recommended from a 20-acre site. Nematode analysis costs $20 per sample, and four samples are recommended by the OSU Plant Pathology Laboratory. If dagger nematodes are detected, soil fumigation may be necessary. An additional $700–$1,200 per acre expense would be required in year 0 if soil fumigation was required on your site (Cross, 1989).

We assume our site requires a lime application of 2 tons per acre. The site is disced again after the lime application. We assume drainage tiles will be required on 10% of the site at a cost of $3,000, or $150 per acre.

A pickup is driven 5,000 miles per year for various tasks relating to the vineyard. Seven hours of general labor is hired for each acre to account for time repairing and maintaining improvements, monitoring and controlling pests, and in later years canopy management, crop adjustment, repairing trellises, and monitoring fruit ripening. It is assumed the general labor is the main driver of the pickup truck, and therefore there is no operator labor cost included with the pickup truck operation. A miscellaneous charge of $25 per acre is included each year to cover office supply, seminar, and bookkeeping expenses.

Operating capital interest is charged on variable costs, based on the dates the expenses were incurred and assuming that all operating credit is repaid in the fall. This charge is included to reflect that money invested in winegrapes could be invested elsewhere and earn interest. The vineyard must provide at least as great a return as your best alternative to be a sound investment. The same reasoning applies to intermediate and long-term capital.

Total variable cash expense in year 0 is $728 per acre. The remaining costs are fixed and include cash and non-cash expenditures. Cash fixed expenses include insurance on machinery, equipment, and site improvements and property taxes. Machinery and equipment needed in year 0 include a 35-horsepower tractor and a pickup. Total fixed cost is $1,650 per acre, and total cost in year 0 is $2,378 per acre.

Establishment: Year 1

The final site improvement is the addition of an irrigation system in year 1. Although many older Oregon vineyards were established without irrigation, it is becoming increasingly important to have the capability to apply water when needed. Closer vine spacing and drought-intolerant rootstocks make irrigation a critical tool for consistent annual production of high-quality fruit. Irrigation materials cost is assumed to be $24,900 and labor costs are $12,450.

Table 5 shows the enterprise budget for establishment costs in year 1, the year in which grapevines are planted. Before planting, decisions about trellis type and spacing must be made. In our vineyard, we chose 7-foot row spacing with 5 feet between vines. This spacing requires 1,245 vines per acre. Vines cost $3.10 each, assuming they are rootstock-grafted vines. Special clones or rootstocks in short supply may be more expensive. Grape phylloxera, a root-feeding, aphid-like insect, is well established in the Willamette Valley and is present in other production regions. The only current method of control for this pest is to plant vines grafted onto resistant rootstock. Self-rooted European winegrapes, although they cost only $1.00 per vine, are susceptible and will die should an infestation occur. Planting of self-rooted vines is not recommended.

Planting costs include the cost of each vine plus $0.05 per vine to prepare it for planting. Approximately 27 vines can be hand planted per hour. This means planting requires 46.1 hours of labor per acre.

The vineyard row middles are rototilled after planting. Cultivation prepares the vineyard for an annual cover crop and also provides weed control. Additional weed control includes strip spraying in the row middles and spot spraying with an herbicide in the vine rows. Hoeing may also be required. A light application of nitrogen fertilizer is applied to encourage growth in the first year. About 1 hour of labor is required per acre for rodent control.

Plastic grow tubes are used to assist vine establishment in the first year. These cost $0.50 per vine and about $109 (120 tubes per hour) of labor per acre to put them on the vines and an additional $109 to take them off at the end of the season. Grow tubes are removed after the first season, and it is assumed they can be resold for half the original purchase price. Therefore, the budgeted price for grow tubes is $0.25 per vine.

An annual cover crop is planted in the fall, with a rented planter. Seed costs about $10 per acre. General labor, miscellaneous, management, and operating capital interest expenses are included as discussed for the year 0 budget. Total variable cost is $5,803 per acre.

Fixed costs are the same as in year 0 for all costs except machinery and equipment depreciation, interest, and insurance and interest on investment. Machinery and equipment expenses increase due to purchases of a sprayer and a tiller.

Our establishment year budgets do not include an interest charge on the previous year's expenditures.

Table 5. Winegrape, Establishment Year 1, $/Acre Economic Costs and Returns

VARIABLE COST				
Description	*Labor*	*Machinery*	*Materials*	*Total*
Plant Vines	484.05	0.00	3921.75	4405.80
Vines (1245 vines x 3.10 = 3859.50)				
Vine prep. (1245 vines x 0.05 = 62.25)				
Cultivate	25.41	12.90	0.00	38.31
Grow Tubes (1245 vines x 0.25 = 311.25)	108.89	0.00	311.25	420.14
Strip Spray	5.08	2.45	20.00	27.53
Herbicide (1 acre x 20.00 = 20.00)				
Fertilizer (nitrogen) (38 lbs (21-0-0) x 0.14 = 5.32)	5.08	2.45	5.32	12.85
Spot Spray	36.58	2.45	10.00	49.03
Herbicide (1 acre x 10.00 = 10.00)				
Rodent Control	10.50	0.00	0.00	10.50
Cultivate	12.71	6.45	0.00	19.16
Plant Annual Cover Crop	10.16	4.38	14.00	28.54
Planter Rent (1 acre x 4.00 = 4.00)				
Seed (1 acre x 10.00 = 10.00)				
Remove Grow Tubes	108.89	0.00	0.00	108.89
General Labor	73.50	0.00	0.00	73.50
Vineyard Management	300.00	0.00	0.00	300.00
Machine Shed	3.38	12.50	0.00	15.88
Pickup	0.00	21.15	0.00	21.15
Miscellaneous	0.00	0.00	25.00	25.00
Operating Capital Interest	0.00	0.00	246.63	246.63
Total VARIABLE COST				5802.91

FIXED COST		
Description	*Unit*	*Total*
CASH Cost		
Insurance	acre	65.84
Property Taxes & License	acre	30.00
Total CASH Cost		95.84
NONCASH Cost		
Land Interest Charge	acre	1200.00
Depreciation & Interest	acre	662.14
Total NONCASH Cost		1862.14
Total FIXED Cost		1957.98
Total of ALL Cost		**7760.89**

But the cost of each year's expenditures must be regarded as an investment for future income that should include interest as an opportunity cost as those expenses accumulate. Table 2 summarizes all the establishment year budgets as well as the full production year expenses and returns. One can see that, for each establishment year, an interest cost is calculated for the accumulated investment as of the end of the preceding year (indicated by the curved arrows). These interest costs are not included in the annual budget tables, but they are included in the calculation of the amortized establishment cost, discussed below.

Total establishment cost in year 1 is $7,761 per acre. The cumulative establishment cost, including opportunity cost interest, for years 0 and 1 is $10,377 per acre.

Establishment: Year 2

The enterprise budget for year 2 is shown in Table 6. Vines are pruned in the spring, using hand labor at 11 hours per acre. A strip spray herbicide is used for weed control. An estimated 1% of the vines are replanted to replace dead and damaged vines.

At this point, the trellis is installed. We included the cost of the trellis, $2,500 per acre including wires and installation, in Table 3. Trellis cost is based on 400-foot rows with 7 feet between the rows. Treated wooden posts are placed every 25 feet, with ground anchors on the end posts. Two permanent high-tensile wires are installed at this time, and 5/16-inch bamboo sticks are included for training vines.

Throughout the spring and summer, rodent control and cultivation operations are performed as necessary. The vines are suckered and trained three times. The first time, disbudding and training is performed by hand labor at 50 vines per hour. The second and third times through the vineyard, 100 vines are disbudded and trained per hour. The total labor requirement for disbudding and training is just under 50 hours per acre. Training ties attach the vines to the trellis. Hoeing, requiring about 20 hours per acre, controls weeds in the rows.

Disease control efforts are begun with a spray program to control powdery mildew. Sulfur is applied three times, and a synthetic fungicide is applied twice.

Remaining operations and variable costs are the same as in year 1, including planting a cover crop, general labor, miscellaneous expenses, management, and operating capital interest. Total variable cost in establishment year 2 is $1,637 per acre.

Fixed costs are calculated the same as in year 1. Insurance, depreciation, and interest on site improvements all increase compared to year 1 due to the added trellis expense of $50,000. Machinery and equipment expenses increase due to purchases of an air-blast sprayer for $5,000. Total fixed cost in year 2 is $2,335 per acre. The total budgeted cost for establishment year 2 is $3,972 per acre. Cumulative year 0, 1, and 2 cost, including interest, is $15,387 per acre.

Establishment: Year 3

In establishment year 3, vines have reached their third leaf. A small grape harvest is expected in year 3, with yields increasing in year 4 and full production in year 5. Table 7 contains the enterprise budget for year 3. Canes are tied after pruning, and the canes removed are flailed with a flail-chop mower.

Seven spray applications are performed with three applications of sulfur, three applications of synthetic fungicide to prevent powdery mildew, and one application of another fungicide to protect the grapes from bunch rot. Vines are sprayed if it appears the grape yield will provide returns greater than spray and harvest expenses. Three applications of a soluble boron fertilizer are made prior to bloom to encourage good fruit set. Boron fertilizer is tank-mixed and applied along with a fungicide.

Vines are trained throughout the summer. Vines are suckered and disbudded twice, requiring 30 hours of labor per acre. Moving the catch wire the first time and tucking the vines takes 4 hours per acre. The catch wire is moved and the vines are tucked a second and third time, requiring 6 and 8 hours per acre, respectively. Total labor required for training is 48 hours per acre.

After training, the vines are trimmed with a hedger. Crop levels are adjusted by cluster thinning to a target yield of 1.5 tons per acre, requiring 15 hours of labor. Installation of bird netting is an expensive practice for bird control. The alternative used in this budget is electronic bird repellers. A 20-acre vineyard would require four units, which entails a capital investment of $1,700 that would have to be replaced every ten years.

All harvest costs in year 3 are based on a yield of 1.5 tons per acre. Grapes are picked by hand, and pickers are paid $200 per ton. This rate is above the typical rate paid in a mature vineyard because of the low yield. Pickers must be paid more per pound because it takes longer to pick each pound of grapes. Grapes are picked into buckets, then placed in totes which are typically supplied by the wineries.

Totes are moved to the loading area with a tractor and loader (fork attachment). A trucking firm is hired to load the totes and deliver them to a winery for $30 per ton. A $25.00 tax must be paid on each ton of grapes sold. Total harvest expenses are $419 per acre. Note that if you expect your yield to be only 0.5 tons of grapes in year 3, it may not make economic sense to harvest your grapes.

Total variable cost in year 3 is $2,553 per acre. The income from 1.5 tons of grapes is $2,325, leaving a difference of -$228 per acre. Thus, 1.5 tons of grapes does not cover the variable costs of establishment in year 3, but to be worth harvesting it need only exceed the variable costs of harvesting the crop and preserving the grapes (disease control).

Fixed costs increase in year 3 due to purchases of equipment and site improvements. The hedger purchased in year 3 is a hydraulically controlled, scissor-cut hedger that trims one side of the vine row at a time. It costs $8,000. A front-end loader, a 3-point fork attachment, and a flail mower cost $3,500, $135, and $3,500, respectively. These four additional pieces of equipment increase the insurance, depreciation, and interest costs of machinery and equipment use. With these additions, fixed cost is $2,416 in year 3.

Total cost in year 3 is $4,969 per acre. This cost is offset by the $2,325 return per acre in grape sales, leaving a net cost of $2,644 per acre. Cumulative establishment cost including year 3 is $19,570 per acre.

Establishment: Year 4

Year 4 is the final year of establishment. Table 8 shows the enterprise budget for year 4. This budget differs from year 3 in only a few respects.

An additional labor operation is required after pruning to pull brush away from the vines (10 hours per acre). Powdery mildew is controlled by five applications of sulfur and three applications of

Table 6. Winegrape, Establishment Year 2, $/Acre Economic Costs and Returns

VARIABLE COST *Description*	*Labor*	*Machinery*	*Materials*	*Total*
Prune	115.50	0.00	0.00	115.50
Strip Spray in Row	5.08	2.45	12.00	19.53
Herbicide (1 acre x 12.00 = 12.00)				
Replant Vines (1%)	48.09	0.00	40.30	88.39
Vines (13 vines x 3.100 = 40.30)				
Rodent Control	10.50	0.00	0.00	10.50
Cultivate (3 times)	38.12	19.35	0.00	57.47
Disbud & Train	522.90	0.00	6.00	528.90
Training Ties (1 acre x 6.00 = 6.00)				
Hoeing	210.00	0.00	0.00	210.00
Mildew Control				
Sulfur Spray (3 times) (12 lbs x .70 = 8.40)	15.25	7.34	8.40	30.98
Synthetic Fungicide (2 appl. x 20.00 = 40.00)	10.16	4.89	40.00	55.06
Plant Annual Cover Crop	10.16	4.38	14.00	28.54
Planter Rent (1 acre x 4.00 = 4.00)				
Seed (1 acre x 10.00 = 10.00)				
Irrigation Maintenance	10.50	0.00	10.00	20.50
Vineyard Management	300.00	0.00	0.00	300.00
General Labor	73.50	0.00	0.00	73.50
Machine Shed	3.38	12.50	0.00	15.88
Pickup	0.00	21.15	0.00	21.15
Miscellaneous	0.00	0.00	25.00	25.00
Operating Capital Interest	0.00	0.00	35.83	35.83
Total VARIABLE COST				1636.73

FIXED COST *Description*	*Unit*	*Total*
CASH Cost		
Insurance	acre	91.97
Property Taxes & License	acre	30.00
Total CASH Cost		121.97
NONCASH Cost		
Land Interest Charge	acre	1200.00
Depreciation & Interest	acre	1013.35
Total NONCASH Cost		2213.35
Total FIXED Cost		2335.32
Total of ALL Cost		**3972.05**

Table 7. Winegrape, Establishment Year 3, $/Acre Economic Costs and Returns

GROSS INCOME				
Description	*Quantity*	*Unit*	*$/Unit*	*Total*
Grapes	1.5	ton	1550.00	2325.00
Total GROSS INCOME				2325.00

VARIABLE COST				
Description	*Labor*	*Machinery*	*Materials*	*Total*
Prune	157.00	0.00	0.00	157.00
Tie Canes	105.00	0.00	6.00	111.00
Training Ties (1 acre x 6.00 = 6.00)				
Flail Canes	16.94	9.31	0.00	26.25
Install Catch Wire	126.00	0.00	0.00	126.00
Rodent Control	10.50	0.00	0.00	10.50
Flail Mow	16.94	9.31	0.00	26.25
Cultivate	38.12	19.35	0.00	57.47
Spot Spray	36.58	2.45	30.00	69.03
Herbicide (1 acre x 30.00 = 30.00)				
Mildew Control (3 times)	15.25	7.34	8.40	30.98
Sulfur (12 lbs x .70 = 8.40)				
Fertilizer (Boron, 3 times)	0.00	0.00	3.75	3.75
(3 appl. tank-mixed with mildew control x 1.25 = 3.75)				
Rot Control	5.08	2.45	37.00	44.53
Fungicide Spray (1 appl. x 37.00 = 37.00)				
Mildew Control	15.25	7.34	60.00	82.58
Synthetic Fungicide Spray (3 appl. x 20.00 - 60.00)				
Plant Annual Cover Crop	10.16	4.38	14.00	28.54
Planter Rent (1 acre x 4.00 = 4.00)				
Seed (1 acre x 10.00 = 10.00)				
Training	504.00	0.00	0.00	504.00
Hedging	12.71	7.76	0.00	20.47
Cluster Thinning	157.50	0.00	0.00	157.50
Bird Harassment	21.00	0.00	55.00	76.00
Pyrotechnics (1 acre x 55.00 = 55.00)				
Drop Wires	63.00	0.00	0.00	63.00
HARVEST				
Picking, Custom	300.00	0.00	0.00	300.00
Tractor & Loader	25.41	10.99	0.00	36.40
Load & Haul, Custom	0.00	0.00	45.00	45.00
Grape Tax	0.00	0.00	37.50	37.50
Total HARVEST				418.90
Miscellaneous	0.00	0.00	25.00	25.00
Irrigation Maintenance	10.50	0.00	10.00	20.50
Vineyard Management	300.00	0.00	0.00	300.00
General Labor	105.00	0.00	0.00	105.00
Machine Shed	3.38	12.50	0.00	15.88
Pickup	0.00	21.15	0.00	21.15
Operating Capital Interest	0.00	0.00	51.59	51.59
Total VARIABLE COST				2552.87
GROSS INCOME minus VARIABLE COST				-227.87

Table continues overleaf

FIXED COST *Description*	*Unit*	*Total*
CASH Cost		
Insurance	acre	97.28
Property Taxes & License	acre	30.00
Total CASH Cost		127.28
NONCASH Cost		
Land Interest Charge	acre	1200.00
Depreciation & Interest	acre	1088.94
Total NONCASH Cost		2288.94
Total FIXED Cost		2416.22
Total of ALL Cost		**4969.09**
NET PROJECTED RETURNS		**-2644.09**

synthetic fungicides. Bunch rot is controlled by one application of a synthetic fungicide. A light application of nitrogen fertilizer is applied to sustain vine growth. A permanent cover crop is planted in year 4, at a cost of $50 per acre for seed. Hedging is performed twice in year 4 as the canopy grows larger. Leaf pulling is required three weeks after bloom at 15 hours per acre.

In year 4, 2.0 tons of grapes are harvested. Pickers are paid $150 per ton compared to $200 per ton in the previous year. Custom hauling expenses remain $30 per ton. Total variable cost in year 4 is $2,831 per acre. Total fixed cost is $2,443 per acre. Total cost in year 4 is therefore $5,274 per acre. This cost is offset by returns of $3,100 per acre, for a net cost of $2,174 per acre. Total establishment cost is $23,700 per acre. This figure represents the total economic cost of taking an unimproved site and establishing a vineyard on it.

Figure 2 shows cumulative establishment costs per acre. This graph highlights the large investment that growing winegrapes requires and provides an estimate of the capital required to establish a vineyard. Figure 3 shows the breakdown of each establishment year cost in variable and fixed cost categories. This graph shows the considerable variable cost incurred during the second establishment year, which was caused by the purchase of the grafted vine plants ($3.10 per plant), as well as increasing fixed costs throughout the establishment period.

Production Years

The final enterprise budget is shown in Table 9. This budget presents the typical annual costs and returns expected from a mature 20-acre vineyard. The long-term yield expected is 2.5 tons per acre, and the expected average price is $1,550 per ton of grapes. The total return expected is $3,875 per acre. Leaf pulling is included in this budget at the same rate as in year 4. This budget assumes that cluster thinning is required at 15 hours per acre. However, the amount of cluster thinning varies because of changes in crop load in response to seasonal weather variation.

Total variable cost is $3,063 per acre. The return over variable cost is $812 per acre. Assuming all variable costs are paid in cash, we expect a return of approximately $812 per acre to land, capital, and risk. This return is the money available to pay fixed costs of insurance, taxes and license, depreciation, and interest.

Total fixed cost is $2,465 per acre. But this does not yet include the amortized cost of vineyard establishment. If the $23,700 cumulative cost of establishing the vineyard were treated as a loan that must be paid off over the 25 year productive life of the vineyard, at 10% interest it would require an annual payment of $2,611. If the establishment costs are paid with an owner's capital, this charge represents repayment of the investment plus the foregone interest in alternative investments the owner could have made instead of establishing grapes.

The net projected return over all costs is -$4,264 per acre. This indicates that a grape yield of 2.5 tons valued at $1,550 per ton is not sufficient to pay the *full* costs of production. In fact, we calculate that a price of $3,256 per ton is required to cover total production costs. Does this indicate that winegrapes will lose money every year? Not necessarily.

Our fixed costs include interest on operating capital and establishment costs at 10%, as well as 10% interest on site improvements, land, machinery, and equipment. In addition, we have included expenses for all labor and capital. If you are willing to accept a lower rate of return on capital or value your labor and management at lower rates, you may achieve an economic profit from your vineyard in the long run.

Also, the prices or yields you achieve may increase your returns. Published average prices were used in this study, but superior vineyard management and a good site could enable better than average fruit quality that brings a higher than average price. Adjusting the mix of varieties grown would also influence the overall vineyard returns.

Table 8. Winegrape, Establishment Year 4, $/Acre Economic Costs and Returns

GROSS INCOME				
Description	*Quantity*	*Unit*	*$/Unit*	*Total*
Grapes	2.0	ton	1550.00	3100.00
Total GROSS INCOME				3100.00

VARIABLE COST				
Description	*Labor*	*Machinery*	*Materials*	*Total*
Prune	189.00	0.00	0.00	189.00
Pull Brush	105.00	0.00	0.00	105.00
Tie Canes	189.00	0.00	6.00	195.00
Training Ties (1 acre x 6.00 = 6.00)				
Flail Canes	16.94	9.31	0.00	26.25
Rodent Control	10.50	0.00	0.00	10.50
Strip Spray in Row	5.08	2.45	20.60	28.13
Herbicide (1 acre x 20.60 = 20.60)				
Fertilizer	5.08	2.45	3.99	11.52
Nitrogen (28.5 lbs (21-0-0) x 0.14 = 3.99)				
Flail Mow	16.94	9.31	0.00	26.25
Cultivate (3 times)	38.12	19.35	0.00	57.47
Spot Spray	36.58	2.45	10.00	49.03
Herbicide (1 acre x 10.00 = 10.00)				
Mildew Control (3 times)	15.25	7.34	8.40	30.98
Sulfur (12 lbs x .70 = 8.40)				
Fertilizer (Boron, 3 times)	0.00	0.00	3.75	3.75
(3 appl. tank-mixed with mildew control x 1.25 = 3.75)				
Mildew Control	12.70	7.62	60.00	80.32
Fungicide Spray (3 appl. x 20.00 - 60.00)				
Rot Control	5.08	2.45	37.00	44.53
Fungicide Spray (1 appl. x 37.00 = 37.00)				
Sucker Control (Leaf Pull)	157.50	0.00	0.00	157.50
Plant Permanent Cover Crop	10.16	4.38	54.00	68.54
Planter Rental (1 acre x 4.00 = 4.00)				
Seed (1 acre x 50.00 = 50.00)				
Training (move catch wires 3 x)	189.00	0.00	0.00	189.00
Leaf Pull (on one side)	157.50	0.00	0.00	157.50
Hedging (2 times)	25.41	15.52	0.00	40.93
Cluster Thinning	157.50	0.00	0.00	157.50
Bird Harassment	21.00	0.00	55.00	76.00
Pyrotechnics (1 acre x 55.00 = 55.00)				
HARVEST				
Picking, Custom	300.00	0.00	0.00	300.00
Tractor & Loader	25.41	10.99	0.00	36.40
Load & Haul, Custom	0.00	0.00	60.00	60.00
Grape Tax	0.00	0.00	50.00	50.00
Total HARVEST				446.40
Drop Wires	63.00	0.00	0.00	63.00
Irrigation Maintenance	21.00	0.00	20.00	41.00
Vineyard Management	300.00	0.00	0.00	300.00
General Labor	157.50	0.00	0.00	157.50
Machine Shed	3.38	12.50	0.00	15.88
Pickup	0.00	21.15	0.00	21.15

Table continues overleaf

Miscellaneous	0.00	0.00	25.00	25.00
Operating Capital Interest	0.00	0.00	56.23	56.23
Total VARIABLE COST				2830.86
GROSS INCOME minus VARIABLE COST				+269.14

FIXED COST

Description	*Unit*	*Total*
CASH Cost		
Insurance	acre	98.85
Property Taxes & License	acre	30.00
Total CASH Cost		128.85
NONCASH Cost		
Land Interest Charge	acre	1200.00
Depreciation & Interest	acre	1114.15
Total NONCASH Cost		2314.15
Total FIXED Cost		2443.00
Total of ALL Cost		**5273.86**
NET PROJECTED RETURNS		**-2173.86**

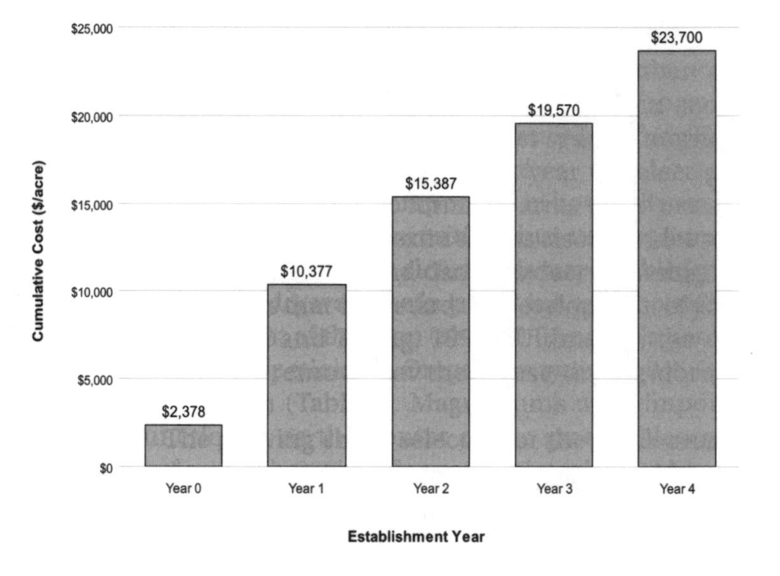

Figure 2. Cumulative economic costs of winegrape establishment.

Throughout this study, we use conservative estimates of costs and include all typical operations. Your costs may be lower, again allowing you to achieve a long-term profit. Furthermore, the size of our example vineyard could be expanded to improve returns. The farm equipment and water well would probably be adequate for at least 10 additional acres. Spreading the equipment and site improvement amortization costs over more acres would increase the vineyard's return.

Finally, this analysis ignores any tax advantages associated with vineyard ownership. It also ignores potential long-term capital gains from appreciation in the value of the vineyard.

One encouraging aspect of this analysis is that it appears that winegrape returns will more than cover cash costs. The breakeven price for grapes to cover cash variable costs is $1,225 per ton at a 2.5 ton yield. The state average price paid by wineries in 2000 was $1,000 per ton of Chardonnay, $1,300 per ton of Pinot gris, and $1,820 per ton of Pinot noir (Rowley et al., 2001). Thus, it appears that for a mix of these grape varieties the vineyard will at least cover the cash costs of producing a crop from year to year.

Cash Flow Analysis

The enterprise budgets in this chapter show the economic costs and returns of establishing and producing winegrapes. These costs include cash costs as well as non-cash costs. Fixed assets costs were spread over their useful lives as depreciation and interest. In analyzing the feasibility of establishing a new vineyard, estimated cash flow requirements are also important for planning cash needs and borrowing. Most lenders require a projected cash flow statement to accompany loan applications.

Table 9. Winegrape, Production Years, $/Acre Economic Costs and Returns

GROSS INCOME				
Description	*Quantity*	*Unit*	*$/Unit*	*Total*
Grapes	2.5	ton	1550.00	3875.00
Total GROSS INCOME				3875.00

VARIABLE COST				
Description	*Labor*	*Machinery*	*Materials*	*Total*
Prune	189.00	0.00	0.00	189.00
Pull Brush	168.00	0.00	0.00	168.00
Tie Canes	189.00	0.00	6.00	195.00
Training Ties (1 acre x 6.00 = 6.00)				
Flail Canes	16.94	9.31	0.00	26.25
Flail Mow (3 times)	50.82	27.92	0.00	78.74
Rodent Control	10.50	0.00	0.00	10.50
Strip Spray	5.08	2.45	20.60	28.13
Herbicide (1 acre x 20.60 = 20.60)				
Fertilizer	5.08	2.45	5.32	12.85
Nitrogen (38 lbs (21-0-0) x 0.14 = 5.32)				
Mildew Control (5 times)	25.41	12.23	14.00	51.64
Sulfur (20 lbs x .70 = 14.00)				
Fertilizer (Boron, 3 times)	0.00	0.00	3.75	3.75
(3 appl. tank-mixed with mildew control x 1.25 = 3.75)				
Mildew Control	12.70	7.62	60.00	80.32
Fungicide Spray (3 appl. x 20.00 - 60.00)				
Rot Control	10.16	4.89	74.00	89.06
Fungicide Spray (2 appl. x 37.00 = 74.00)				
Spot Spray	18.29	1.22	5.00	24.51
Herbicide (0.5 acre x 10.00 = 5.00)				
Sucker Control (Leaf Pull)	157.50	0.00	0.00	157.50
Training (move catch wires 3 x)	189.00	0.00	0.00	189.00
Cluster Thinning	157.50	0.00	0.00	157.50
Leaf Pull (on one side)	157.50	0.00	0.00	157.50
Hedging	38.12	23.28	0.00	61.40
Bird Harassment	21.00	0.00	55.00	76.00
Pyrotechnics (1 acre x 55.00 = 55.00)				
HARVEST				
Picking, Custom	375.00	0.00	0.00	375.00
Tractor & Loader	25.41	10.99	0.00	36.40
Load & Haul, Custom	0.00	0.00	75.00	75.00
Grape Tax	0.00	0.00	62.50	62.50
Total HARVEST				548.90
Drop Wires	63.00	0.00	0.00	63.00
Irrigation Maintenance	42.00	0.00	10.00	52.00
Trellis Maintenance	42.00	0.00	20.00	62.00
Vineyard Management	300.00	0.00	0.00	300.00
General Labor	157.50	0.00	0.00	157.50
Machine Shed	3.38	12.50	0.00	15.88
Pickup	0.00	21.15	0.00	21.15
Miscellaneous	0.00	0.00	25.00	25.00
Operating Capital Interest	0.00	0.00	61.30	61.30
Total VARIABLE COST				3063.38
GROSS INCOME minus VARIABLE COST				+811.62

Table continues overleaf

FIXED COST Description	*Unit*	*Total*
CASH Cost		
Insurance	acre	100.19
Property Taxes & License	acre	30.00
Total CASH Cost		130.19
NONCASH Cost		
Amortized Establishment Cost	acre	2611.04
Land Interest Charge	acre	1200.00
Depreciation & Interest	acre	1134.75
Total NONCASH Cost		4945.79
Total FIXED Cost		5075.98
Total of ALL Cost		**8139.36**
NET PROJECTED RETURNS		**-4264.36**
Breakeven Price, Total Variable Cost	3063.38/2.5 tons = 1225 per ton	
Breakeven Price, Total Cost	8139.36/2.5 tons = 3255.74 per ton	

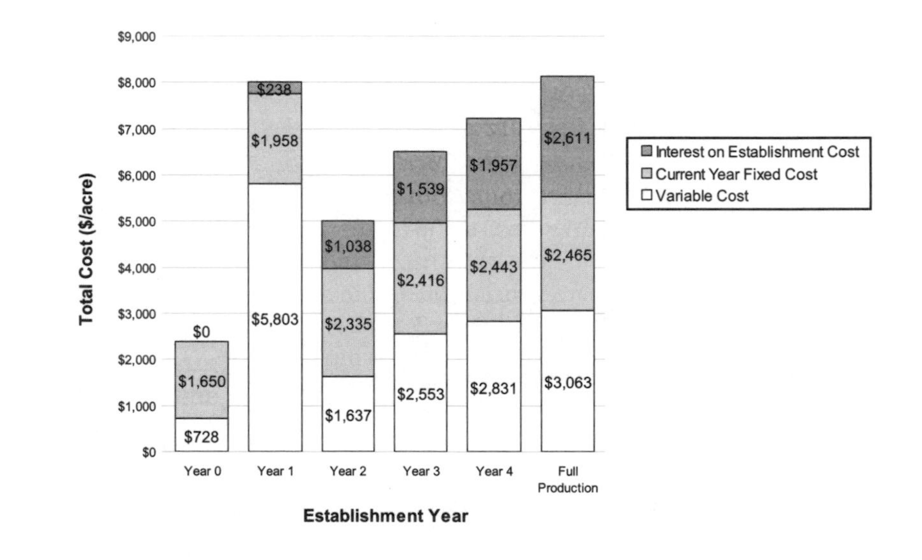

Figure 3. Annual winegrape establishment costs per acre.

Table 10 shows a summary of the total cash flows required to establish 20 acres of winegrapes. This cash flow projection summarizes the cash receipts and expenditures throughout the establishment period. Returns are received only in years 3 and 4, when 1.5 and 2.0 tons of grapes are harvested, respectively. Returns per acre are multiplied by 20 acres to estimate total returns.

Operating costs are taken from the enterprise budgets in Tables 4–9. We assume that all variable costs, including labor, are cash expenses. Your variable cash expenses may be less than the values shown if you provide all or part of the labor. Variable costs for the vineyard are calculated by multiplying the variable costs per acre for each year by 20.

Cash fixed costs are also taken from the enterprise budgets in Tables 4–9. These are the costs incurred for insurance, property taxes, and licenses. Cash fixed costs per acre are multiplied by 20 to calculate total cash fixed costs for the vineyard.

The remaining cash expenditures pay for site improvements and purchase machinery and equipment. The majority of site improvement expenditures occur in year 0 (machine shed, road and utility development, and well), year 1 (irrigation system), and year 2 (trellis), while the machinery and equipment purchases are spread fairly evenly over the the first four years of the establishment period.

Annual cash flows are calculated by subtracting operating costs, site improvement expenditures, and machinery and equipment purchases from total returns in each year. The negative values indicate that a cash deficit occurs in each year until year 4, and outside cash must be provided. The cumulative cash flow shows the total cash flow in each year, including all previous cash flows.

Table 10 shows that over the 5-year establishment period, just under $600,000 of cash is used in establishing this 20-acre vineyard (assuming the land needs to be purchased). Obviously, anyone considering

Table 10. Winegrape Establishment Cash Flow Analysis, 20 Acres

Item	*Year 0*	*Year 1*	*Year 2*	*Year 3*	*Year 4*	*Total*
RECEIPTS	0	0	0	46,500	62,000	108,500
OPERATING COSTS						
Variable Costs	14,560	116,058	32,735	51,057	56,617	271,027
Cash Fixed Costs	1,495	1,917	2,439	2,546	2,577	10,974
SITE IMPROVEMENTS						
Land	240,000					240,000
Machine Shed	10,000					10,000
Road Development	3,300					3,300
Utilities (wires & poles)	6,600					6,600
Well	10,000					10,000
Irrigation System		37350				37,350
Trellis and Wire			50,000			50,000
Electronic Bird Repeller				1700		1,700
MACHINERY & EQUIPMENT						
Small Tools	3,000					3,000
Tractor, 35 hp	24,000					24,000
Pickup	15,000					15,000
Sprayer		1,500				1,500
Tiller		2,000				2,000
Air-blast Sprayer (100G, 3 pt.)			5,000			5,000
Hedger				8,000		8,000
Fork Attachment				135		135
Flail Mower				3,500		3,500
Front End Loader				3,500		3,500
ANNUAL CASH FLOW	-327,955	-158,825	-90,174	-23,938	2,806	
CUMULATIVE CASH FLOW	-327,955	-486,780	-576,953	-600,892	-598,086	-598,086

Table 11. Costs and Returns ($/acre) at Alternative Price Levels, Assuming Yield is 2.5 Tons/Acre

	Price ($/Ton)					
	1,050	*1,300*	*1,550*	*1,800*	*2,050*	*2,300*
Returns	2,625	3,250	3,875	4,500	5,125	5,750
Variable Cost	3,063	3,063	3,063	3,063	3,063	3,063
Returns over Variable Cost	-438	187	812	1,437	2,062	2,687
Fixed Cost	5,075	5,075	5,075	5,075	5,075	5,075
Total Cost	8,139	8,139	8,139	8,139	8,139	8,139
Net Returns	-5,514	-4,889	-4,264	-3,639	-3,014	-2,389

Table 12. Costs and Returns ($/acre) at Alternative Yield Levels Assuming Price is $1,550/Ton

	Yield (Tons/Acre)					
	1.5	*2.0*	*2.5*	*3.0*	*3.5*	*4.0*
Returns	2,325	3,100	3,875	4,650	5,425	6,200
Variable Cost	2,844	2,954	3,063	3,173	3,283	3,393
Returns over Variable Cost	-519	146	812	1,477	2,142	2,807
Fixed Cost	5,075	5,075	5,075	5,075	5,075	5,075
Total Cost	7,919	8,029	8,139	8,248	8,358	8,468
Net Returns	-5,594	-4,929	-4,264	-3,598	-2,933	-2,268

Table 13. Net Returns ($/acre) at Alternative Price and Yield Levels

	Yield (Tons/Acre)					
Price ($/Ton)	*1.5*	*2.0*	*2.5*	*3.0*	*3.5*	*4.0*
$1,050	-6,344	-5,929	-5,514	-5,098	-4,683	-4,268
$1,300	-5,969	-5,429	-4,889	-4,348	-3,808	-3,268
$1,550	-5,594	-4,929	-4,264	-3,598	-2,933	-2,268
$1,800	-5,219	-4,429	-3,639	-2,848	-2,058	-1,268
$2,050	-4,844	-3,929	-3,014	-2,098	-1,183	-268
$2,300	-4,469	-3,429	-2,389	-1,348	-308	732

establishment of a new vineyard needs access to a considerable amount of cash.

Sensitivity Analysis

In this final section we evaluate the impacts of varying prices and yields on economic profitability.[1]

Table 11 examines costs and returns when yield is held constant at 2.5 tons and price ranges from $1,050 per ton to $2,300 per ton. Returns over variable cost are positive for all prices greater than $1,225 per ton (obtained from the bottom of Table 9). This suggests that, in the short run, growers cover their variable production costs at prices greater than $1,225 per ton. Over the long run, fixed costs must also be paid. According to the full production year budget, $3,256 per ton or greater results in positive net returns over total cost.[2] This is a much higher price than has been experienced historically in winegrape production.

The results of varying yield when price is held constant at $1,550 per ton are shown in Table 12. This table shows that returns over variable costs are positive even when yield drops slightly below 2 tons per acre. Net returns over total cost are not, however, positive in the range of our table. In fact, a 5.8 ton/acre yield would be required. This unrealistic yield level suggests that vineyards would need to reduce costs or receive higher grape prices to be profitable in the long run.

Table 13 shows net returns over total costs when both price and yield vary. Large losses occur at low yields and prices, and a slight profit finally occurs at a combination of the highest yields and prices in the table. Still, even a combination of higher prices and higher yields than are commonly experienced in the Willamette Valley do not provide positive returns over total costs for the study's scenario of production methods. Again, growers will need to find ways to lower costs or receive higher prices to be profitable on a total cost basis.

Summary

Establishing a vineyard requires a large investment of time and money. We estimate that economic establishment costs amount to $23,700 per acre for a hypothetical 20-acre vineyard. Slightly under $600,000 in cash may be required during a 5-year establishment period.

After the vineyard reaches full production, winegrape sales should cover variable costs of production. However, at a price of $1,550 per ton for grapes, net projected returns over all economic costs are negative. The estimated breakeven price to cover all costs is $3,256 per ton. At grape prices below this level, not all factors of production are being paid. These "negative results" suggest that there are other motivations besides profit maximization at work in the Oregon winegrape industry.

A major limitation of this study is that income taxes are ignored. All costs and returns are estimated on a before-tax basis. Anyone considering the establishment of a vineyard should check with tax accountants and legal counsel to determine potential tax incentives or disincentives associated with a vineyard investment and incorporate these factors into their own investment analysis.

It should be reemphasized that we use assumptions in our projections that may not apply to all circumstances. Profitability is restricted in the sample scenario by the relative inefficiency of equipment usage and site improvements on a relatively small, 20-acre vineyard. Grape prices received were assumed to be average, although a good site and superior management could improve the price received. Additional assumptions that reduced profitability are the 10% interest charge on capital, establishment costs, site improvements, land, machinery, and equipment. Also, expenses were charged for all labor and capital. If vineyard owners are willing to accept a lower rate of return on capital or place a lower value on their own labor and management inputs, the vineyard could be profitable in the long run.

References

Cross, Tim. 1989. *Custom Rates for Oregon Agriculture, 1988.* Publication SR 835, Oregon State University Agricultural Experiment Station, Corvallis. (No charge for single copy; order from Dept. of Agricultural and Resource Economics, OSU, Ballard Extension Hall 213, Corvallis, OR 97331-3601).

Cross, Tim, and Bart Eleveld. 1988 *Understanding and Using Enterprise Budgets.* Publication EM 8354, Oregon State University Extension Service, Corvallis. (No charge for single copy; order from Dept. of Agricultural and Resource Economics, OSU, Ballard Extension Hall 213, Corvallis, OR 97331-3601.)

McGrann, James, et al. 1986. *Microcomputer Budget Management System,* Texas A&M University, College Station.

Rowley, H., B. Eklund, L. Burgess, and R. F. Kriesel. 2001. *2000 Oregon Vineyard Report.* Oregon Agricultural Statistics Service, Portland (and earlier editions).

Satouf, Lionel, Tim Cross, Bernadine Strik, and Brenda Turner. 1993. *Vineyard Economics: The Costs of Establishing and Producing Wine Grapes in the Willamette Valley.* Publication EM 8533, Oregon State University Extension Service, Corvallis.

Notes

1. The summary enterprise figures shown in Table 2 are also available as an Excel spreadsheet from the first author of this chapter. The spreadsheet is designed to allow users to enter their own costs of establishment and production and then to solve numerically for long- and short-run breakeven prices and yields. A long-run breakeven would be the price or yield that allows the recovery of all costs of establishment and production, while the short-run breakeven would indicate the price or yield needed to cover just the variable costs of production during a full production year.
2. This breakeven price is accurate only if the price received in "baby crop" years 3 and 4 is fixed at the assumed price of $1,550 per ton. Varying price or yield of these early cropping years affects the amortized establishment cost. The $3,255 breakeven price is obtained by a simplified procedure of dividing net returns in the full production years by assumed yield (see bottom of Table 9).

2

Wine Production Costs and Returns

Esteban Vega-H, Robin Cross, and Catherine Durham

This chapter provides a summary of the costs of producing 1999 vintage premium Pinot noir and Pinot gris, from crush pad to cased goods, at one Oregon winery. Just as one vineyard block does not represent all sites and soils, one winery cannot represent the many talents that contribute to Oregon's wine industry. This study provides, in as much detail as possible, one example for comparison or planning purposes.

During the 1999 harvest season, 105 Oregon wineries crushed fruit. Three out of four of those wineries reported producing 10,000 or fewer cases; the median reported 5,000 cases (Northwest Farm Credit Services, 1999, p. 31). At 40%, Pinot noir represented the largest share of the crush. Chardonnay and Pinot gris followed at 19 and 15%, respectively. Table 1 characterizes our study winery and compares it to 1999 Oregon averages.

Several assumptions are built into our study winery. The Pinot noir was produced without pressing, fermented in stainless steel, aged 29 months in oak of 7-year barrel life, without fine filtration. The Pinot gris was pressed, fermented in stainless steel, filtered, and received no oak aging. The owner wage rate is $20 per hour. Cost-of-capital is 7.0%, based on an average return on assets for Oregon wineries (Northwest Farm Credit Services, 1999, p. 23). Production level and corresponding business accounts are treated as constant over the production cycle (1999–2002), so there is no meaningful distinction between cash and accrual.

Table 1. Comparing the Subject Winery to Oregon Averages

	Winery	*Oregon*
1999 PN Production (cases)	3,000	4,381
1999 PG Production (cases)	3,000	1,689
1999 All Wines Production (cases)	8,000	11,252
Reported Cooperage (cases)	8,000	17,706
PN Direct Case Sales (cases)	3,000	Not Available
PG Direct Case Sales (cases)	500	Not Available
PN Price Received[a](per case)	$194	Not Available
PG Price Received[b] (per case)	$78	Not Available
PN Share of Estate Grown[c]	31%	41%
PG Share of Estate Grown	0%	35%
Share of Owner Labor[d]	43%	Not Available
PN Juice Yield (gallons/ton)	165	Not Available
PG Juice Yield (gallons/ton)	176	Not Available

Source: Oregon volume information provided by tonnage data from the Oregon Agricultural Statistics Service's 1999 Winery Report. The author has imposed a juice yield on the Oregon tonnage data equal to subject winery's juice yield of 165 gallons per ton for Pinot noir, 176 gallons per ton for Pinot gris, and 170 gallons per ton average over all wines. This is in contrast to the 1999 Winery Report, which imposes a yield of 150 gallons per ton for all wines.

PN = Pinot noir; PG = Pinot gris.

[a] Price received for Pinot noir is understated, representing sales of the 1999 vintage as of June 1, 2002, which represents the first 25% of the 1999 vintage.

[b] Price received for Pinot gris represents all Pinot gris sales from May 1, 2000, to April 30, 2001, the main marketing period for the 1999 vintage.

[c] All grapes treated as purchased at $2,000 and $1,600 per ton for Pinot noir and Pinot gris, respectively.

[d] Owner labor treated as hired at total hourly cost of $20.

Profitability

Table 2 outlines the cost per bottle for several items. Owner labor and cost-of-capital are included to illustrate the opportunity cost of wine production. These two items typically would not appear in an accountant-prepared income statement. Profit, excluding these two items, is recalculated at the bottom of the table to illustrate the impact of owner labor and cost-of-capital on winery profits.

The $16.20 per bottle price received by the winery for Pinot noir covers all costs, with $3.41 remaining. The high level of profit is due, in part, to the high proportion of wine sold directly to the consumer.

The winery received an average of $6.50 per bottle for Pinot gris, which is less than the average cost per bottle at $8.28. Before owner wages and cost-of-capital, Pinot gris sales resulted in a profit of $0.39 per bottle, 5% above costs.

Table 2. Cost Summary By Bottle, 1999 Vintage

	Pinot noir		*Pinot gris*	
	$	%	$	%
Production Costs				
Grape Purchases	2.40	19	1.80	22
General Supplies	0.11	1	0.08	1
Crush Labor	0.10	1	0.09	1
Racking Labor	0.08	1		
Bottling Labor	0.45	4	0.45	5
Bottling Supplies	1.05	8	1.05	13
Palletizing Labor	0.01	0.04	0.01	0.1
Production Labor, Non-owner	0.10	1	0.06	1
Production Labor, Owner	0.35	3	0.24	3
Repairs and Maintenance	0.07	1	0.05	1
Small Tools and Lab Fees	0.01	0.1	0.01	0.1
Barrel Depreciation[a]	0.40	3		
Related Utilities	0.03	0.2	0.02	0.2
Total Production Costs	5.15	40	3.86	47
Selling Costs				
Related Utilities	0.03	0.2	0.02	0.2
Automobile	0.05	0.4	0.03	0.4
Promotion and Supplies	1.54	12	0.66	8
Selling Labor	0.76	6	0.28	3
Selling Labor, Owner	1.01	8	0.49	6
Total Selling Costs	3.39	27	1.48	18
Administrative Costs				
Bookkeeping	0.08	1	0.06	1
Professional Services	0.27	2	0.18	2
Other Administrative	0.49	4	0.34	4
Administrative Labor, Owner	0.64	5	0.44	5
Insurance and Taxes	0.46	4	0.31	4
Total Administrative Costs	1.94	15	1.34	16
Other Noncash Costs				
Production Depreciation	0.28	2	0.20	2
Administrative Depreciation	0.45	3	0.31	4
Cost-of-capital	1.58	12	1.09	13
Total Other Noncash Costs	2.30	18	1.60	19
Total Costs	12.79	100	8.28	100
Price Received by Winery[b]	16.20	127	6.50	79
Net Profit	3.41	27	(1.78)	(21)
Return On Assets (Peak)	39		(28)	
Internal Rate of Return	16.4		n/a	
Net Profit, before owner wages and cost-of-capital[c]	6.84	54	0.39	5

[a] Based on average return on assets for Oregon wineries, provided by Northwest Farm Credit Services (1999, p. 23).

[b] Price received by winery based on actual weighted average prices received for wines by variety in 1999.

[c] Eliminates the consideration of cost-of-capital and owner labor.

Table 3. Cost Summary by Production Year and Case, 1999 Vintage

Category	*Production Year or Item*[a]	*Unit*[b]	*Unit Cost*	*Pinot noir*	*Sub-total*	*Pinot gris*	*Sub-total*
Grape Purchases							
$2,000/ton Pinot noir							
$1,600/ton Pinot gris	Grapes			28.82	28.82	21.61	21.61
Supplies General	Year 1			0.13		0.13	
	Year 2			0.50		0.50	
	Year 3			0.44		0.33	
	Year 4			0.31	1.38		0.96
Crush Labor		Hours	11.6	1.18	1.18	1.10	1.10
Racking Labor	Year 1	Hours	11.6	0.11			
	Year 2	Hours	11.6	0.44			
	Year 3	Hours	11.6	0.44			
	Year 4	Hours	11.6	0.00	0.99		
Bottling Labor		Hours	11.6	5.44	5.44	5.44	5.44
Bottling Supplies	Bottles	Bottles	5.8	5.79		5.79	
	Capsules	Bottles	1.2	1.19		1.19	
	Corks	Bottles	2.6	2.58		2.58	
	Labels	Bottles	2.4	2.39		2.39	
	Supplies	Bottles	0.6	0.62	12.57	0.62	12.57
Palletizing Labor		Hours	11.6	0.06	0.06	0.06	0.06
Production Labor, Other	Year 1	Hours	11.6	0.00		0.00	
	Year 2	Hours	11.6	0.46		0.46	
	Year 3	Hours	11.6	0.40		0.30	
	Year 4	Hours	11.6	0.29	1.15		0.76
Production Labor, Owner	Year 1	Hours	20.0	0.40		0.40	
	Year 2	Hours	20.0	1.53		1.53	
	Year 3	Hours	20.0	1.34		1.00	
	Year 4	Hours	20.0	0.96	4.23		2.94
Repair and Maintenance	Year 1			0.08		0.08	
	Year 2			0.30		0.30	
	Year 3			0.26		0.20	
	Year 4			0.19	0.83		0.58
Small Tools and Lab Fees				0.09	0.09	0.06	0.06
Depreciation Barrels	Year 1	Barrels	0.5	0.53			
	Year 2	Barrels	3.2	3.16			
	Year 3	Barrels	1.1	1.05			
	Year 4	Barrels	0.0	0.00	4.74		
Production Utilities	Year 1			0.03		0.03	
	Year 2			0.13		0.13	
	Year 3			0.11		0.08	
	Year 4			0.08	0.36		0.25
Selling and Administrative Utilities	Year 1			0.03		0.03	
	Year 2			0.13		0.13	
	Year 3			0.11		0.08	
	Year 4			0.08	0.36		0.25
Automobile	Year 1			0.06		0.06	
	Year 2			0.21		0.21	
	Year 3			0.18		0.14	
	Year 4			0.13	0.58		0.40
Promotion and Supplies				18.45	18.45	7.88	7.88
Selling Labor		Hours	11.6	9.16	9.16	3.42	3.42
Selling Labor, Owner		Hours	20.0	12.11	12.11	5.83	5.83

Category	Production Year or Item[a]	Unit[b]	Unit Cost	Pinot noir	Sub-total	Pinot gris	Sub-total
Bookkeeping	Year 1			0.10		0.10	
	Year 2			0.37		0.37	
	Year 3			0.32		0.24	
	Year 4			0.23	1.01		0.70
Professional Services	Year 1			0.30		0.30	
	Year 2			1.15		1.15	
	Year 3			1.01		0.76	
	Year 4			0.72	3.18		2.21
Other Administrative Costs	Year 1			0.56		0.56	
	Year 2			2.13		2.13	
	Year 3			1.87		1.40	
	Year 4			1.33	5.89		4.09
Administrative Labor, Owner	Year 1	Hours	20.0	0.73		0.73	
	Year 2	Hours	20.0	2.77		2.77	
	Year 3	Hours	20.0	2.42		1.82	
	Year 4	Hours	20.0	1.73	7.65		5.31
Other Fixed Costs	Insurance			3.51		2.34	
	Taxes			2.04	5.55	1.36	3.70
Production Depreciation	Year 1			0.32		0.32	
	Year 2			1.22		1.22	
	Year 3			1.07		0.80	
	Year 4			0.76	3.37		2.34
Administrative Depreciation	Year 1			0.51		0.51	
	Year 2			1.94		1.94	
	Year 3			1.70		1.27	
	Year 4			1.21	5.36		3.72
Cost-of-capital	Year 1	Rate	7.0%	1.79		1.79	
	Year 2	Rate	7.0%	6.85		6.85	
	Year 3	Rate	7.0%	5.99		4.49	
	Year 4	Rate	7.0%	4.27	18.91		13.14
Total					153.43		99.34

[a] Production Year 1 corresponds to calendar year 1999, Year 2 to 2000, and so on. [b] Unit type not applicable if blank.

The traditional return on assets (ROA) measure of profitability is calculated here as net profit divided by the peak level of assets used during the production process. ROA is relatively high for Pinot noir at 39%, and low for Pinot gris at -28%. These calculations assume that winery capital is rented at a rate of 7%. Higher rental rates would, of course, result in lower returns.

ROA may not, however, portray an accurate picture of profitability, because the wine production process typically spans more than one fiscal period. A more accurate measure might be the internal rate of return (IRR), which is an estimate of the rate of return associated with a stream of payments. In this case, payments are calculated as quarterly revenues less expenses. The IRR is more reasonable at 16.4% for Pinot noir and not applicable for Pinot gris because of the negative return. Both the ROA and IRR calculations assume that wine sales are received during the last two quarters of production for each wine.

Cost Allocation

Several allocation methods are used in this study and tabulated in Table 3. Costs incurred in readily countable units, such as bottles and corks, are reported in those units. Labor hours, where possible, are calculated for each activity and multiplied by the average wage. Selling costs are allocated by number of cases sold. Certain production costs, such as repairs and utilities, which are not readily assignable, are allocated to each wine lot using a "gallon-day" rate, that is, a rate that reflects the volume of the wine lot and the number of days each lot spends in the winery. A more familiar version of this type of allocation might be the "ton-mile" rate, commonly used to bill for trucking services.

Sensitivity

Input costs affect profitability in proportion to the input's share of the total expense. Labor represents the largest single input, at $3.50 per bottle for Pinot noir, or 27% of the total cost. A 10% increase in the average wage, holding other factors constant, reduces net profit by $0.35, or 10%. Grapes are the second largest input, at $2.40 per bottle for Pinot noir, or 19%. Holding other factors constant, a 10% higher grape price lowers net profit by $0.24, or 7%. A different grape price may, however, be associated with different quality or a different supply situation, each of which may affect the wine price as well. Thus, the assumption of holding other factors constant is particularly unlikely in this instance.

Limitations and Extensions

Validity of Assumptions. A key factor that is often contemplated but rarely quantified is the intangible value of participation in the Oregon wine industry. Industry members might be willing to accept a return of more or less than the 7.0% real cost-of-capital and $20 per hour owner wage rate utilized in this study. Also not addressed is the premium deserved for the risks associated with weather, fermentation, competition, and policy shifts.

Estimated Prices Received by the Winery. The price received for the 1999 Pinot gris is calculated as the average price received for all Pinot gris over the marketing period from May 1, 2000, to April 30, 2001. Thus, a small amount of older vintage Pinot gris sales are included. The price received for Pinot noir, in contrast, represents sales of the 1999 vintage as of June 1, 2002. As of that date, only 25% of the total 1999 vintage had been marketed. As a result, the Pinot noir price received may be under- or overstated.

Labor Cost by Activity. The process of tracking hours spent on specific tasks, such as crushing, racking, and bottling, is costly and subject to inaccuracy. Efforts have been taken to recreate and account for these hours, but the final numbers represent approximations only.

Market Value. Opinions about the market value of winery assets continue to evolve. Estimated total capital employed by the winery is in line with appraised value. Depreciation costs were derived from cost-basis records for depreciable assets.

References

Northwest Farm Credit Services. 1999. *Wine Industry Study.* December. Spokane, Wash.

Acknowledgments

The authors would like to thank the Oregon Wine Industry, Northwest Farm Credit Services, Bart Eleveld, and Steven T. Buccola. Any remaining errors or omissions are solely those of the authors.

Notes to Appendix Table 1

All numbers represent simple averages. The Oregon 1997 sample consists of eight wineries. Case production varies from 3,000 to 70,000 cases. The California 1997 sample consists of nineteen wineries, with case production up to 50,000, and was provided by Deloitte and Touche.

Cost Basis: Numbers are typically presented on a cost basis. Market basis was utilized for two Oregon wineries but does not significantly alter the results.

Consolidations: Financials are included on a consolidated basis when two conditions exist: (a) the winery entity alone does not constitute a "viable economic unit" (an entity that both owns the full complement of assets controlled by the operation and retains sufficient liquidity and solvency to conduct operations on an ongoing basis); and (b) non-wine assets do not significantly dilute winery assets. In general, Northwest winery owners with significant non-wine interests "capitalize" the winery entity to a level that allows the winery to operate on its own capital.

Appendix Table 1. Financial Benchmarks, 1997

Benchmarks 1997	*Oregon*	*California*	*California*
	All	*High ROA*	*Low ROA*
Cases Sold (000's)	30	24	29
Total Revenue (000's)	2,282	3,445	3,288
Total Revenue/Case	77	146	115
Solvency			
Current Ratio	2.6	2.8	1.8
Acid Test	0.6	0.5	0.4
D:E	0.8	-2.4	0.7
D:A	44%	170%	42%
A:E	1.8	-1.4	1.7
CL/Case	28	31	64
NCL/Case	36	366	42
L/Case	64	397	106
Efficiency			
Gross Margin	49%	63%	51%
Total Revenue:Inv	1.35	2.00	1.24
Total Revenue:NCA	1.09	1.01	0.84
Total Revenue:A	0.54	0.63	0.46
Interest:Total Liabilities	5.4%	0.4%	5.3%
Assets Required to Sell a Case of Wine			
CA/Case	73	89	116
NCA/Case	70	145	136
A/Case	143	234	252
Profitability			
EBIT:Total Revenue	12%	34%	17%
EBIT:A	7%	21%	8%
EBT:E	8%	-29%	9%
Inventory			
Inventory in Gallons	142,887	72,700	122,301
Inventory Gallon Years	2.0	1.3	1.8
Inventory Dollar Years	1.5	1.4	1.6
Gallon Years:Dollar Years	1.4	1.0	1.1
$$$/Case			
Total Revenue/Case	77	146	115
COGS/Case	39	54	56
Gross Mar/Case	38	93	59
S,G&A/Case	28	43	39
EBIT/Case	9	49	19
Int/Case	3	2	6
EBT/Case	6	48	14
Income%			
Total Revenue	100%	100%	100%
COGS	51%	37%	49%
Gross Margin	49%	63%	51%
S,G&A	37%	30%	34%
EBIT	12%	34%	17%
Interest	5%	1%	5%
EBT	8%	33%	12%

Source: Reprinted here with permission from the Northwest Farm Credit Services (1999).

3

Site Assessment

Gregory V. Jones and Edward W. Hellman

Choosing a location to grow grapes can be a challenge within such a diverse landscape as the state of Oregon. There is a wide range of scales on which grape-growing issues should be considered, but successful winegrape production is grounded on a thorough understanding of each vineyard site's characteristics. Whether you have an established vineyard, are evaluating a potential vineyard site, or are seeking a suitable site, you must learn the characteristics of the site to know its capabilities, deficiencies, and unique qualities. Understand that the site characteristics described individually below are often related. For example, temperature and sunlight are related, and both are influenced by geographic latitude, elevation, slope, aspect, and topography. Consider the sum total of all factors influencing the site when evaluating a site's characteristics.

Matching the site with appropriate grape varieties, clones, and rootstocks helps establish the potential productivity, wine quality, and profitability of the vineyard. Good management practices, and of course cooperation from Mother Nature, enables realization of the vineyard's potential. It is a continuous learning process to understand how grapevines interact with your vineyard's site, and optimization of your vineyard's potential requires careful application of this knowledge to your planning and management practices.

In this chapter, for simplicity, we refer to a vineyard site as if it has uniform characteristics. In fact, most vineyard sites larger than a few acres are not uniform but have variable topography, soil, and possibly mesoclimate. It is important to recognize and assess these variations within a site in order to make sound decisions on varieties, clones, rootstocks, vineyard design, and management practices. Moreover, most vineyard sites are not ideal in every respect, so adjustments and compromises are made, and methods of mitigating deficiencies can be employed.

This chapter examines the various factors that influence a site, provides guidelines for site evaluations, discusses the resources available to growers, and considers some new technologies that are becoming important components in successful grape growing. Specific site assessment overviews for the major grape-growing regions in Oregon are given in Chapter 4.

Assessment Criteria and Evaluation

Climate and Weather

The average climatic conditions of a given region determine to a large degree the grape varieties and styles of wine produced. Furthermore, wine production and quality are chiefly influenced by site-specific factors, management decisions, and short-term climate variability. Evaluating a vineyard's climate requires distinguishing between three scales of influence: the regional climate, or *macroclimate*, the local climate, or *mesoclimate*, and the climate within and between the soil and grapevine canopy, or *microclimate*.

The macroclimate of a region describes the broad weather and climate patterns of a relatively large area, roughly from 100 to 1,000 miles or more. In most regions of the world, including Oregon, the dominant viticultural climate types are the Mediterranean and marine west-coast climates. These climates are synonymous with mild wet winters and warm dry summers, which are ideal for the cultivation of *Vitis vinifera* grapes.

The mesoclimate, sometimes called the site climate, is intermediate between the macroclimate and the microclimate of the vineyard. This term is often confused with "microclimate," but given the importance of climate on this scale, they should be differentiated by grape growers. Mesoclimatic influences operate on a scale of hundreds to thousands of feet horizontally and tens of feet vertically, and they are largely influenced by the local topography. Regional variations in mesoclimate can be drastic, and it is at this scale that careful site evaluation can have the biggest impact on grapevine growth and fruit quality. The major influences of mesoclimate can be evaluated in terms of elevation, slope, aspect, and air drainage barriers.

The microclimate of a vineyard is the climate from the soil upward into the vine canopy, and it plays a significant role in wine quality. The climate within this region can be quite different from outside the canopy and is mostly due to the influence and interaction between the vines and factors such as quantity and quality of sunlight, air and soil temperatures, wind speed, and humidity. The microclimate of a vineyard is largely controlled by management practices; thus, a grower can have a great influence on fruit and wine quality by optimizing the canopy climate (see Chapter 22).

Evaluation of a site is often hampered by a lack of site-specific weather and climate information. Observations should be collected firsthand by temporarily instrumenting the site or, second best, interpolated from nearby public weather stations. The grape grower should recognize that understanding a site's weather and climate characteristics is not a short-term activity and should consider investing in weather monitoring equipment. Portable weather stations, as described below under "Technological Assessment Tools," enable a user to monitor the site for a few weeks or a season. This information should then be used to make comparisons to other sites within the region and similar sites worldwide.

Temperature. Temperature is a critical aspect of site assessment in Oregon, particularly the influence of summer heat on grape maturity and quality, the potential for frost in late spring or early fall, and midwinter low temperature injury. Recognize that topography, elevation, slope, and aspect interact to influence temperature variations within a site.

The amount of heat captured is probably the single most important site influence on the success of an Oregon vineyard. Oregon is considered a viticultural cool climate, falling within regions I and II of Winkler's heat summation index (Winkler et al., 1974). The relative coolness of western Oregon's climate limits the varieties and clones and the size of the crop that can be ripened successfully. For some regions in Oregon (the Walla Walla and Columbia Valleys), minimum temperatures during the winter can be even more critical as a site characteristic. See Chapter 25 for a more detailed discussion of this topic.

The heat summation index, or growing degree-days, is a rough measure of the cumulative amount of functional heat experienced by grapevines during a growing season defined as April 1 through October 31 in the Northern Hemisphere. Degree-days are determined by subtracting 50°F from the mean daily temperature (maximum + minimum ÷ 2) and calculating the cumulative sum through the growing season. The base temperature of 50°F is used in calculations because almost no shoot growth occurs below this temperature. Using this index, five climatic regions for wine production are defined; region I has less than 2,500 degree-days, and region V has more than 4,000 degree-days. A comparison of average heat summation for locations within selected grape-growing regions of the world is shown in Table 1. The Willamette Valley ranges from 1,900 to 2,200 degree-days, and the southern Oregon valleys range from 2,200 to 3,000 degree-days. Average degree-day accumulations for Oregon's growing regions are shown in the color section following page 72.

Table 1. Heat Summation Given as Degree-Days above 50°F

Location	*Heat Summation*[a]
Salem, OR	2021
McMinnville, OR	2066
Coonawarra, AU	2170
Forest Grove, OR	2205
Beaune, France	2300
Bordeaux, France	2390
Roseburg, OR	2445
Sunnyside, WA	2570
Yakima, WA	2600
Grants Pass, OR	2870
Medford, OR	2815
Napa, CA	2880
Sonoma, CA	2950

Source: Most sites from Winkler et al. (1974) or Oregon Climate Service (1993); Medford, Grants Pass, and Roseburg derived from temperature data by the author.

[a] For the Period April 1 to October 31.

Because of annual weather variations, western Oregon can experience insufficient degree-days in some years. Fruit may develop to a mature stage with insufficient heat but not develop all of the fully ripe characteristics desired by wineries. Therefore, mesoclimate site characteristics, combined with good canopy management and crop control, are critical to capturing sufficient heat to ripen the fruit. In summary, only the warmer vineyard sites produce fully ripe fruit in less than ideal weather years.

Frost events, known as radiation or ground frosts, occur as the ground and the air in the lower layers of the atmosphere within and just above a grapevine canopy give off heat, warming the air in successive layers upward. As the ground and lower layers of the atmosphere cool, the heat energy lost is conveyed upward to form what is called a radiation inversion. On nights when inversions form, a warmer thermal zone develops upslope that provides a measure of protection from the coolest air in the valley floor sites (Figure 2). The thermal zone varies from region to region but is generally found 100–500 feet above the valley floor, depending on the width of the valley and degree of slope. Inversions are common in most of the grape-growing regions in Oregon and occur most

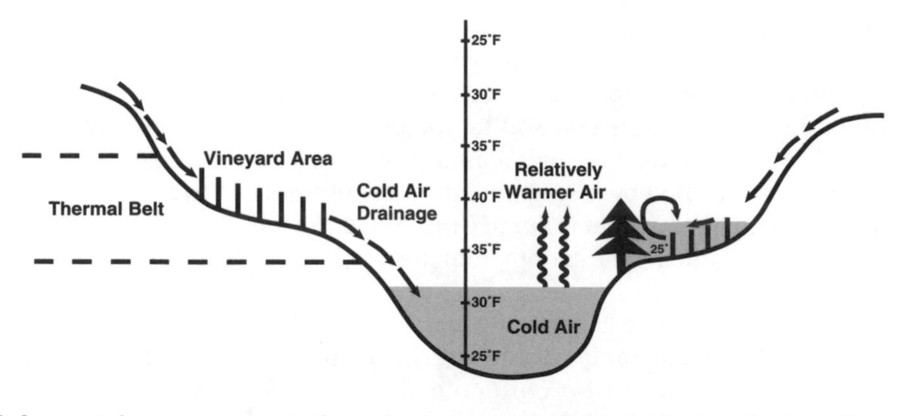

Figure 2. Patterns of cold air drainage in a valley or hillside setting.

frequently on long, calm, cloud-free nights; see Chapter 26 for more information.

Along with inversions, cold air drainage can greatly affect the relative amount of frost on a given site. Cold air drainage occurs because cold air is heavier than warm air and tends to flow downhill as regions upslope cool more quickly. Cold air drains downhill until impeded by an obstruction large enough to pool the cold air or until the topography flattens. If a slope is open to proper airflow, cold air pooling is generally not a problem, but any obstruction of airflow, such as fences, treelines, windbreaks, should be avoided or removed.

Another type of frost event, an advection frost or freeze, occurs as cold air masses are brought into a region with the passage of a cold front. These freeze events occur sporadically during the spring, fall, and winter and can cause problems over most of a region. Oregon has occasionally experienced winter freeze events that damaged grapevine canes and trunks. Site characteristics have much less influence on advection freezes than on radiation frosts.

Evaluation of a site's temperature characteristics can be conducted with many types of instrumentation. It is recommended that a site be instrumented with a properly housed temperature sensor, a thermister, which is calibrated and positioned midslope. Monitor the daily maximum and minimum temperatures over the course of the season for calculation of accumulated degree-days. It is also important to record the dates when the first and last frosts occur. Keep in mind that low temperatures are much more influenced by local site conditions. If a weather station is not available, the site can be observed by high-quality maximum/minimum recording thermometers, small electronic temperature sensors, or hand-held instant thermometers. Inexpensive thermometers, including the U-shaped maximum/minimum variety, should be avoided because instrument errors are large.

Solar Radiation. Sunlight, referred to as *solar radiation* or *insolation,* provides the energy through the photosynthetic process necessary for grapevine growth and maturation of the fruit. Site factors that influence the amount of insolation include geographic latitude, seasonal variation in the angle of incidence of the sun's rays, daylength, cloud cover, the reflective nature of the soil surface, and topographical variation.

Of all of the insolation factors, the only one that can be managed to any beneficial degree is topographical variation. The amount of insolation received by a site depends on the elevation of the sun combined with the inclination of the slope toward the sun. Therefore, a site should be evaluated in terms of how the slope and aspect influence sunlight reception.

Insolation values do not change drastically from similar sites in a region unless factors such as aspect, topography, or surrounding vegetation affect the site. Often one of the best methods for evaluating insolation is visual observation over the course of many days or seasons, which gives insight into shadow patterns produced by obstructions.

Water. Water availability is a major factor influencing vine vigor and capacity; therefore, a reliable, adequate supply of water is critical to the success of a vineyard. Total water supply assessment should consider local rainfall, soil water-holding capacity, and the availability of a suitable water supply for irrigation. Oregon's production regions have annual rainfall ranging from less than 10 inches to over 50 inches per year. Even in the higher rainfall regions, irrigation is a valuable tool that makes available many other site management options such as the use of drought-intolerant rootstocks or development of a site with shallow soil. If the site has a well, it should be tested for flow rate, seasonality, and water quality. The site should also be evaluated for suitable pond sites, surface water rights, and availability of irrigation district water.

Annual rainfall can be assessed by acquiring long-term precipitation data from the nearest reporting weather station or the Oregon Climate Service. Although topographical variations influence precipitation variability from site to site, local long-term data can give the grower an adequate assessment. Greater accuracy can be achieved with a standard tipping bucket rain gauge connected to an electronic weather station located on-site. Soil water-holding capacity information is provided in county soil surveys, as described later in this chapter.

Wind. Winds can have both positive and negative effects on grapevines. New shoots can be broken off by high winds, and ripening fruit can be desiccated, reducing berry volume and possibly quality. On the other hand, drying winds at night and in the early morning can help reduce the occurrence of fungal diseases by limiting the formation of dew on leaves and berries. Night winds can also be beneficial by limiting radiation frosts. Undesirable wind can be mitigated somewhat by location, topography, and the use of natural and human-made windbreaks. Topographical variations, trees, and buildings can create windbreaks. Be careful to avoid establishing a windbreak where it might serve as an obstruction for cold air pooling during frost or freeze conditions.

When evaluating a site's wind climate the best methods are through historical data and direct observation. Historical data come from wind atlases that are available in some libraries, and long-term data can be obtained from the regional weather reporting site. General wind direction, average wind speed, and gust speeds are usually available and give insights into possible locations for windbreaks. Direct observation on the site also provides good evidence of wind directions. Site-specific wind information generally does not provide enough significant information to justify adding it to a weather station.

Topography

The effects of topography on grape and wine quality can be evaluated in terms of how it influences temperature variability both at the soil level and within the canopy. Topographic factors that have the most influence on mesoclimate include the relative elevation, the slope, the aspect, the relative isolation of the hill and how this affects air drainage, and the proximity to bodies of water. Gladstones (1992) has found that, after carefully examining topographies of vineyards worldwide, the best vineyards usually have two or more of the following features:

- They are on slopes with excellent air drainage and are usually situated above the fog level in a thermal zone.
- The best are usually on the slopes of projecting or isolated hills that enhance air drainage.
- The slopes directly face the sun during at least some part of the day. South-facing is best, with easterly aspects receiving morning sun and westerly aspects receiving the afternoon sun (Northern Hemisphere).
- If inland, they tend to be close to substantial rivers or lakes. Additionally, mountain/valley breeze locations during the summer are important.

When inspecting a site's topography, components of elevation, slope, aspect, and hill isolation should be examined individually but evaluated collectively since it is their interactive effect on grape production that is crucial. Very few sites have the perfect blend of each component; some general guidelines for assessment are provided below.

Elevation. The effects of a site's elevation can be assessed from both absolute and relative standpoints. Overall, absolute elevation above sea level determines the general climatological characteristics of a site's temperature regime. On the average, temperature changes 1.1°F per 330 feet of elevation. This effect can leave higher-altitude sites with fewer growing degree-days, which can retard vine growth and fruit maturation. Frost probabilities also increase at higher elevations. Relative elevation of a site, the local relief from a valley bottom to the site's elevation, largely determines the air drainage and slope temperature variations.

Slope. Slope is the degree of inclination of the land, which is measured as the angle of the drop in elevation over a horizontal distance. Both slope and aspect are important in sunlight reception, cold air drainage, and frost and wind protection. Cool climate viticulture often takes advantage of sloping sites, which alter the angle of incidence of the sun's rays that strike the surface. This effect can be substantial; a vineyard with a 10° south-facing slope can receive as much as 25% more insolation than a flat site. Greater insolation increases the growing degree-days, so a south-facing slope is warmer, promoting earlier or more complete fruit ripening.

A sloped site also enables cold air to drain away, reducing the risk of frost damage. Hillside sites, however, have increased risk of soil erosion, higher vineyard management costs, and a greater hazard for operating equipment. Erosional forces increase in direct proportion to increases in slope, and many of the soils in Oregon experience slow downhill creep unless preventive practices are employed.

Aspect. A site's aspect describes the compass direction in which the slope faces, and in general southeast to west aspects are favored for maximum sunlight receipt (Table 2). Depending on numerous other factors such as obstructing trees, other hills, and rock outcroppings, a properly situated slope can enhance growth and maturation or limit disease problems (Jackson and Schuster, 1987). For any site in the Northern Hemisphere, northwest-, north-, and northeast-tending sites experience delayed grape growth stages, lower sunlight and heat receipt, and slower soil and canopy evaporation). Southeast-, south-, southwest-, and west-tending slopes exhibit earlier grape growth stages and show varying increases in insolation, heat, and evaporation (Wolf, 1997).

Hill Isolation. Isolated or projecting hills or ranges of hills are of special interest because they create the most dominant thermal zones within the area. The Dundee Hills and Eola Hills are examples of isolated ranges of hills in the Willamette Valley. Isolated hills provide a temperature-modifying effect compared to the surrounding valley floor. This occurs as cold air is efficiently drained away, and with no new source of cold air the isolated or projecting hill is in the thermal zone above surface inversions.

Soil

Soil characteristics define the baseline of water and nutrient availability, which in combination provide the greatest initial influence on grapevine vigor. Because soil is the starting point for determining vigor, vineyard plans and management practices should give careful consideration to soil influences. Vineyard planning decisions such as choice of rootstock, vine spacing, and trellis and training system are greatly influenced by the inherent vigor-promoting capacity of a soil. Similarly, vineyard management practices such as irrigation, fertilization, canopy management, and use of cover crops can be used to compensate for soil-related deficiencies or excesses that influence vigor. Soil is also considered by some people to contribute to the *terroir*[1] of a region's wine (Wilson, 1999). There is no single soil type that is ideal for growing grapes and making high-quality wines. Successful vineyards in Oregon are located on a variety of sandy, gravely, and clay-loam soils. The most critical criterion for a vineyard soil is good internal drainage. Grapevines grow poorly under water-logged conditions. A soil that is saturated during the growing season creates a low oxygen environment that inhibits root activity. Soils that are somewhat poorly drained can be improved by the installation of drain tile systems. Consult a soils professional for recommendations on design and installation of a drain tile system.

Soil depth is defined by the proximity to the surface of an impenetrable layer, usually of unbroken rock but in some cases a hardpan layer. Vineyard soil recommendations often call for a deep, well-drained soil. But how deep is deep enough? Recommendations vary from a minimum 24 inches to 40 inches, but the important relationship to consider is that, within a given soil type, the deeper the soil, the greater the total water-holding capacity. Water availability can be the most important factor influencing vine vigor and capacity. Shallow soils, in the absence of irrigation, may not have the capacity to store enough rainwater to sustain vines and ripen the fruit.

Some viticultural regions schedule irrigation to induce controlled water deficits as a method to restrict grapevine vigor. This approach is easier to implement in low rainfall regions, on a shallow soil with relatively limited rooting area, or on a sandy soil with low water-holding capacity. Be aware that some shallow soils may also have the opposite problem of becoming waterlogged. The restriction layer of some shallow soils may prevent water drainage, causing them to become saturated after a rain. Very shallow soils also can present difficulties in installation of trellis posts.

Generally, soils of moderate fertility are preferred for vineyards. Grapevines are not heavy feeders of nutrients, and very fertile soils encourage high vigor, which makes canopy management difficult. Soil nutrient problems that may need to be ameliorated include deficiencies in nitrogen, phosphorus, potassium, boron, zinc, and magnesium. Grapes are successfully grown in the varying production regions of Oregon on a wide range of soil pH values. Soil

Table 2. Relative Effects of Site Aspect (compass direction of slope) on Climate Characteristics and Grapevine Phenology

	Aspect							
Parameter	North	Northeast	East	Southeast	South	Southwest	West	Northwest
Initial growth in spring	Retarded	Retarded	Retarded	Advanced	Earliest	Earliest	Advanced	Retarded
Daily maximum canopy temperatures	Minimum	Less	Less	Less	Maximum	Greater	Greater	Less
Speed of evaporation in the morning	Slow	Moderate	Rapid	Moderate	Slow	Slow	Very Slow	Slow
Radiant heating of fruit in summer	Minimum	Less	Less	Less	Maximum	Greater	Greater	Moderate
Radiant heating of vines in winter	Minimum	Less	Less	Moderate	Maximum	Greater	Greater	Less

nutrient availability is affected by pH, so familiarize yourself with the potential nutrient problems in your region.

Soils should be assessed on a site-by-site and within-site basis. Hillside soils are notoriously variable and can contain several soil types within a relatively small area. A preliminary assessment of soil types at a site can be obtained from the soil surveys published by the USDA Soil Conservation Service, an agency now called the Natural Resources Conservation Service (NRCS). Soil surveys have been prepared for most counties and are available at the local USDA-NRCS office, libraries, or the county Agricultural Extension office (individual soil surveys are discussed in Chapter 4). The soil surveys contain information on soil types, their properties including drainage and water-holding capacity, and detailed maps of soil locations. However, soil survey maps tend to be spatially limited in their accuracy. More detailed soil information is best obtained by using a backhoe to dig exploratory trenches at strategic locations throughout the site. Soil drainage can often be evaluated visually; a well-drained soil has a distinctive uniform brown or red color. Poorly drained soils have topsoil or subsoil that is gray or displays mottling (alternating areas of reddish brown and gray color). Representative soil samples should be collected for laboratory analysis to provide baseline information on essential nutrient content, soil pH, and organic matter content. Soil test interpretation guidelines are available from the Agricultural Extension Service. See Chapter 18 for more detailed soils information.

Other Site Assessment Factors

Site assessments for a potential vineyard should also consider the size, shape, and possible varitions in layout of the vineyard. Size considerations are influenced by the available land, regional zoning criteria, and clearing or preparation decisions. A site should be divisible into blocks to create reasonably uniform management units that are appropriate for the characteristics of the site and the variety/rootstock combinations to be grown. Adequate space should be available between blocks for farm equipment passageways.

Site selection should also include an assessment of the property and the surrounding area for potential pest problems. Consider possible damage from wildlife, insects, and diseases. Deer, elk, gophers and other burrowing animals, and birds can cause extensive damage to vines, consume large amounts of fruit, and create dangerous conditions for vineyard workers and machinery. Sites situated near woodlands receive the greatest pressure from large mammals and birds.

It is also important to consider the proximity and type of development near your site. Homeowners often do not understand or appreciate the realities of standard farming practices. Conflicts can arise from noise generated by field equipment or bird control devices, and pesticide spray drift from your vineyard can create serious concern. The problem of spray drift can be two-sided. Be aware that neighboring farms growing other crops may utilize phenoxy-type herbicides as part of their production practice, and homeowners use them to clear out poison oak, wild blackberry, and thistle. Small quantities of phenoxy herbicides can damage grapevines if spray droplets or volatilized material contact the vines. Ideally, these issues would be discussed with your neighbors prior to development of the vineyard.

Technological Assessment Tools

Many of the new assessment tools and methods are potentially useful for grape production, but the usefulness of the technology must be weighed against the costs involved. One of the more valuable tools is an on-site weather station. Weather stations can range from simple devices to complex multi-instrument stations. The complexity and concomitant costs should be evaluated in relation to need. There are two broad categories of weather stations: manually recorded stations, in which daily averages are recorded by hand; and electronic stations that send information to a database. Electronic stations are popular because they allow continuous and user-absent recordings of climate data. Numerous companies sell electronic weather stations that can meet the needs of grape growers. National Weather Service offices, Extension offices, and consultants can assist in selecting the appropriate device. Several factors are relevant in choosing a device:

- The number and type of instruments on the weather station. For most needs, the recording of temperature, humidity, and precipitation are sufficient. Some vineyards may also want to measure atmospheric pressure, insolation, wind speed and direction, and soil moisture.
- The type of computing infrastructure needed to download and process the data. The weather station manufacturer can supply this information.
- The type of software needed to analyze the data. Often the software comes with the weather station, but other programs may be needed.
- Location in the vineyard. The weather station should be placed far away from structures and devices, which bias the recorded data if too close, and should be situated where it gives an average vineyard response to the individual parameters.

• The power source. Most weather stations will operate with a minimal amount of power that can come from direct electrical connections, batteries, or solar panels. Since most stations are remote and direct connections unlikely; a solar panel connected to batteries is recommended.
• Data retrieval. The retrieval of recorded data usually occurs through direct download via cable or telephone hook-up, or with a floppy disc or laptop computer. Cost and ease of retrieval vary with the retrieval method.
• The ease of use and time commitment needed to operate the weather station. The weather station and software should be easy to use and the time spent retrieving data should be minimal.

Other site assessment and vineyard management tools include Geographic Information Systems (GIS), Global Positioning Systems (GPS), aerial photography, and satellite imagery. These tools can enhance the ability to evaluate a vineyard site and can be used to track spatial and temporal variability of grapevine growth and development, management inputs, and production history. Sources for these technologies include state, county, and city planning departments, private companies, consultants, and research universities.

Climate and Weather Resources

Oregon Climate Service, Office of the State Climatologist, Oregon State University, Corvallis, Ore. http://www.ocs.orst.edu/

Western Regional Climate Center, National Oceanic and Atmospheric Administration (NOAA), Reno, Nev. http://www.wrcc.sage.dri.edu/

Climate Diagnostics Center, NOAA-Cooperative Institute for Research in Environmental Sciences, University of Colorado, Boulder, Colo. http://www.cdc.noaa.gov/

The Pacific Northwest Cooperative Agricultural Weather Network (Agrimet), U.S. Department of the Interior, Bureau of Reclamation. http://www.pn.usbr.gov/agrimet

National Weather Service, NOAA, Portland, Ore. http://www.wrh.noaa.gov/Portland/climate/

National Weather Service, NOAA, Medford, Ore. http://nimbo.wrh.noaa.gov/Medford/climo/

Soil Resources

USDA-Natural Resources Conservation Service local offices

Soil Quality Institute, USDA-Natural Resources Conservation Service. http://soils.usda.gov/sqi

Department of Crop and Soil Science, Oregon State University, Corvallis, Ore. http://www.css.orst.edu/

Central Analytical Lab, Oregon State University, Corvallis, Ore. http://www.css.orst.edu/Services/Plntanal/CAL/calhome.htm

Oregon State University Extension Service. http://osu.orst.edu/extension/

References

Gladstones, J. 1992. *Viticulture and Environment.* Winetitles, Adelaide, Australia.

Jackson, D., and D. Schuster. 1987. *The Production of Grapes and Wine in Cool Climates,* Butterworths Horticultural Books, Wellington, New Zealand.

Wilson, J. E. 1999. *Terroir: The Role of Geology, Climate, and Culture in the Making of French Wines.* University of California Press, Berkeley.

Winkler, A. J., J. A. Cook, W. M. Kliewer, and L. A. Lider. 1974. *General Viticulture.* University of California Press, Berkeley.

Wolf, T. K. 1997. *Site Selection for Commercial Vineyards.* Publication 463-016, Virginia Agricultural Experiment Station, Winchester.

Note

1. *Terroir* is a French term used to describe all the ecological factors that make a particular type of wine special to the region of its origin, of which soil is thought to be a major contributor.

4

Winegrowing Regions

Gregory V. Jones, Michael McLain, and Scott Hendricks

Oregon contains a diverse array of lands and climates for grape growing, ranging from dry, high-elevation regions such as the Columbia and Walla Walla Valleys to the intermountain valleys of the Rogue and Umpqua Rivers and the broad agricultural belt of the Willamette Valley. These regions grow a wide range of varieties including many cool-climate (e.g., Pinot noir) and warm-climate (e.g., Merlot) varieties and are regionally, nationally, and internationally known for their characteristic wine styles.

The regional designation of winegrowing regions in the United States is controlled by the Bureau of Alcohol, Tobacco, and Firearms (BATF, U.S. Department of the Treasury). Each American Viticultural Area (AVA), as they are known, is defined as "a delimited, grape-growing region distinguishable by geographical features, the boundaries of which have been delineated by an approved map" (Code of Federal Regulations, 2001). Started in 1978, this designation system is meant to provide a "comprehensive scheme for appellation of origin labeling," representing a widely used geographic designation, wine style, quality assurance, and marketing tool to winegrape growers and wine producers. Currently there are 145 approved AVAs in the United States and six in Oregon (map in color section following page 72). Below are overviews of each of the Oregon AVAs.

Willamette Valley AVA

The Willamette Valley AVA (map in color section following page 72) was established in 1983 and consists of portions of Lane, Benton, Polk, Yamhill, Washington, Clackamas, Marion, and Linn counties (Code of Federal Regulations, 2000). The Willamette Valley is the largest wine-producing region in Oregon, with more than 7,600 planted acres of grapes and 113 wineries, 73% and 81%, respectively, of the state totals in 2000 (Oregon Vineyard and Winery Reports, 1987–2000). Pinot noir is the flagship variety of the region, comprising more than 58% of the valley's acreage and 91% of the state's total Pinot noir acreage. The other principal varieties are Pinot gris (16%), Chardonnay (13%), Riesling (5%), Pinot blanc (1%), and Gewürztraminer (1%).

The valley is oriented north–south, and the Willamette River flows 127 miles from south of Cottage Grove to where it merges with the Columbia River at Portland. In some places the valley reaches 60 miles in width between the Coast Range on the west and the Cascade Range on the east. From an elevation of about 320 feet at Cottage Grove, the river drops to approximately 20 feet above sea level, where it meets the Columbia River. The borders of the AVA roughly follow an elevation line of approximately 900 feet on the west, south, and east walls of the valley. The north portion of the AVA is bounded by the Columbia River. Latitude coordinates are 43° 50' at the south to 45° 40' at the north.

Climate

The Willamette Valley is Oregon's coolest winegrape region. Heat accumulation is probably the single most important site selection criterion, with average degree-day accumulations at favorable sites generally ranging from 2,000 to 2,200 (Table 1). Overall, the climate is maritime with relatively mild, cool, wet winters and dry, warm summers. Most precipitation occurs in winter in the form of rain, with 50% falling from December through February. The remaining rainfall occurs during the spring and fall months, with July through September mostly dry. Temperatures infrequently rise into the 90s in summer and almost as infrequently fall below freezing in the winter.

The Coast Range protects the Valley from the predominant east-flowing storm systems arriving from over the Pacific Ocean, which bring our renowned rain. As the moisture-laden clouds reach the coast and begin to rise and condense over the Coast Range, rainfall levels rise from 60–70 inches per year at the beaches to as high as 200 inches per year at the highest peaks. Then, as the air rapidly drops into the Willamette Valley, the opposite occurs, and within 10 miles annual rainfall totals can drop by over 100 inches per year. On the valley floor rainfall averages about 44 inches, but the condensation/rainfall effect occurs again as elevation rises on the isolated hills of the valley. The upper reaches of the Salem and Eola Hills receive 52 inches per year, and over 60 inches falls on the Chehalem Hills at Bald Peak. Rainfall levels also begin to rise again as elevation increases in the

Table 1. Average Climate Characteristics for Representative Stations in the Willamette Valley AVA

Station (Elevation)	Average July Maximum Temperature (°F)	Average January Minimum Temperature (°F)	Average Mean Growing Season[a] Temperature (°F)	Growing Degree Days (Apr-Oct, 50°F base)	Precipitation (inches)	Frost-free Period (# of days last to first. 32°F)
Beaverton (270 ft.)	78.9	32.5	59.3	2097	39.8	190
Corvallis (230 ft.)	80.1	32.9	58.8	2005	42.7	NA[b]
Corvallis Water Bureau (590 ft.)	78.4	31.7	57.6	1796	66.1	191
Cottage Grove 1 S (650 ft.)	81.2	32.4	57.9	1820	45.5	131
Dallas 2 NE (290 ft.)	82.6	32.8	59.0	2022	48.4	168
Eugene WSO AP (360 ft.)	82.0	33.2	59.3	2099	49.3	NA
Forest Grove (180 ft.)	81.6	32.3	59.9	2205	43.9	177
Hillsboro (160 ft.)	80.1	33.0	59.0	2038	37.6	173
Lacomb (520 ft.)	77.8	31.5	57.9	1810	55.9	NA
McMinnville (150 ft.)	81.9	33.8	59.1	2066	41.9	NA
Oregon City (170 ft.)	81.9	34.6	61.3	2495	47.1	215
Portland WSFO AP (20 ft.)	79.8	33.6	60.8	2386	36.3	223
Salem WSO AP (200 ft.)	81.5	32.6	58.9	2021	39.2	165
Silverton (410 ft.)	78.4	32.4	59.1	2034	45.9	207
Stayton (430 ft.)	80.1	32.5	58.9	2030	51.5	185

Source: All data are from the 1961-1990 climate normals for that station (OCS, 2000)

[a] April through October. [b] NA=data not available.

Cascade foothills to the east, reaching over 100 inches at the peaks. Winter rains usually provide adequate water to recharge soil moisture.

Because the majority of the prevailing "weather" comes from the west, the Coast Range is the primary topographical feature that establishes the Willamette Valley macroclimate. There are, however, three major low-elevation "passes" from the ocean into the valley, and proximity to these passes greatly influences a site's specific characteristics, or mesoclimate. These passes, over which flow the main highways to the coast, are as follows: in the south, from Eugene to Florence (State Highway 126); in midvalley, from Corvallis to Newport (U.S. Highway 20, known as the Philomath Gap); and in the north, from Sheridan to Lincoln City (State Highway 18, also known as the Van Duzer Corridor). A fourth, less influential pass follows U.S. Highway 26 into the valley at Forest Grove/Hillsboro. During the growing season, as stagnant air heats up in the valley, cooler air moves through these corridors from the coast and fans out as it enters the valley. Exposed sites closer to the fan can have significantly lower temperatures and therefore fewer total degree-days. Overall, the airflow contributes to cool nights late in the growing season, which promotes the desired slow ripening of fruit that enables full flavor development in Pinot noir. But cool winds also rob an exposed site of the heat necessary to mature the fruit. The cooling effect is reduced by greater distance from the corridor or the presence of ridges to block the wind.

It would be ideal if weather-reporting stations were scattered among the vineyards up and down the valley. Unfortunately, most weather stations are in cities, away from the farms. However, data collected from the 30-year climate normal period of 1961–1990 show that from north to south through the valley there is not a large variation in degree-day accumulation numbers (Table 1). Although there is a gradual rise in degree-days from the south to the north, a significant portion of the differences may be attributable to the elevation of the reporting station or its proximity to a city center. As an example, Corvallis has two stations, one at 590 feet elevation reporting 1,796 degree-days and the other at 230 feet reporting 2005 degree-days. Successful grape production in the Willamette Valley AVA requires at least 2,000 degree-days, which is attainable at sites throughout the valley. Cooler sites and cooler portions of a site can be more suited to white varieties and earlier ripening clones of Pinot noir. Warmer areas allow more choices and increased odds of ripening Pinot noir.

Topography

From Eugene, where the Willamette Valley broadens in the south, to Albany, there are a few scattered low buttes and hills, and few vineyards. A combination of high clay soils and potential lowland frost has discouraged vineyard development on the valley floor. The most prominent midvalley landforms begin just south of Salem and include the Salem Hills, with numerous vineyards planted on the lower south and east slopes; the Eola/Amity Hills, with west-slope vineyards clustered north of State Highway 22, south of Holmes Gap, and on south and east slopes north of Bethel Heights Road (most below 800 feet elevation); the Dundee Hills, with most non-north-facing slopes below 900 feet elevation planted to vineyards; and the Chehalem Hills, with vineyards dotting its abundant south slope and south-facing portions of the north slope in Washington County, and the low hills to the south and west that are becoming known as the Willakenzie district. Numerous vineyards are scattered on the east slopes of the Coast Range along its entire length, with the lower and more protected sites enjoying greater heat accumulation. The Cascade foothills on the east side of the valley have enormous amounts of undeveloped acreage with soils, slopes, and elevations apparently suited to production.

Soils

Geological History. The Willamette Valley floor was the ocean floor 35 million years ago; therefore, much of the parent material is of varied marine sediments. About 20–30 million years ago, basalt lava flows spilled down the Columbia Valley and into the Willamette Valley, capping existing sandstone bluffs. These caps are the underlying material of all the isolated mid-valley hills.

About 15 million years ago, the valley was a much different place, to which the deep red soils of the Salem, Eola, and Dundee Hills give witness. These "laterite" soils, including Jory and Nekia, developed from weathering of the basalt, combined with decayed vegetation produced in a warm, wet environment. The Salem Hills contain abundant bauxite, an extreme type of laterite formed only in hot tropical climates, which is so rich in aluminum that it is used for ore.

Subsequent to this hot, moist climate period, there was a dry period 3–10 million years ago during which the valley and mountains were a desert. During this period, extensive sand, gravel, and gravelly sediments washed into the valley's low-lands. As faults opened, they were filled above the brim with sediments. Roughly a million years ago, as the Willamette River beds filled with deposits, the river meandered about the broad valley and formed a large, deep freshwater lake stretching from the Oswego/Oregon City gaps to Eugene. Resultant "buff lake" silts were deposited as high as 200 feet. This lake existed during the last interglacial stage but drained as rivers cut through the silt in advance of the next glaciers.

Then came one of the truly amazing events in the valley's history: the Missoula Floods. In the latter stages of the last ice age about 15,000 years ago, glaciers repeatedly formed ice dams across the Clark Fork of the Columbia River, where it currently crosses into Canada. The dams backed up about 500 cubic miles of water as the glaciers melted and retreated. Finally, the dams gave way and a titanic wall of water washed across the Columbia Plateau, down the Gorge and into the Willamette Valley, filling it to a depth of 400 feet and laying down a thin layer of silt. Pieces of glacial ice-encapsulated rocks that originated in Montana also floated in with the floodwaters. The icebergs subsequently melted and deposited the rocks all over the valley. A good example of these "erratic" rocks is easy to see at the Erratic Rock Wayside off State Highway 18 southwest of McMinnville. It is generally recognized that multiple cataclysmic floods (as many as forty) occurred between 15,000 and 12,800 years ago as glaciers came and went (Allen and Burns, 1986).

Soil Types. In general, there are two basic soil types upon which vineyards are commonly grown in the Willamette Valley: those formed from or over the volcanic basalt, and those formed from or over sedimentary parent material. Vineyard soils of volcanic origin include Jory, Nekia, Laurelwood, Saum, and Yamhill. Soils of sedimentary origin or bedrock include Willakenzie, Cornelius (with Kinton in a complex), Bellpine, Melbourne, Steiwer, and Woodburn. Bellpine is interesting because it consists of a red, silty-clay loam much like the volcanic Jory, but it overlays a sedimentary base. Consult Soil Conservation Service soil surveys for complete descriptions of soil characteristics.

The soils are primarily volcanic at the mid- and higher elevations of the isolated valley hills. Jory, arguably the best known soil in the Willamette Valley, is a red, silty-clay loam 4–6 feet deep over rotted and weathered basalt that is found all over the valley. The Eola, Amity, and Salem Hills and the Cascade foothills commonly contain Jory and an essentially shallower version of it called Nekia, which is typically 3 feet over fractured and weathered basalt. Lower on these hills are alluvial deposits from the various lakes that have filled the valley. A typical soil found at lower elevations is Woodburn, a dark grayish-brown silt loam generally over 65 inches deep.

Nonalluvial sedimentary soils are found in the valley hills west and south of Eugene, the foothills of the Coast Range, the low hills west and north of the Dundee Hills, and the western portion of the Chehalem Hills. These soil types include Willakenzie, a dark brown to yellowish-brown silty-clay loam, typically 3–4 feet deep over siltstone, and Bellpine, a red silty-clay loam 3–4 feet over siltstone and sandstone. Currently, a reassessment of sedimentary soils is being conducted by the Natural Resources Conservation Service (NRCS) in Yamhill County with its focus primarily upon Willakenzie and Peavine soil types. The Willakenzie soils in this area are not consistent and will likely be reclassified as Goodin (20–40 inches deep), Melbourne (48–60 inches), and Wellsdale (>60 inches) when not conforming to the typical Willakenzie soil profile. Recent coring has shown that most of the Ribbon Ridge area will be reclassified as Wellsdale. The majority of Peavine found at lower elevations will be reclassified as Jory, Bellpine, or Windy Gap, which is described as a cross between Jory and Bellpine.

Water

Obtaining water rights for irrigation requires application to the Oregon Water Resources Department, a process that can take as long as two years. Be aware that the state has designated "restricted" areas where no new irrigation rights are being granted because of dropping well water levels. These include Parrot Mountain and portions of the Eola and Salem Hills.

Despite its wet reputation, wells in the Willamette Valley can be stingy. There is much variation, even from parcel to parcel, but in general wells on upland sites in the valley average about 10 gallons per minute with depths averaging about 250 feet (unpublished data, author's average of vineyard site wells from 20 years of data). Water quality also can be an issue, with problems caused by hard or alkaline water, or high levels of salt and iron. Research the depth and production of wells on adjacent tax lots at the State Water Resources Department. Existing wells should have a "well log" showing depth and production, drawn up by the drilling contractor.

Location Factors

With few exceptions, land suited to winegrape production is zoned Farm/Forest (FF), Exclusive Farm Use (EFU), or Forest (F); therefore, homesite approvals are restricted. The zoning criteria state that, if there is no existing habitable residential structure on the property, a new home cannot be built on the site until the land has produced over $80,000 (in 1994 dollars) in gross farm gate income in two successive years, or three of the last five years. At current crop prices, it would be difficult to produce this income on a 20-acre vineyard. Winery buildings can be constructed as a state-approved "outright allowed" use with 15 acres of planted vines, subject to county/state site approval restrictions. Less than 15 acres requires a conditional use permit through the County Planning Department and involves public hearings to determine the permit issuance. Surrounding property owners have significant input on the approval or denial, and on conditions. They also have the right to appeal and can therefore delay a project.

Development Areas

There are no formal subregions to the Willamette Valley, but some areas that have been developed for vineyards or have potential to be developed are briefly described below:

- The smaller, narrow valleys in the hills south and west of Eugene have considerable vineyard development and potential, but frost is a concern. The primary soil association in this area (and in the Coast Range foothills all the way to the Van Duzer corridor) is Jory and Bellpine, with spots of Willakenzie turning up occasionally, such as in the Briggs Hill area.
- Nearing the Highway 126 corridor in the Coast Range foothills, frost and cool airflow are of concern and can cause problems in the side valleys to the corridor.
- North of Highway 36 begins an area that could be one of the warmest in the valley, with fruit maturation consistently ahead of other areas. This would seem to hold true north to where the Highway 20 corridor begins its influence just south of Corvallis. Again, side valleys open to the corridor should be avoided except where they are interrupted by a ridge or are sufficient distance from the corridor.
- Beginning at the McDonald Forest "jut" just north of Corvallis and ending at Dallas is a huge bowl open into the Willamette Valley with numerous undeveloped sites, some quite large. Reporting weather stations are in Corvallis and Dallas, but both are outside the "bowl." A vineyard in the bowl near Monmouth shows maturity data quite similar to Dundee. The soil association from Eugene to Dallas in the foothills is primarily Bellpine, with some Jory.
- After the hills north of Dallas begins the Van Duzer corridor, and caution should rule on sites exposed to the "fan."
- On the other side of the Van Duzer north and west of the Highway 18 community of Bellvue, vineyards begin to dot the foothills and then continue all the way north to Gaston, with primarily the Willakenzie and Jory soil associations.
- North of Patton Valley the soils shift to the Laurelwood and Melbourne associations and continue on the hillsides to the Sunset Highway at Banks. There is little development north of Highway 26.
- In the Willamette Valley proper, with one exception of a valley floor vineyard near Harrisburg, vineyards have been planted on low hills north of Corvallis and begin to become more concentrated in the Salem Hills. The south faces

of the Salem Hills and Eola Hills west of Salem bear the brunt of the dissipated Van Duzer airflow and can be quite windy. Protection from that airflow should increase heat.

• The west side of the Eola Hills, though exposed to the Van Duzer, is a fair distance from the "mouth," and sites with some protection are producing fine wines. The east, or lee side, of the Eola Hills is quite warm, especially on the lower slopes. Heat dissipates rapidly as elevation nears 800 feet.

• The Amity Hills are quite similar to the Eola Hills.

• The Dundee Hills are the most developed in the Willamette Valley, with good sites commanding the highest prices per acre in the AVA. Vineyards are currently found on almost all non-north slope sites.

• The Chehalem Hills, beginning on Parrot Mountain at the southeast, are seeing more development. The south face of Parrot Mountain is notoriously rocky with shallow soils, which are obvious and easily avoided. About midway northwest on the south face of the Chehalem Hills there is a shift from volcanic to sedimentary soils, with some sites showing both types.

• South and west of the Chehalem Hills are low rolling hills east of Carlton and Yamhill. This region is becoming known as the Willakenzie district, named after the soil type. Concerns with this soil type center around inconsistency, with some sites needing tiling to enhance drainage and others needing irrigation. The reclassification of this soil mentioned above may clarify the situation.

• Over the top of the Chehalem Mountain, on south-facing portions of the north slope, the soil is volcanic again, with much Laurelwood. This protected area is quite warm and is minutes from Hillsboro/Beaverton/Portland, with resultant development pressure and higher land prices.

• The east side of the Willamette Valley is relatively undeveloped but contains huge amounts of gentle sloping sites on the lower slopes of the Cascades. Several commercial vineyards are located near Oregon City, Mt. Angel, and Silverton.

• With the exception of two small vineyards at Sweet Home, there are few vineyards on the east side of the valley from south of Highway 22 at Salem to where the valley closes at Cottage Grove.

In summary, high-quality wines have been produced throughout the Willamette Valley, with the best sites sharing certain characteristics:

1. Hillsides above frost-pooling lowlands and generally below 900 feet in elevation.
2. Slopes in all directions other than northerly, although, as in Burgundy, some fine wines have been produced from north/northeast-sloping sites.
3. Soils that are neither too deep and fertile nor too shallow and austere.
4. Sites protected from cool flowing air through the coastal corridors, accumulating adequate heat for ripening.

In Europe, many centuries of grape cultivation have led to the knowledge of the best varieties to grow in specific regions. The Willamette valley has less than forty years of experience, so the industry is still in the experimental phase. Still, some of our variety decisions have resulted in exceptional wines that have received worldwide recognition and a constantly increasing interest and investment in the industry. With less than 8,000 acres currently planted in the Willamette Valley, and likely more than 50,000 additional plantable acres currently in other uses, the potential for growth is large.

Umpqua Valley AVA

The Umpqua Valley AVA (map in color section following page 72) was established in 1984 and lies wholly within Douglas County (Code of Federal Regulations, 2000). The main delimiting boundary is the 1,000-foot contour, because most good sites are found below this elevation. A few sites are situated above 1,000 feet, but they are limited. The AVA limits extend to the towns and surrounding valleys of Riddle, Canyonville, and Days Creek to the south, the Callahan Escarpment and the mountains west of the Umpqua River, the mountains north of Scottsburg to the divide southwest of Cottage Grove, and then southward along the mountains to the east.

Wine making in the Umpqua Valley started in the 1880s when two German brothers (the Doerner family) planted vines and made wine near Roseburg. Post-prohibition winegrape growing in Oregon got its start in 1961 when Richard Sommer planted vines, ushering in a new age of viticulture in Oregon. Today, the region has over 750,000 agriculture acres and, as of 2000, was planted with 618 acres of vines on 36 vineyards, which are increases of 97% and 17%, respectively, from 1987 (Oregon Vineyard and Winery Reports, 1987–2000). Production has increased 82% over the same period, with 1,316 tons of grapes crushed at eight wineries in 2000. The dominant varieties are Pinot noir, Chardonnay, Riesling, Cabernet Sauvignon, Pinot gris, Merlot, and Gewürztraminer. Although the varieties listed above make up over 80% of the crop in the Umpqua Valley, the region is proving to be well suited for other varieties such as Tempranillo, Syrah, Malbec, and Dolcetto.

Climate

Often called the "Hundred Valleys of the Umpqua," the valleys in this AVA generally tend in an east–west direction producing, a "cross valley" effect of allowing cool Pacific breezes inland. The region is overall warmer and drier than the Willamette Valley AVA to the north but generally cooler and wetter than the Rogue Valley AVA to the south. Precipitation averages 30–60 inches per year (Table 2) but varies greatly from south to north in the region. Depending on elevation, the growing season averages 180 to over 215 days, with the average last and first frosts occurring on April 10 and October 31, respectively (median frost dates defined for 32°F; Oregon Climate Service and Western Regional Climate Center, 2000). Growing degree-days sum from 2,200 to over 2,500 in most of the lowlands, varying from south to north.

Topography

The diversity in the landscape of the Umpqua Valley AVA comes from the joining of the three mountain formations in the region: the Klamath Mountains, the Coastal Range, and the Cascades, each with distinct geological origins. The Klamath Mountains, which extend into the south-southwestern portion of the AVA, consist of folded and faulted metamorphic and igneous rocks that formed in an oceanic setting and were accreted to the North American continent about 150 million years ago. The Coastal Range makes up the majority of the western component of the Umpqua's landscape and has a varied geologic history. This mountain province was also formed by volcanic island chains colliding with North America, roughly 50 million years ago. Much of the sediments in the Coastal Range have accumulated from the erosion of the ancient volcanoes and contain marine deposits that indicate their oceanic origin. The Cascade Mountains were developed during two distinct volcanic regimes: the older, more deeply eroded Western Cascades province that makes up the eastern portion of the AVA, and the younger, easterly volcanoes of the High Cascades.

The majority of the Umpqua AVA consists of valley lowlands broken up by isolated hilltops and ridges. The lowlands are drained by numerous small streams and by the meandering North and South Umpqua Rivers, which exit the AVA west of Elkton and ultimately empty into the Pacific Ocean. The best sites in the region are on isolated hillsides, benches, and footslopes ranging from 500 to 1,000 feet in elevation. Some viable lowland areas with gradual slopes to the rivers can, however, be found. Narrow valley stretches are located in the southern (south and west of the Winston/Dillard area) and eastern (east of Interstate 5) portions of the AVA, with the best sites in the wider stretches and along the footslopes. The wider valley stretches are found in the Winston/Dillard area, Lookingglass Valley, the greater Garden Valley area, Coles Valley, Calapooya Creek Valley, Camas Swale (east of Sutherlin), and south and west of Elkton. These regions contain a greater diversity of potential site options with isolated hills, terraces, and gradual slopes to the stream or river systems.

Soils

Because of the complex joining of the three mountain formations, the general soil structure in the Umpqua Valley AVA is derived from a widely varying mix of metamorphic, sedimentary, and volcanic parent material. Most valley floor locations consist of deep alluvial material washed downstream from either the Western Cascades or the Klamath Mountains. Hillside and bench locations generally have mixed alluvial deposits but can possess a wide arrange of soil types caused by complex faulting, especially in the southern part of the AVA. The more frequent soil types found in most potential vineyard locations include, but are not limited to, Bellpine, Camas, Coburg, Dixonville, Evans, Newberg, Nonpareil, Oakland, Philomath, Roseburg, Sutherlin, and Windy Gap (NRCS, 1997). The soils are categorized below by the type of location where they are typically found:

Flat to Gradual Slopes

- The Camas–Newberg complex, a widely distributed suite of soils found in flood plains in western Oregon. They are generally very deep and excessively drained soils that formed in gravelly and very gravelly coarse textured alluvium.
- The Coburg series of soils, which are generally deep and poor to moderately drained soils that formed in mixed alluvium. The texture varies from silt to clay loam, with some areas experiencing frequent flooding or having high water tables.
- The Roseburg and Evans soil types formed from mixed alluvium. These soils are moderate and well drained, and fine and coarse loamy, respectively. They are limited in extent but are found along the edges of flood plains and up gradual slopes off the valley floors.

Gradual to Steep Slopes

- The Windy Gap–Bellpine complex, which is mostly deep, well-drained soils on hills, terraces, and plateaus. These soils formed in colluvium and residuum derived from sedimentary rock sandstone and siltstone from the Roseburg,

Lookingglass, Flournoy, and Tyee Formations. Both are common on smooth convex-shaped foothills.

- The Nonpareil–Oakland–Sutherlin complex. The Sutherlin series are fine loams generally overlying clay; they are deep and moderately well drained. They formed in mixed alluvium and colluvium over residuum weathered from sandstone and siltstone. The Nonpareil series is similar but shallower. The Oakland series is similar, but it is also formed from mixed shale deposits and is finer. The commonality among the series is that they are found on broadly convex footslopes and moderately steep hillslopes and are mostly found as improved pastures.
- The Philomath–Dixonville complex are shallow to moderately deep, well-drained soils that formed in colluvium weathered from basic igneous rock. Philomath soils are on convex foothills adjacent to the valley floors. The Dixonville series is mostly colluvium and residuum weathered from basalt and silt to clay loam in texture. Both soil types are common in both natural and improved pasturelands.

Development Areas

The Umpqua's numerous valleys suggest that many sites would be suitable to winegrape growing. Site assessment should consider, however, the width of the valley and its ability to drain away cold air during the winter and spring. Subvalleys within the region exhibit a wide range of site exposures and soil characteristics. Careful evaluation of site characteristics is required in this highly variable region. To facilitate discussion, we divide the Umpqua AVA into five areas (from south to north, and loosely defined by the local topography).

South Douglas Area. The South Douglas area contains the valley extensions south of Roseburg that drain the South Umpqua River and its tributaries. The area currently has three wineries and over ten vineyards of varying sizes. The climate in this area is slightly warmer during the growing season than the rest of the Umpqua Valley AVA and is generally drier due to the intervening mountains to the west (Table 2). Most of the area receives beneficial breezes from the Pacific Ocean through the Camas Valley Divide to the southwest. During the winter and spring, a fog line develops near Roberts Mountain, with fog to the north and clear conditions to the south. The clear mornings in the southern valleys should benefit vines with ample sunshine to dry out dew and increased heat accumulation. Water is generally available from reservoir systems but should be critically assessed from site to site. The valleys range from broad open reaches to narrow widths with some isolated hills, benches, or terraces and gentle slopes off the surrounding ridges.

The area has numerous valley extensions with widely varying grape-growing conditions. Careful site assessment is needed to identify the few optimum sites in each area. Southwest of Riddle, the Cow Creek extension is wide, has good water sources, and provides some marginal bench areas with south to northwest exposures. The Canyonville to Days Creek extension meanders west to east, providing some good southerly exposed bench areas at 800–1,000 feet. Proceeding north, the main valley stem starts out wide and narrows from Tri City through Myrtle Creek and toward Roberts Mountain. Although there are some good benches, there is little south-sloping land and they may be more prone to frosts. To the east of Myrtle Creek, two small valleys have some potential, but their narrow width and the growth of the town eastward limits the number of possible sites. A few vineyards have been successful just south of Round Prairie along the South Umpqua River, but frost risks are greater and the deeper soils encourage higher vigor.

From Dillard through Winston and the Green district, the valley widens and offers some promising isolated hills and benches, with sites at 400–800 feet. Currently, two wineries and three vineyards are found in this area. Heading northeast up the Roberts Creek valley to Dixonville, there are some beautiful isolated hills and gentle slopes dropping off the surrounding ridges that offer many potential south-facing sites. The main drawback to this area is water availability, which must be assessed on a site-by-site basis.

Southwest of Winston, the Olalla and Tenmile Valleys offer potential growing sites that experience more of a Pacific Ocean effect due to their proximity to the Camas Valley Divide. Both valleys offer a mix of isolated hills, benches, and gently sloping sites off the surrounding ridges. Frost danger is high in the narrow valleys, so good air drainage is a critical site feature. Of the two valleys, the Olalla has better water availability from Ben Irving Reservoir. Tenmile Valley is more dependent on natural stream or site water. The two valleys currently have a few successful vineyards and one winery. Northwest, up Tenmile Creek, the area around the community of Reston offers some potential sites, but elevations over 1,000 feet may preclude a viable vineyard.

Just south of Roseburg, from Interstate 5 and the South Umpqua River westward, three valleys contain potential grape-growing sites: Happy Valley, Lookingglass Valley, and Flournoy Valley. Happy Valley is oriented east–west and has some wonderful south-facing exposures that gently slope off the ridge to the north. Although these sites seem promising, the major limitation is water. Surface water is seasonal in Happy

Table 2. Average Climate Characteristics for Representative Stations in the Umpqua Valley AVA

Station (Elevation)	*Average July Maximum Temperature (°F)*	*Average January Minimum Temperature (°F)*	*Average Mean Growing Season[a] Temperature (°F)*	*Growing Degree Days (Apr-Oct, 50°F base)*	*Precipitation (inches)*	*Frost-free Period (# of days last to first. 32°F)*
Drain (292 ft.)	83.0	33.7	59.8	2185	45.7	175
Elkton (122 ft.)	83.9	35.9	61.1	2278	53.8	220
Flournoy Valley (700 ft.)	NA[b]	NA	NA	NA	45.2	NA
Winchester (460 ft.)	85.2	36.2	61.4	2436	34.3	227
Roseburg (465 ft.)	83.5	34.6	61.1	2445	32.4	207
Riddle (680 ft.)	83.4	33.9	60.8	2378	30.7	183

Source: All data are from the 1961-1990 climate normals for that station, except for Elkton and Flournoy Valley, which are from monthly climate summaries for 1948-1998 and 1978-1998, respectively (OCS, 2000; WRCC, 2000).

[a] April through October. [b] NA=data not available.

Valley, making well water the major source for irrigation. In the Lookingglass Valley to the west, water availability is better, but sites are mostly limited to valley floor locations with deep, rich soils. Although some gentle slopes are found, there are few if any isolated hills or benches in the valley. The Flournoy Valley is a broad, gently sloping area with east to southeast slopes and fair water sources. There is no history of successful vineyards in the area; development may be limited by elevations over 900 feet.

Garden Valley/Melrose Area. The broad open area with gently rolling hills northwest of Roseburg is the historical center of Umpqua Valley grape growing. A commercial vineyard existed here in the 1880s, and Richard Sommer planted the first post-Prohibition vineyard in 1961. The area currently has three wineries and contains the most planted acreage in the Umpqua Valley AVA. The area is nearly as warm as the South Douglas area, but precipitation and number of days with fog are higher. In the vicinity are two areas, Garden Valley and Melrose, which are bounded on the west by the Callahan Escarpment and a broad mountainous region to the north. Garden Valley extends from west of Winchester along the North Umpqua River, past the confluence with South Umpqua River, and northwest to Woodruff Mountain. Although there are some isolated hills and slope lands, Garden Valley mostly offers valley floor sites at 400–600 feet. The valley floor has deep, rich soils and water availability is generally not a problem, but it has some potential for frost. The best sites have some air drainage to minimize frosts. An example of a successful valley floor site in this region is Henry Estate, where Scott Henry developed a training system to fit the circumstances of his site. The Melrose area extends from a broad bend in the South Umpqua River, through the town of Melrose, southwest to the ridge that separates it from the Flournoy Valley. The region offers more isolated hills and gentle slopes off ridges than does Garden Valley, with elevations ranging from 500 to 1,000 feet. Water is less readily available than in Garden Valley but is generally sufficient for grapes.

Although not necessarily a part of the Garden Valley/Melrose area, the extension of the North Umpqua from Winchester to the AVA boundary near Glide could be considered. This valley offers some southern exposures, bench lands, and isolated hills that seem promising. The major drawback is the high frost potential from cold air draining off the uplands of the Cascades down the North Umpqua River valley.

Calapooya/Umpqua Area. The next development area to the north starts in the Coles Valley vicinity, including the town of Umpqua, and extends northeasterly along the Calapooya Valley extension, through the cities of Oakland and Sutherlin, and into many subvalleys, including the Camas Swale. The area is somewhat varied in its grape-growing potential due to its broad east–west size and wide rather than narrow valley landscapes. The climate is generally similar to the Garden Valley/Melrose area but offers the benefits of cooling winds from the Pacific Ocean during the hot days of summer. The area has one winery and a few successful vineyards, although they are mostly confined to the Coles Valley area and the lower reaches of the Calapooya Valley. The Coles Valley area has many north-facing slopes and benches coming off Woodruff Mountain that may present good sites for some cooler climate varieties. Coles Ridge is an isolated topographical feature in the valley that may have good grape-growing potential. On the northern end of the valley, flat valley floor sites with deep, rich soils predominate. Water availability is generally not a problem if the site is relatively close to the Umpqua River.

Northeast along the Calapooya Creek, the most promising sites are northwest of the creek, with a few sites on good south to southeast slopes coming off Tyee Mountain. A couple of vineyards have been developed in the drainages of Tyee Mountain. From Interstate 5, the Calapooya Valley branches off into many subvalleys that have had little, if any, grapevines planted. Many of these valleys may not be suitable for vineyards because of higher elevation, narrow valleys, and high frost potential. Extremely careful site selection is needed to find the few sites that are suitable. One valley that presents some potential is the Camas Swale from Sutherlin to Nonpareil, where there are some benches, isolated hills, and gentle slopes off the ridge to the north. Unfortunately, some of the best locations in Camas Swale are being developed for housing as the city of Sutherlin grows eastward. To the north, Rice Hill provides the drainage divide for the area; the more isolated Rice Valley area to the south shows some promise as a grape-growing area. The main limitation in this area is water, with elevation considerations also important.

Northwest Douglas Area. The Northwest Douglas area consists of the main stem of the Umpqua River from the broad meander near the town of Kellogg, extending toward Elkton, and downstream to Scottsburg. The landscape in this area provides some good bench sites above the river, some south-facing sites sloping down to the river, and a few isolated hills. The region has moderate temperatures and heat accumulation, but the most limiting factor to grape growth is high rainfall (Table 2). Rainfall at Elkton averages over 50 inches per year. The saturated soils take longer to warm up and dry out in the spring, delaying vine growth and increasing humidity and the accompanying risk of fungal diseases. The most promising areas in Northwest Douglas can be found to the south of the town of Elkton (with three vineyards along State Highway 138), where there are gentle sloping sites and a few isolated hills. Westward from Elkton the valley narrows and rainfall amounts increase.

North Douglas Area. A series of ridges from Elkton southeast to Rice Hill separates the North Douglas area from the Northwest Douglas area. This upland region ranges from 1,000 to over 2,400 feet in elevation and is the largest, but mostly unsuitable, area of the Umpqua Valley AVA. Surrounding this upland area, the North Douglas area consists of the areas surrounding the towns of Drain, Yoncalla, and Curtain. The area also has the broad Pleasant Valley area, roughly from Rice Hill to east of Yoncalla, and a few other isolated valleys. The climate is similar to that found to the west, but the ridges that separate the locales act to reduce the average annual precipitation from nearly 54 inches at Elkton to just under 46 inches at Drain. A few vineyards have done well in this area, but high annual rainfall is a concern.

Putnam Valley, mostly to the north of State Highway 38, contains some beautiful south facing slopes currently used for orchards. To the south and west of Mount Yoncalla there are some gently sloping lands, a few benches, and an isolated hill or two. This area has the potential to offer more heat accumulation and frost protection than the rest of the area. The broad Yoncalla Valley and the greater Pleasant Valley offer the largest area of potential sites in this area. The area is roughly 300–600 feet in elevation with deep, moderately rich soils and a few sloping sites. To the northeast, Scotts Valley is a higher-elevation extension of Pleasant Valley with some potential south-facing sites at 500–800 feet. From Drain along State Highway 99, Scotts Valley northward along Interstate 5 to the town of Curtain and the northern AVA boundary a few miles to the northeast, the valleys are narrow, the elevations higher, and plantable area scarce.

Rogue Valley AVA

The Rogue Valley AVA (map in color section following page 72) is the southernmost winegrape-growing region of Oregon and is located entirely within Jackson and Josephine Counties (Code of Federal Regulations, 2000). The Klamath and Siskiyou Moun-tains form the western and southern boundaries of the region, and the Cascade Mountains are the eastern boundary. Established in 1991, the Rogue Valley AVA consists of three main areas: the Bear Creek Valley, the Applegate Valley, and the Illinois Valley, which together have over 15,000 potential acres for grape growing. In 2000, the Applegate Valley was approved as an American Viticultural Area lying wholly within the Rogue Valley AVA (Federal Register, 2000). The Applegate Valley is discussed in general in this section and in more detail in the section that follows.

Although this AVA takes its name from the Rogue Valley, it is better thought of as the Rogue "region," because there are diverse areas that lie within its boundary. Drained by the Rogue River and its tributaries, the region presents an array of lands and climates for grape growing. The AVA northern limits start at the intersection of Interstate 5 and the Josephine and Douglas County line and extends west and then southward to the border with California by following the Siskiyou National Forest boundary. The southern border of the AVA follows the boundaries of the Siskiyou and Rogue River National Forests to just southeast of Ashland. After encompassing the upper end of Bear Creek and some of its tributaries, the eastern border extends to just north of Shady Cove. The northern margin of the AVA then proceeds west and north back to the Josephine/Douglas County intersection along Interstate 5.

Grape growing and wine making in the Rogue Valley AVA has its origin in the middle of the nineteenth century, but after years of little interest a substantial growth in vineyards has occurred in the past two decades. The Rogue Valley AVA currently has 78 vineyards with 1,334 acres, which are increases of 105% and 342%, respectively, from 1987 (Oregon Vineyard and Winery Reports, 1987–2000). The number of wineries has grown from five in 1987 to twelve in 2000, with the total amount of grapes crushed increasing from 514 to 1,645 tons (220% increase). Because of a diverse climate, the region offers the various conditions needed to produce both cool- and warm-climate grape varieties. Grape varieties commonly grown in the Rogue Valley AVA are Merlot, Pinot noir, Cabernet Sauvignon, Chardonnay, Pinot gris, Riesling, and Gewürztraminer. In addition, plantings of Cabernet Franc, Pinot blanc, Early Muscat, Gamay, Malbec, and Syrah are increasing.

Climate

The Rogue Valley AVA has the highest elevations, but is overall the warmest and driest, of the western Oregon AVAs. The general north–south-tending valleys, with their proximity to the Pacific Ocean and intervening topographical barriers, create a climate transect of wetter and cooler conditions in the western parts of the region to the warmer and drier eastern areas. Precipitation varies from 12 to 60 inches across the region, declining in amount from west to east (Table 3). Overall, less than 15% of the total precipitation occurs during the growing season of April through October. The growing season averages 155–185 days, with the average last and first frosts on May 10 and October 10, respectively (median frost dates defined for 32°F: Oregon Climate Service and Western Regional Climate Center, 2000). The growing season is shorter in the Rogue Valley AVA than in the other AVAs due to higher elevations that bring later and earlier frost potential. Growing degree-days exhibit a similar spatial trend, with values ranging from 2,200 to nearly 2,900 from west to east.

Topography

The landscape of the Rogue Valley AVA is extremely diverse, with many vineyards on flat to very steep slopes (up to 30%) that are distributed along isolated hills, stream terraces or benches, and at the foot of alluvial fans. Elevation of potential and existing vineyard sites ranges from near 900 feet in the lower Grants Pass area to over 2,200 feet in the upper Bear Creek Valley. Although isolated hills can be found throughout the region, they are more common in the broader stretches of the valleys where they are separated from the surrounding mountains. Some good examples can be found in the mid to upper Bear Creek Valley and the broad region of the Illinois Valley. Terraces are found along each of the stream systems, with varying elevations above the valley floor depending on age and depth of stream downcutting. One of the more common vineyard site types can be found on footslopes at the base of the surrounding mountains. These sites are often 100–300 feet above the valley floor, in the thermal zone, and provide adequate slopes for maximizing solar radiation receipt. Footslopes can be found throughout the region with varying exposures (aspects); some good examples are along the southern fringe of the Bear Creek Valley near the towns of Talent and Phoenix. The general north–south-tending valleys produce many south to westerly and north to easterly aspects, which should be carefully considered in reference to the surrounding landscape.

Table 3. Average Climate Characteristics for Representative Stations in the Rogue Valley AVA

Station (Elevation)	*Average July Maximum Temperature (°F)*	*Average January Minimum Temperature (°F)*	*Average Mean Growing Season[a] Temperature (°F)*	*Growing Degree Days (Apr-Oct, 50°F base)*	*Precipitation (inches)*	*Frost-free Period (# of days last to first. 32°F)*
Applegate (1276 ft.)	NA[b]	NA	NA	NA	25.6	NA
Ashland (1750 ft.)	86.8	29.6	60.2	2338	19.2	154
Cave Junction (1280 ft.)	88.5	31.9	60.7	2403	59.8	153
Grants Pass (960 ft.)	90.1	32.7	63.1	2870	31.1	169
Medford Airport (1300 ft.)	90.5	30.1	62.6	2815	18.9	166
Medford Exp. Station (1457 ft.)	88.8	30.1	61.2	2490	21.2	140
Ruch (1549 ft.)	89.3	29.7	61.3	2531	26.0	137
Williams (1450 ft.)	NA	NA	NA	NA	33.7	NA

Source: All data are from the 1961-1990 climate normals for that station, except for Applegate and Williams, which are from monthly climate summaries for 1979-1998 and 1900-1998, respectively (OCS, 2000; WRCC, 2000).

[a] April through October. [b] NA=data not available.

Soils

The broad valleys of the Rogue AVA contain thick beds of rock and gravel derived from a mix of alluvial and glacial origins. Numerous isolated hills and terraces with a variety of soil types can be found within the AVA. Soil type varies greatly over the region, but there is a general pattern of dominant textures. Throughout most of the region, rich, deep loams are found on the flatter sites and heavier clay soils are found on eastern slopes, which have higher and more prolonged water availability. South- and west-facing slopes tend to have more granitic soil formations. Furthermore, although most soils in the region have the moisture-holding capacity needed for winegrape production, water availability is a principal consideration for any site here. Some water is generally needed for irrigation on the best hillside vineyards, but the greatest water need may be for protection against frosts in the early spring.

Another soil characteristic that varies from area to area is soil pH. In the Illinois Valley, with its greater rainfall, leaching of the soil has produced a slightly more acidic suite of soils. In the Bear Creek Valley there tend to be less acidic soils, and the Applegate Valley is intermediate between the two.

The main soil series found in many of the vineyards of the Rogue Valley AVA include the following: the Central Point, Kerby, Manita, Ruch, and Shefflein series in the Applegate Valley; the Agate–Winlow Complex, Brockman, Carney, Central Point, Coleman, Darrow, Evans, Holland, Medford, Provig–Agate Complex, Ruch, Selmac, Shefflein, Vannoy, and Wapato series in the Bear Creek Valley; and the Brockman, Cornutt–Dubakella Complex, Foehlin, Kerby, Pollard, and Takilma series in the Illinois Valley (Federal Register, 2000; Soil Conservation Service, 1983 and 1993).[1] Overall, there are four common vineyard soil series found in the different areas of the AVA, and they have the following general characteristics:

- Central Point—mostly coarse-loamy, deep, and well-drained soils that formed in alluvium weathered from granitic and metamorphic rocks. Most Central Point soils are found on low stream terraces and alluvial fans and have slopes of 0–3%.
- Ruch—widespread fine-loamy, generally very deep and well-drained soils that formed in mixed alluvium. Much of the Ruch soils can be found on high stream terraces, alluvial fans and footslopes with slopes of 2–60%.
- Shefflein—generally fine-loamy, deep, and well-drained soils that formed mostly in alluvium and colluvium (fine material deposited at the base of mountains) settings from granitic rocks. Shefflein soils are typically found along mountain slopes, ridges, and alluvial fans and can have slopes of 2–35%.
- Kerby—mostly fine-loamy, deep, and moderately to well-drained soils that formed in a mixed stream alluvium environment. The majority of the Kerby soils can be found on low stream terraces and have slopes of 0–3%.

Development Areas

The Rogue region can roughly be divided into three main areas that have successfully produced wine-grapes—the Illinois Valley, the Applegate Valley (discussed separately in the next section), and the Bear Creek Valley. Three additional areas that are relatively untested for commercial grape production are also discussed.

Illinois Valley Area. The Illinois Valley extends along U.S. Highway 199 from just south of the town of O'Brien to just north of Cave Junction. The main growing area is the broad valley southeast of Cave Junction, which contains three wineries and numerous successful vineyards. Elevations range from near 1,200 feet to over 1,800 feet, although most of the best sites are at 1,200–1,400 feet. The area is the wettest and coolest in the Rogue Valley AVA because of the greater maritime effect (Table 3). Most of the valley gets 2,200–2,600 degree-days. Cave Junction accumulates up to 2,400 growing degree-days. Average growing season temperatures are lower than most locations in the greater Rogue Valley AVA. Spring and fall frosts are frequent, and water for frost protection by sprinklers is generally necessary. Early ripening varieties have done well in this area, including Pinot noir, Pinot gris, Riesling, Gewürztraminer, and Chardonnay. Varietal experimentation is ongoing in the Illinois Valley

The main site limitations are land availability and water. There are few available benches, isolated hills, and sloping sites. Most successful vineyards are found on flatter lands. The hillside sites that are available generally have insufficient water. Water availability is best at the lower elevations. Soils in the Illinois Valley are varied, with little research done to identify the best soil and variety combinations.

Associated with the main stem of the Illinois Valley, the east–west-oriented valley area around Selma has many similar growing characteristics but is slightly warmer than the rest of the valley. Chardonnay and Pinot noir have done well in this area. Elevations are roughly 1,200–1,600 feet, although the narrowness of the valley extensions can be problematic with cold air drainage.

Bear Creek Valley Area. The Bear Creek Valley area is a large subregion that stretches from Blackwell Hill east of the town of Gold Hill and Lower Table Rock, southeast through the towns of Central Point, Medford, Jacksonville, Phoenix, Talent, and just past Ashland to the southern boundary of the AVA.

Elevations rise from north to south; the valley floor and surrounding hillsides are roughly 1,200–1,600 feet in elevation, and the southern portion near Ashland ranges from 1,700 to just over 2,000 feet. Historically this area has housed a substantial pear industry, but it has seen increased development of vineyards in response to recognition of its capacity for high-quality grape and wine production and the desire to diversify the agricultural industry. Today, three wineries and numerous vineyards call the Bear Creek Valley home.

The climate in this area is drier than most of the Rogue AVA, with annual rainfall averaging 18–22 inches. Days are sufficiently warm during the growing season, but nighttime lows can be dangerous during the spring. Sites should maximize cold air drainage and have adequate frost protection. Average growing season degree-days sum from 2,400 to over 2,800, with the highest values found around Medford and decreasing toward the surrounding hillsides. Within the Bear Creek Valley there are a wide array of possible grape-growing sites with numerous benches, south- and west-facing hillsides, and some ideal isolated hills. In addition, many north- and east-facing slopes can be found along the southern valley margin.

Soils tend to vary from deep, rich loams on the flatter sites to heavier clay soils on eastern slopes, which have greater and more prolonged water availability. The south- and west-facing slopes have soils that are more granitic. Water availability varies from site to site, but most vineyards depend to some degree on irrigation district water. Varieties that have done well in this area include Cabernet Sauvignon, Merlot, Chardonnay, Cabernet Franc, Sémillon, Sauvignon blanc, Pinot gris, Gewürztraminer, and Pinot noir. Other varieties receiving some attention include Malbec, Syrah, and Tempranillo. Some specific areas within Bear Creek Valley are worthy of note:

- The Willow Creek area northwest of Central Point; although few commercial vineyards are found here, it may offer promise if good water sources are available.
- The hillsides along the western valley extension toward Jacksonville offer a diverse collection of potential sites. Most of these are to the south and west of the Stage and Old Stage Roads.
- The Murphy and Griffin Creek area south of Medford offers some potential west- and east-facing sites.
- The Coleman Creek and Pioneer Road area southwest of Phoenix is a small area with few sites, but it offers good potential.
- The broad Anderson and Wagner Creek valley extension south and west of Talent has potential sites that vary from relatively flat to western, northern, and eastern exposures.
- The south- and west-facing slopes, benches, and isolated hills from east of Medford near Roxy Ann Peak and Hillcrest Road southeast to Emigrant Lake. There are many potential sites along this northeast valley margin; much of the land is currently planted to pears. The higher elevations probably require more frost protection.

Northwestern Rogue Valley Area. Along the northwestern portion of the Rogue Valley AVA are the lowest elevations and some of the warmest sites in the region. This area roughly consists of the greater Grants Pass area, extending north to the towns of Merlin, Hugo, and Sunny Valley. Rainfall at Grants Pass is nearly half that of Cave Junction in the Illinois Valley. The growing season average temperatures are the highest of the AVA, and average degree-day summation is 2,870 in Grants Pass. Although this area has adequate warmth to grow grapes, a climatic concern is increased fog during the fruit-ripening period of late summer and early fall. There are some promising benches, south-facing slopes, and isolated hills with predominately granitic soils throughout the area, but there has been little experience with grapes here.

Valley of the Rogue Area. Named after the state park near the center of the area, this region covers the main stem of the Rogue River and Interstate 5 from Grants Pass to the east of Gold Hill. It includes the Evans Valley north of the town of Rogue River. Although the river valley is narrow over much of its length, a few sites with south-facing slopes can be found north of Interstate 5 between Rogue River and Gold Hill. Potential site elevations along this east–west outlet of the Rogue River vary from just under 1,000 feet to over 1,200 feet. Some of these sites have historically supported fruit orchards and have begun to attract attention as grape-growing areas. Northward up Evans Creek, the Evans Valley has supported a few successful vineyards on elevations of 1,100–1,300 feet. Along the Rogue River, water availability, both for irrigation and for frost protection, is paramount. The narrow valleys produce strong dams of cold air that accumulate along the riverside vegetation. In addition, this outlet to the Pacific Ocean acts as the main cold air drainage outflow of the upper Rogue and Bear Creek Valleys. Site-specific climate, soil, and varietal data are limited in the area.

Northeastern Rogue Valley Area. Mentioned separately from the main Bear Creek Valley because of its dryness, this area consists of the upper reaches of the Rogue River Valley. Starting north of Medford and continuing along State Highway 62, this region consists of the Antelope and Yankee Creek extensions to the southeast, the Butte Creek Valley southeast to the town of Lakecreek, north past Eagle Point and up the Beagle Sky Ranch and Sams Valley extension to the northwest, and finally to the northern AVA boundary near Shady Cove. This area is the driest of the subregions, with annual rainfall usually less than 16 inches. Elevations range from 1,200 to 1,700 feet, generally increasing toward the upper Rogue River and Shady Cove. Numerous benches, south- and west-facing slopes, and isolated hills can be found throughout the subregion. There are a few commercial vineyards in the area. Water for irrigation and frost protection can be a limiting factor in this area. Much of the Antelope, Yankee, and Butte Creek Valleys have ample water from the Medford Irrigation District, but some areas may not have access to water. The other valleys generally have seasonal streams, making development of other water sources more important.

Applegate Valley AVA

Oregon's newest AVA was established in 2000 (Federal Register, 2000) and is wholly contained within the Rogue Valley AVA, nestled between the Illinois and Bear Creek Valleys. The Applegate Valley AVA (map in color section following page 72) is considered a subappellation of the Rogue Valley AVA. The Applegate Valley is approximately 50 miles long, oriented southeast–northwest from near the California border to where it joins the Rogue River west of Grants Pass.

Grapes have been grown in the Applegate since 1870, and currently there are 26 vineyards, over 360 acres of vines planted, and four wineries. Varieties that have proved successful here are Cabernet Sauvignon, Chardonnay, and Merlot. Other varieties, such as Syrah and Dolcetto, are under evaluation.

Climate

Surrounded by the Siskiyou Mountains, which accentuate climatic differences in the region, the area is intermediate to the Illinois and Bear Creek Valleys in all climate parameters. It is warmer and drier than the Illinois Valley and cooler and wetter than the Bear Creek Valley (Table 3). The only long-term reporting weather station in the AVA is Ruch, which exhibits degree-day values of over 2,500 for the growing season and 26 inches of rainfall annually. The location of Ruch is not, however, completely representative of the valley climate, and a potential grower should discuss variations with other growers to assess sites within the AVA. It must be noted that the frost-free period is short in the Applegate (137 days), indicating that site characteristics that maximize heat accumulation and facilitate cold air drainage are important for vineyard success.

Topography

The Applegate is a long valley of varying widths, averaging 3,000–10,000 feet. Narrower valley stretches may receive less solar radiation due to obstruction of late day sun by the surrounding mountains. Elevations range from 1,200 to 1,600 feet, with the most promising sites in the AVA on south- and west-facing slopes found along, and generally north of, State Highway 238 from the Missouri Flat benchland to Ruch and just southward. There are many potential sites along the valley floor, but the width of the valley and the potential cold air drainage and frost problems should preclude large-scale plantings on the flatter sites.

One area of relatively unexplored grape-growing potential is the Williams Valley, which extends south from the town of Provolt. This area has not had grapes planted to any large degree, but it may be a viable area for some cool-climate varieties. This north-facing, narrow valley has some east and west slopes, but higher elevations could be problematic with frost problems.

Soils

Soils across the region vary greatly, but most soil types in the Applegate Valley originate from granitic parent material. The majority of the vineyards in the Applegate Valley are planted on stream terraces or benches cut by prior locations of the Applegate River or on alluvial fans draining off the surrounding mountains. Both types of sites generally provide deep, well-drained soils. The dominant soil series found throughout most vineyards here are Central Point, Kerby, Manita, Ruch, and Shefflein (Federal Register, 2000; Soil Conservation Service, 1983, 1993). The Central Point, Kerby, Ruch, and Shefflein soils are described in the Rogue Valley AVA section. The Manita soils are mostly deep, well-drained soils on alluvial fans and hillsides of the valley that formed from colluvium and alluvium in altered sedimentary and extrusive igneous rocks.

Development Areas

The Applegate Valley offers potential sites that range from broad terraces to gently sloping ground approaching both the Applegate and Little Applegate Rivers. From the northeast on Highway 238, potential sites can be found to the east of the highway and the town of Ruch. The main concern in this area is water availability, with most existing sites relying on well water. West of Ruch, the valley opens to its broadest expanse, with moderate to flat land sloping southwest to the Applegate River. Much of this land is currently used for hay production, but it would provide adequate exposure and heat accumulation for winegrapes.

Heading south along the Applegate River and the Little Applegate River, the valley narrows, but there are many potential sites along alluvial terraces north and east of the rivers. Heading northwest on Highway 238 from Ruch, nearly the entire northeastern stretch to southeast of the town of Murphy contains potential sites. Most of this area consists of narrow to broad terraces including China Gulch, along Humbug Creek, and the Missouri Flat area, with numerous existing vineyards. Also along this stretch of the valley are a few promising sites along the river, but careful selection is needed to maximize air flow and solar radiation receipt due to the leveling out of the landscape and the obstruction of the mountains to the southwest.

A region of the AVA that has not been fully explored is the Williams Valley, which experiences more rainfall than other parts of the AVA and has limited southerly to westerly exposures. Reasonable sloping land and heat accumulation should, however, be achievable.

Columbia Valley AVA and Hood River Valley

The Columbia Valley AVA (map in color section following page 72) was established in 1984 and amended to include new areas in 1987 and 1993 (Code of Federal Regulations, 2000). A recent realignment, to expand the Walla Walla Valley AVA and extend the Columbia Valley AVA to encompass all of the Walla Walla, was done to correct a mistake in the 1984 boundary (Federal Register, 2001). The Columbia Valley is the largest AVA in the Pacific Northwest, encompassing over 18,000 square miles straddling the Washington and Oregon border. Although the AVA is mostly synonymous with the state of Washington, it includes grape-growing areas in parts of Wasco, Sherman, Gilliam, Morrow, and Umatilla Counties in Oregon. The region is divided by the Columbia River, which drains the vast upland plain created from large basalt flows that covered the region 17–12 million years ago.

Grapes have been grown in the region since before 1900, including over 4,000 acres of vines reported in The Dalles region in the early 1900s. As of 1999, Oregon had 527 of a total of 14,777 acres of winegrapes in the AVA (Washington Winegrape Acreage Survey, 1999. This survey is not produced annually, therefore no report was produced for 2000). On the Oregon side of the AVA, the most common varieties are Cabernet Sauvignon, Merlot, and Riesling, with smaller acreage of Syrah, Pinot noir, and Chardonnay (Oregon Vineyard and Winery Reports, 1987–2000). Varieties being tested include Cabernet Franc, Tempranillo, Sauvignon blanc, and Viognier.

Currently there are no commercial wineries operating on the Oregon side of the AVA. Production from the Oregon vineyards has two main markets: Washington's Columbia Valley AVA wineries near Yakima, and wineries throughout Oregon. Most of the grapes in The Dalles area tend to have market support from Oregon wineries, whereas those farther east in the AVA find their market potential to be greater in Washington (Lonnie Wright, Columbia County Vineyards, personal communication).

Climate

The climate of the Columbia Valley region is defined as semi-arid, a product of the rain shadow effect of the Cascade Mountains to the west. The region is generally warmer and drier than other Oregon AVAs, except the Walla Walla Valley, averaging less than 15 inches of precipitation (Table 4). Irrigation is required in this region, and water quality can be a problem. Many wells contain higher than desired salinity levels, although better water from the local irrigation district is available in some areas.

Elevation and aspect are the critical variables that determine a site's climate in this region. The length of the growing season is strongly dependent on elevation, ranging from 150 to over 215 days. Average last and first frosts are on April 19 and October 17 (median frost dates defined for 32°F: Oregon Climate Service and Western Regional Climate Center, 2000). Although many sites in the region experience shorter growing seasons than in other Oregon AVAs, the long day-lengths and cloudless days provide suitable grape-growing conditions.

Growing degree-day totals range from near 2,000 at higher elevations to over 3,300, which is the highest heat accumulation in the state. Heat accumulation varies greatly by location in the region. Weather stations from low-elevation locations near the Columbia River report 100–600 degree-days more than sites not far from the river. The region experiences higher daytime and lower nighttime temperatures than other Oregon AVAs, providing good ripening conditions for many warm climate varieties. Low temperatures can, however, be problematic; most areas are prone to vine damage in cold winters. Sites should maximize cold air drainage and solar radiation receipt.

Wind is the other major climate determinant; the Columbia River Gorge and Plateau experience some of the strongest and most persistent winds in Oregon. Dropping off the plateau and funneling into the gorge, these winds most often come from the east or southeast, and can average 10–20 miles per hour during the growing season. Local topographical variations and planted windbreaks give some protection from the wind. End rows that face the predominant wind directions are sometimes constructed with taller trellises, which provide some wind blockage for the rest of the vineyard.

Table 4. Average Climate Characteristics for Representative Stations in the Columbia Valley AVA

Station (Elevation)	*Average July Maximum Temperature (°F)*	*Average January Minimum Temperature (°F)*	*Average Mean Growing Season[a] Temperature (°F)*	*Growing Degree Days (Apr-Oct, 50°F base)*	*Precipitation (inches)*	*Frost-free Period (# of days last to first 32°F)*
Arlington (285 ft.)	91.6	28.8	65.2	3303	8.8	182
Condon (2861 ft.)	82.3	23.6	56.5	1610	14.1	127
Dufur (1330 ft.)	85.1	24.4	58.0	1800	12.5	123
Heppner (1883 ft.)	85.7	25.9	59.5	2100	14.0	151
Hermiston 2 S (620 ft.)	88.4	25.7	62.4	2936	9.1	164
Hood River Exp. Station (500 ft.)	80.1	28.2	58.7	2082	31.1	157
Pendleton Airport WSO (1482 ft.)	87.8	27.3	62.1	2729	12.0	187
The Dalles (102 ft.)	87.8	29.9	64.6	3175	14.0	218

Source: All data are from the 1961-1990 climate normals for that station (OCS, 2000; WRCC, 2000).

[a] April through October.

Topography

The expansive Columbia Valley AVA is found mostly within the Columbia Basin and Plateau physiographic regions, with the western portion including some of the Eastern Cascade Mountain province. The majority of the region is made up of a vast basalt plateau that has been uplifted and deeply eroded by the Columbia River and its many tributaries. Along the Columbia River, steep escarpments lead to upland terraces that can rise 100–1,000 feet above the river. Along most of the tributaries that flow into the Columbia are towering vertical basalt cliffs or sandy terraces. Both situations have few adequate sites for vineyards. In the western portion of the AVA, to the south and west of The Dalles, the landscape consists of wide, nearly level ridgetops bisected by steep V-shaped canyons with alluvial deposits and volcanic material. The best sites are found where the canyons are wide enough, providing adequate slopes and air drainage.

Soils

Soil composition and depth vary greatly over this large region, but the underlying rock material over most of the plateau is basalt. Soil types vary from sandy, well-drained soils to silt and clay loams that range from shallow to very deep. Common soil types in the vineyard areas to the south, southwest, and west of The Dalles include the Cherryhill–Chenowith association (Soil Conservation Service, 1982). These soils are found on slopes and terraces that consist mostly of silt loam and other loam types with rooting depths of 60 inches or more. The most common limitations are shallow depths on moderate to steep slopes and water erosion, which can be minimized by using a proper cover crop. Grapevines grown on shallow soils can be prone to drought stress even with adequate irrigation. In this region, soil depth also influences the cold hardiness of grapevines. Shallow soils tend to be drier, and dry soil results in greater root system damage during hard freezes. Deeper soils are generally found in the lower areas, which have poor drainage and an increased risk of frost damage. Deeper soils also encourage excessive vegetative growth that may not harden off adequately prior to cold winter temperatures.

The Columbia Valley AVA experiences many of the soil-related problems associated with drier climates. Soils commonly have high calcium or salinity levels. These soil types often have a hardpan of accumulated salts located a few inches to over a foot or more below the surface. The hardpan acts as a barrier to root growth and water penetration. Hardpans generally require deep ripping to break up the barrier, although this may not be possible in all instances. Calcium-rich soils are also generally high in carbonate and have relatively high pH values.

Development Areas

The Columbia Valley AVA in Oregon is vast, but it currently has only a few areas with established vineyards to indicate a potential for growth. Site limitations are the availability of water, competition for good land with other agricultural interests, and potential damage from phenoxy-type herbicide drift from wheat farms. The majority of existing vineyards are concentrated in and around The Dalles. Between 150 and 250 acres of grapevines are planted south, southwest, and west of The Dalles, mostly in the Threemile, Chenowith, and Mill Creek drainages. The topography ranges from 400–800 feet in elevation, with valley and canyon settings. The drainages are oriented southwest–northeast, with some southerly aspects along benches and slopes. The three drainage areas accumulate 2,500–2,900 degree-days (Lonnie

Wright, Columbia County Vineyards, personal communication).

Vineyards have been established with varying success in isolated areas south and east of The Dalles. Vineyards near the towns of Boardman and Clarke, just west of Hermiston, were killed in the winter of 1990/91 when temperatures dropped as low as -20ºF. Other small vineyard operations are active along the Umatilla River near the town of Echo and in the area around Hermiston and Umatilla. Farther west, potential vineyard sites exist near the Columbia River and a mile or two upstream along Willow Creek, Alkali Canyon, the John Day River, and the Deschutes River. New vineyards are being established near the town of Dufur, along Fifteenmile Creek, just outside the AVA boundary.

Hood River Valley

The Hood River area is not within the Columbia River AVA, but it is home to two wineries and eleven vineyards with 137 planted acres (Oregon Vineyard and Winery Reports, 2000). The Hood River Valley is between the town of Hood River and The Dalles (map in color section following page 72), where the semiarid air of eastern Oregon collides with the marine air of western Oregon. Consequently, a dramatic east–west change in climate occurs over a relatively short distance here. Precip-itation levels in the Hood River Valley can be double those found in the Columbia Valley AVA, depending on elevation. Growing season temperatures and degree-days are lower than in the Columbia Valley AVA, and winter temperatures are milder due to lower elevation and proximity to the coast.

The main agricultural area of the Hood River Valley is oriented north–south, running from the Columbia River southward. Much of the prime land is planted to pear, apple, and cherry orchards. Irrigation water is plentiful from five farmer-owned irrigation districts that obtain water from the Hood River and a couple of reservoirs. Soil types vary considerably. The major agricultural soils series are Parkdale loam, formed from volcanic ash deposits; Van Horn fine sandy loam, formed from alluvial deposits; Hood loam soils, formed in lake deposits; and Oak Grove loam, formed in deep clayey outwash and alluvial materials.

The greatest limitation is finding a site with a good aspect and proper elevation. The north–south valley has increasing elevation toward the south. Most locations have aspects ranging from north to east or west. Few southerly exposures are available.

Walla Walla Valley AVA

The easternmost AVA in Oregon, the Walla Walla Valley (map in color section following page 72), was approved as a winegrowing area in 1984 and amended in 1987 and 2001 (Code of Federal Regulations, 2000; Federal Register, 2001). The region consists of a broad upland plain drained by the Walla Walla River and its tributaries, encompassing parts of Oregon and Washington. Although records of grape growing date back to the earliest settlements in the valley, it was immigrants from Italy who brought to the region the skills of wine making and knowledge of grape growing that are still used today. The oldest vineyard, which is still producing a commercial harvest today, has its roots in the Italian home wine-making tradition. Planted on the Harry Feigner farm just north of Milton Freewater in 1926, Black Prince (Cinsault) vines still provide grapes that are blended in small proportion with Syrah at a local winery.

The Walla Walla AVA has developed a single identity with little attention being paid to the state line that runs directly through its middle. Currently it contains approximately 1,000–1,100 acres of grapes on 36 vineyards plus another 17 estate vineyards, a nearly tenfold increase over the past decade (unpublished data collected by the Walla Walla Valley Wine Alliance). On the Oregon side, the AVA lies wholly within Umatilla County and consists of ten vineyards and 367 acres (Oregon Vineyard and Winery Reports, 1987–2000). The AVA is home to 34 licensed wineries with several more in the planning stages, all located in Washington and relatively close to the town of Walla Walla. The main red varieties grown in the Walla Walla AVA are Cabernet Sauvignon, Merlot, Syrah, and Cabernet Franc. White varieties make up only 5% of the total plantings in the AVA and include Chardonnay, Sémillon, Viognier, and Gewürztraminer.

Geographically, the Walla Walla AVA is east of the broad westward turn of the Columbia River and is bounded by the Blue Mountains to the south and east, opening up to the expanse of the Columbia Plateau to the southwest and northeast. Water for irrigation comes largely from the Walla Walla River and its tributaries, but many shallow to deep aquifers provide well water sources. The Walla Walla River drainage includes the major tributaries of Cottonwood Creek and Mill Creek, each influencing the flow of air through the valley and the varied mesoclimates of the region. The Walla Walla River starts on the Oregon side of the Blue Mountains, winds its way through Milton Freewater across to the Washington side of the AVA, and ultimately to the Columbia River past the western end of the appellation. Cottonwood Creek bisects the AVA and, along with Mill Creek, the most northern of the subbasin watersheds, collects many smaller streams and creeks, eventually combining with the Walla Walla as it heads west to the Columbia River. Each of the main drainages creates separate meso-climates, influencing the grape-growing methods used within the AVA.

Climate

The ability to grow grapes in the Walla Walla region is largely determined by the bowl-shaped valley, which produces a mixing of several weather patterns that ultimately influence the mesoclimates of existing and potential vineyard locations. The Blue Mountains to the south and east, which rise to just over 4,500 feet, effectively block or rechannel many of the major weather systems through the various drainages of the Walla Walla River. Dominant airflow is from the southwest, funneling from the Columbia Plateau into the mountains and spilling back to the valley floor. Add to the mix an occasional polar front with cold air from the north, and the region can experience the broadest temperature range of any Oregon AVA, with extremes from above 110°F in the summer to as low as -30°F in the winter.

The unique climate of the Walla Walla produces a variety of growing season conditions. Precipitation averages 10–18 inches over the AVA, with larger amounts in the southeast along the forks of the Walla Walla as they drain from the Blue Mountains (Table 5). Nearly 45% of the annual rainfall comes during the growing season and is a result of a greater summer thunderstorm occurrence than most of the western Oregon AVAs. Of prime consideration for viticulture in this appellation is the length of the growing season. The valley has areas with frost-free days as few as 140, while at other locations the season may be 220 days. Again, the bowl shape of the valley comes in to play, creating regular inversions as cold air drains from the mountains to the valley floor and becomes trapped at the lower elevations. There are few places on the west end of the valley (where the elevation in some places is less than 400 feet) where the cold air can drain, thus backing it up against the rising terrain to the west, near what is known as Nine Mile Hill. In fact, the only drainage for air is the narrow channel created by the Walla Walla River as it heads the last few miles to empty into the Columbia River, near Wallula Gap. Therefore, the south side of the valley, which contains this small outlet, generally drains the cold air faster and is one reason why tree fruit has had a regular place in agriculture near Milton Freewater. On the north side of the appellation, in the Mill Creek watershed, cold air from that drainage is able to empty through a series of canyons that drain into Dry Creek, leaving warm air in its place and creating areas for growing grapes north and east of Walla Walla. However, Dry Creek eventually returns to the Walla Walla River near Touchet and deposits its cold air to the lower elevations on the west end of the AVA. This unique situation confuses many people who study the valley, because some of the warmest summer temperature locations are also the places that get the coldest, both during the frost seasons and during the most severe winter weather. This means that degree days and heat units for particular sites many not indicate the ability to grow winegrapes there, and length of growing season must be factored into site assessment.

Arguably the most important weather consideration in the Walla Walla is the extreme cold temp-

Table 5. Average Climate Characteristics for Representative Stations in the Walla Walla Valley AVA

Station (Elevation)	*Average July Maximum Temperature (°F)*	*Average January Minimum Temperature (°F)*	*Average Mean Growing Season[a] Temperature (°F)*	*Growing Degree Days[b] (Apr-Oct, 50°F base)*	*Precipitation (inches)*	*Frost-free Period (# of days last to first. 32°F)*
Milton Freewater, OR (971 ft.)	88.6	27.8	63.5	3030	14.4	186
College Place, WA (691 ft.)	88.7	28.5	62.1	2603	13.0	157
Touchet, WA (492 ft.)	88.6	29.5	63.0	2790	8.4	161
Whitman Mission, WA (630 ft.)	89.4	25.7	60.9	2499	13.8	133
Walla Walla, WA (1186 ft.)	86.7	29.4	62.4	2652	20.3	201
Walla Walla Airport, WA (1210 ft.)	89.4	28.4	64.0	3049	19.5	206
Garden City Heights, WA (1050 ft.)	90.7	NA[c]	63.6	3031	NA	172

Source: Milton-Freewater, Oregon Climate Service (1993, from 1961-1990 means);

Whitman Mission, Walla Walla Airport, and Garden City Heights, WRCC (2000, period of record);

College Place, Touchet, and Walla Walla, PAWS, WSU IAREC (2000, period of record).

[a] April through October.

[b] Growing Degree Days are calculated from the station's complete period of record, which varies by data source.

[c] NA=data not available.

eratures that occasionally sweep across the area during the winter. The spatial variability in these extreme events causes some places to seldom drop below zero in the coldest winters, while other areas have recorded lows well below -20°F. In any given year, one or more polar weather events occurs, and every few years the low temperatures kill vines back to the ground. Long-term probability of killing frosts or freezes versus the economics of grape production should be carefully assessed for every site.

Topography

The terrain of the Walla Walla region results from the intermingling of four land resource areas: the Columbia Basin, the Columbia Plateau, the Palouse and Nez Perce Prairies (making up the Blue Mountain foothills), and the Northern Rocky Mountains (which make up the Blue Mountains) foothills (Soil Conservation Service, 1988). The region provides many potential winegrape-growing sites, of which few have been fully explored. Most of the best sites are on flat to gently sloping lands, but some isolated hills and benches can be found in the region. Current and potential growing sites in the region vary from 600 to 1,500 feet in elevation, with the best sites generally at 800–900 feet. The alluvial terraces, fans, and plains of the foothills of the Blue Mountains, which spread out from the south and east toward Milton Freewater, provide much of the best potential winegrape acreage. Sites on flat lands and at low elevations experience greater frost, or freeze potential, as do sites at higher elevations. Finding an optimum site requires maximizing relative elevation, slope, aspect, and cold air drainage on terraces and gently sloping hillsides.

Soils

The soils of the region are a wide range of types, from deep, rich silt loams to rocky riverbeds with boulders the size of grapefruit. General soil types vary by land resource area, with mantled wind-blown silts on most of the Columbia Plateau, alluvial deposits reworked by the wind in the Columbia Basin, and uplifted and folded volcanic basalts mixed with alluvial deposits in the Blue Mountains and its foothills (Soil Conservation Service, 1988). Specific soils found in these areas include these:

- The Powder–Umapine–Pedigo and Freewater–Hermiston–Xerofluvents soil complexes, which formed from alluvial deposits on floodplains and terraces. These soils are generally deep but range from poorly drained and containing high amounts of sodium (Pedigo) to excessively drained (Freewater). The Powder, Umapine, and Hermiston soils have few limitations in terms of agriculture. These complexes are generally found along the Walla Walla and Hermiston drainage systems.
- The Quincy–Winchester–Burbank, Adkins–Sagehill–Quincy, Shano–Burke, Oliphant–Ellisforde, and Walla Walla soil complexes. These soil types formed from wind-transported sand, loess (loam deposit), alluvium, and lacustrine sediments (lake). Most can be found laterally from the main stems of the rivers on the lower plains and terraces of the region. The soils in these complexes vary in suitability, with the most common limitations being wind and water erosion, excessive drainage, hardpans, and low natural fertility.
- The Waha–Palouse–Gwin and Athena soil complexes, which formed in loess and colluvium along the foothills of the Blue Mountains. These soils range from shallow (Gwin) to deep and well drained (Palouse and Athena), with the main limitation of increased water erosion on the steeper slopes.

Vineyards are currently planted in a variety of soil types, but limited experience leaves much of the potential unknown. It is recommended that soil variations be discussed with a local Extension agent, other farmers, and consultants.

Development Areas

The Walla Walla Valley presents a wide variety of sites, slopes, soils, and heat accumulation. The area around Milton Freewater offers great potential, with a long growing season and parts of the area protected from extreme cold temperatures. The best area extends up the drainage of the Walla Walla River into the canyon where the river forks. Orchards line this canyon from about 1,500 feet in elevation to the river at Milton Freewater. There are many orchards and some vineyards northeast and northwest of this small town.

In the middle of the valley is the Cottonwood Creek drainage, with numerous existing vineyards and potential sites. Although there are many interesting and suitable sites in this area, the region generally has colder temperatures and shorter growing seasons. Variety, slope, and elevation are the most critical determinants in site selection through the middle portion of the Walla Walla Valley.

On the north side of the Valley is Mill Creek, a watershed that influences a wide area generally outlining the northern parts of the appellation. There are several new vineyards here, especially on the north and east sides of the town of Walla Walla and near the Walla Walla airport. This area has great potential, with many sites having favorable slope, heat accumulation, and a long growing season.

Other areas being explored include some of the smaller stream drainages to the north of the town of Walla Walla and on Nine Mile Hill. Each of these has at least one vineyard planted, with more acreage planned.

Additional limiting factors in the region include the restrictive land use zoning found in some of the farming areas. Many of the best potential winegrape growing sites are in areas where the smallest tract of land that can be purchased is 160 acres. In addition, much of the potential vineyard land of the valley does not have irrigation water and has little hope of obtaining an existing water right. Overall, there are some wonderful sites still to be explored, and no one has yet measured the potential acreage that could grow grapes in the Walla Walla Valley.

The Walla Walla region is new enough, and has attracted such a diverse range of wine makers and grape growers, that new sites are constantly being explored. Where the growing season climate, the slope, and the soil combine to provide outstanding conditions, classic reds have done well. No matter the grape, the major challenge of growing in this valley is finding those spots where all the variables come together. Currently, almost every vineyard is located in an area of "uniques"—a combination of soils, growing season, slope, and varieties that makes them truly individual in nature.

Note

1. This list is meant only as a comparison for potential soil types; it does not imply that these soils are the only ones that should be considered for vineyard suitability.

References and Additional Resources

Allen, J. E., and M. Burns. 1986. *Cataclysms on the Columbia.* Timber Press, Portland, Ore.

Code of Federal Regulations. 2000. Alcohol, Tobacco Products, and Firearms (Title 27, Parts 1-199), Part 9: American Viticultural Areas.

Federal Register. 2000. Rules and Regulations (Vol. 65, No. 241/Thursday, December 14, 2000)

Federal Register. 2001. Rules and Regulations (Vol. 66, No. 38/Monday, February 26, 2001).

Natural Resources Conservation Service. 1997. Soil Survey Geographic (SSURGO) Data Base for Douglas County Area, Oregon. http://www.ftw.nrcs.usda.gov/ssur_data.html

Natural Resources Conservation Service. 2002. Soils. http://soils.usda.gov

Oregon Vineyard and Winery Reports. 1987–2000. Oregon Agricultural Statistics Service, Portland. http://www.nass.usda.gov/or/vinewine.htm

Oregon Climate Service (OCS). 2000. Office of the State Climatologist, Oregon State University, Corvallis. http://www.ocs.orst.edu/

Soil Conservation Service. 1982. Soil Survey of Wasco County, Oregon—Northern Part. G. L. Green (ed.). USDA and Oregon Agricultural Experiment Station, March 1982.

Soil Conservation Service. 1983. Soil Survey of Josephine County, Oregon. R. Borine (ed.). USDA, December 1983.

Soil Conservation Service. 1988. Soil Survey of Umatilla County Area, Oregon. D. R. Johnson and A. J. Makinson (eds.). USDA, November 1988.

Soil Conservation Service. 1993. Soil Survey of Jackson County Area, Oregon. D. R. Johnson (ed.). USDA and Oregon Agricultural Experiment Station, August 1993.

Soil Conservation Service: Soil Surveys of Benton* (1975); Clackamas Area* (1985); Lane (1987); Linn Area* (1987); Multnomah* (1983); Polk* (1982); Washington* (1982); and Yamhill Area*(1974). Various editors. USDA. (* Denotes survey is available online at http://soils.usda.gov)

Washington State University, Public Agriculture Weather System (PAWS). 2000. Irrigated Agriculture Research and Extension Center (IAREC).

Western Regional Climate Center (WRCC). 2000. National Oceanic and Atmospheric Administration (NOAA), Reno, Nev. http://www.wrcc.sage.dri.edu/

5

Varieties and Clones

David R. Lett and Edward W. Hellman

The unique flavors and other quality characteristics of grape varieties are the essence of fine wine. But simply growing a classic winegrape variety does not guarantee that a characteristically high-quality wine will result. Varieties must be appropriately matched to the climate and the site to have any reasonable chance for success. Furthermore, variety selection is not based solely on what will grow well and produce good-quality grapes. The fruit must be marketable to wineries at a profitable price. This chapter discusses variety and clone selection in the major production regions of Oregon. The selection process is the same everywhere; only the local climatic conditions vary, which influence the suitability of varieties. Many options are available to the grower, but this chapter focuses on climatically adapted varieties that have an established market and on varieties that are likely to be adapted but may require additional testing or improved market development. We discuss the performance of most of the principal grape varieties grown in Oregon. Descriptions of the less prominent varieties, along with hundreds of other grape varieties, can be found in the reference books by Robinson (1996) and Galet (1998).

Utilization of specific clones of grapevine varieties has the potential for fine-tuning the matching of grape variety to site conditions and winemaking objectives. A clone is defined (Hartman et al., 1990) as a genetically uniform group of individuals derived originally from a single individual by asexual propagation (cuttings, grafting, etc.). One reference book lists forty distinct clones of Pinot noir (ENTAV-INRA, 1995) and describes their individual characteristics. Clonal evaluations in Oregon have been limited up to this point; only Pinot noir and Chardonnay clones have undergone systematic testing by Oregon State University (Price and Watson, 1995; Oregon Wine Advisory Board, 1993–2000). Recently, an evaluation of Merlot clones was initiated at the Southern Oregon Research and Extension Center. Readers are referred to the recent books by ENTAV-INRA (1995) and Caldwell (1998) for discussions of the characteristics of specific grapevine clones.

Climate and Maturity Classification

The principal site characteristic that should influence variety selection is climate, and temperature is the single most critical factor. Amerine and Winkler (1944) developed the commonly used degree-days system for grapes and employed it as the basis for segregating California into five winegrape-producing regions. Grape varieties were recommended for the climatic regions based on degree-days, ripening season, performance in the vineyard, and subsequent wine quality. Several more elaborate prediction methods (Gladstones, 1992; Jackson and Cherry, 1988; Smart and Dry, 1980) have been developed that are intended to assist growers in characterizing their location compared to known winegrape production regions as universal guides to selection of adapted varieties. However, general climatic characteristics in Oregon's winegrowing regions are well known and are described in Chapter 4. We now also have the benefit of more than thirty-five years' experience of growing and testing varieties. Selection of adapted varieties can, therefore, be based on demonstrated performance in neighboring areas, or by experimentation with untested varieties that have a similar ripening period.

The time of the season at which a grape variety matures its fruit is obviously of critical importance in a cool climate, or where there is a short growing season. Maturity period, therefore, has long been used to match varieties to climates, and also as a descriptor for grapevine variety identification *(ampelography)*. The Pulliat maturity classification system, developed in the nineteenth century and described in Galet (1998), is based primarily on the early-maturing grape variety Chasselas, which is used as the basic Period I indicator vine. Grape varieties are classified into one of five periods (Early, I, II, III, IV) based on the number of days longer than Chasselas required for the fruit to reach maturity. Within each period, varieties are further categorized into one of three groups: early, middle, and late. Chasselas continues to be the standard variety by which maturity time is measured in French viticulture (ENTAV-INRA, 1995), although it is simply expressed as the number of weeks after Chasselas rather than grouping varieties of similar maturity into Periods. The maturity ranking system of Gladstones (1992) places varieties within eight maturity groups, at intervals of 50 biologically effective degree-days.

A comparison of the three maturity classifications for selected grape varieties is presented in Table 1. There is general agreement in the relative ranking of varieties, although there are also some apparent discrepancies. For example, Pinot noir ripens before Pinot gris according to ENTAV-INRA, but the order is reversed in the Gladstone classification, and Pulliat places both varieties in Period I (middle). Similarly, the varieties Sémillon and Viognier ripen 2.5 weeks after Chasselas by the ENTAV-INRA system but are placed in Gladstones' maturity groups 4 and 5, respectively.

Some discrepancy is to be expected considering the natural variability of grapevines and the inherent precision of the classification systems. Pulliat separates varieties into five groups, each with three subdivisions, whereas the Gladstones system has eight groups. The ENTAV-INRA system rounds off the relative ripening time to the closest half-week. It is probable that some discrepancies result from differences in the interaction of variety and climate. Two of the systems are based on French experience, and the third is based on Australian growing conditions. McIntyre et al. (1987) reported considerable variation in phenology of identical varieties growing in the five winegrape climatic regions of California. Clonal differences could also account for differences of as much as one week or more in some cases, and variations in local wine-making practices may modify the definition of fruit maturity. None of these issues is serious enough to preclude the judicious use of maturity classifications as a general guide to selecting varieties. Again, experimentation with varieties not previously grown in a region has a greater likelihood of success if they have a similar maturity time as established varieties in the same or a similar region.

Table 1. Ripening Period Classifications of Selected Grape Varieties

	Pulliat Ripening Period[a]	*ENTAV-INRA Time of Maturity*[b] *(weeks)*	*Gladstones Maturity Group*[c]
Muscat Ottonel	-	0	2
Pinot Meunier	I	+0.5	3
Pinot noir	I	+0.5–1	3
Müller-Thurgau	II early	+1	2
Pinot gris	I	+1.5	2
Pinot blanc	I	+1.5	-
Chardonnay	I late[d]	+1.5	3
Gamay	I late	+1.5	3
Dolcetto	-	-	3
Gewürztraminer	II early	+1.5	3
Tempranillo	-	+2	4
Sylvaner	II early	+2.5	3
Sauvignon blanc	II	+2.5	3
Malbec	-	+2.5	4
Sémillon	II	+2.5	4
Merlot	II	+2.5	5
Sangiovese	-	-	5
Viognier	II	+2.5	5
Syrah	II	+2.5	5
Cabernet Franc	II	+2.5–3	5
Riesling	II	+3	4
Cabernet Sauvignon	II early	+3–3.5	6
Grenache	III	+4	7

[a] Five periods, based on ripening order in relation to the variety Chasselas (Galet, 1998).

[b] Maturity time in relation to maturity date (August 14, 38-year average) of the variety Chasselas at INRA Vassal Station-Languedoc (ENTAV-INRA, 1995).

[c] Based on Australian experience with ripening order, eight groups separated by 50 biologically effective degree-days (Gladstones, 1992).

[d] Oregon experience has been that ripening time depends on the clone.

Variety Selection in the Willamette Valley

Grape growing in western Oregon is an adventure. Like any real adventure, it is fraught with risk and peril, but it can be richly rewarding, spiritually if not financially. The climate constitutes both the risk and the reason for this adventure.

The macroclimate of a region and the mesoclimate of a particular site are the most powerful influences on the maturation of grapes and the quality of the wines that come from them. The Willamette Valley has a marginal climate for growing vinifera grape varieties, and the word "marginal" means exactly that—there is little margin for error, in choosing a site, or in choosing varieties to fit that site. It is vital to pay attention to the basics. Winemaking "wizardry" is far less important in Oregon than producing top-quality grapes, because no amount of enological expertise can correct limitations resulting from the choice of site or variety for the site.

Of course, the question follows: If it's so marginal, why would anyone in their right mind grow grapes here? Why go so close to the edge? The answer, at least, is easy...*flavor.*

Matching Variety with Climate

Where grape maturity coincides with the end of the region's growing season, the fruit has the greatest delicacy and complexity of flavor. So do the wines. Edward Hyams, the English author of *Dionysius: A Social History of the Wine Vine,* summarizes this concept well:

> *It is a curious fact that the most exquisite, although not the most generous, wines always come from near the periphery of the viticultural region; or else from such an altitude within that region that, climatically, the vineyard in question might be in a much higher latitude. Grapes exposed to too much sun, too high summer temperatures, become excessively sugary and wanting in acidity; the wine made from them will be heavy and heady and coarse. It is the grapes which are sweet enough to yield an adequacy of alcohol, yet acid enough to give the wine character, which yield those wines of real delicacy which the modern palate prefers. (1965, pp. 135-136)*

It is well established that winegrape varieties, when grown with just enough heat to ripen the fruit, retain their best flavors and varietal characteristics. These flavors are lost when the fruit is grown in a climate that is too warm for its requirements. It was this concept, in the search for a climate suitable for Pinot noir and related varieties, that guided the first serious plantings of European winegrapes *(Vitis vinifera)* in the Willamette Valley in 1965. The pioneering vineyards in the Willamette Valley in the mid-1960s constituted the first conscious effort in the United States to actually match varieties by ripening period to the climate.

Over many centuries of European winegrowing, the selection of grape varieties to fit the climatic regime in which they are grown has evolved empirically. This is why Pinot noir, Chardonnay, Aligoté, Gamay, and Pinot Beurot (Pinot gris) are grown in the "iffy" climate of Burgundy, where the vintages vary and the grapes just barely make it to maturity before the end of the growing season. By the same token, the varieties of Bordeaux (Cabernet Sauvignon, Cabernet Franc, Merlot, Malbec, etc.) are planted in a climate which is marginal for maturation, and where they, too, are subject to the same yearly maturity variations. If fine wine were simply a matter of getting grapes ripe, then the varieties of Burgundy, Bordeaux, Alsace, and Champagne could all be grown with ease in much-warmer Provence, with essentially no variation in the wines from vintage to vintage.

So why do the European grape growers plant varieties in climatic regimes where maturity (and subsequent wine quality) is always in question? The first author has been asked this question many times when traveling in Europe, and the answer has usually been "It's the tradition." Still, tradition always has a beginning, and A. J. Winkler of the University of California, Davis, answered the question succinctly: "The fact that the variety or varieties grown in each of these regions attains almost perfect development in that environment has made it possible for the wines produced to establish a reputation for their quality that is worldwide" (1938, p. 14).

The French have provided another term to help explain how location influences wine quality—*terroir.* Not so many years ago, many people simply equated terroir with soil, the literal translation of the word. The concept, extensively treated by Wilson (1998), has now evolved to encompass the "complete vineyard habitat": the site, soil, drainage, climate, the grapevines and other organisms sharing the vineyard environment, and even a spiritual aspect. Within the vineyard habitat, soil is certainly a factor, but a far greater influence on grape quality is the interplay of the climate and the ripening period of the variety. Winkler explained the concept clearly way back in the 1930s, and throughout his career at the University of California: "Yet, if environment in the broad sense of zones limits grape growing generally on the surface of the earth, it should equally as definitely in a more restricted application limit the adaptation of the individual variety to regions" (1938, p. 14).

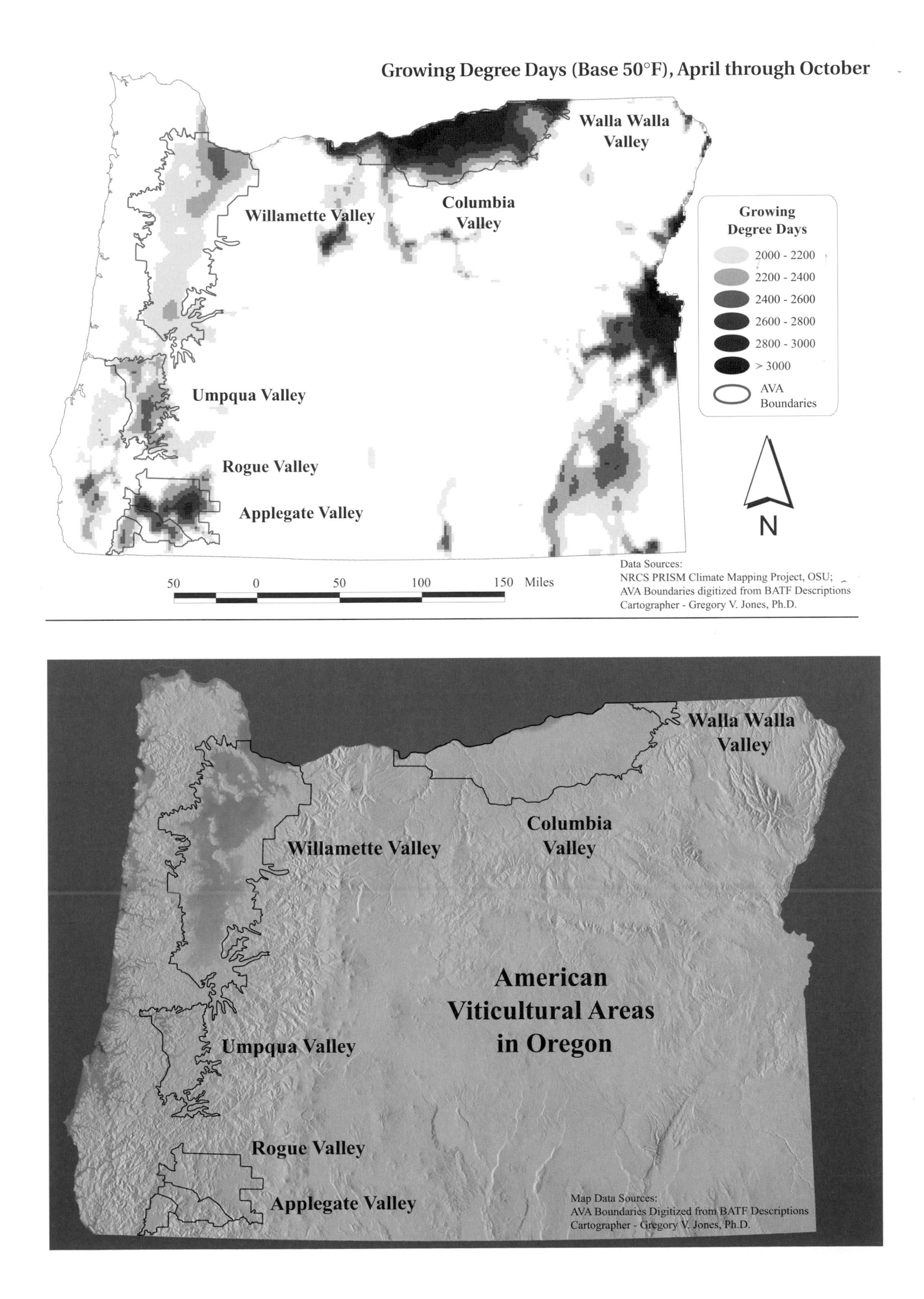
Growing Degree Days (Base 50°F), April through October
Walla Walla Valley
Columbia Valley
Willamette Valley
Umpqua Valley
Rogue Valley
Applegate Valley
Growing Degree Days
2000 - 2200
2200 - 2400
2400 - 2600
2600 - 2800
2800 - 3000
> 3000
AVA Boundaries
N
50 0 50 100 150 Miles
Data Sources:
NRCS PRISM Climate Mapping Project, OSU;
AVA Boundaries digitized from BATF Descriptions
Cartographer - Gregory V. Jones, Ph.D.
Walla Walla Valley
Columbia Valley
Willamette Valley
American Viticultural Areas in Oregon
Umpqua Valley
Rogue Valley
Applegate Valley
Map Data Sources:
AVA Boundaries Digitized from BATF Descriptions
Cartographer - Gregory V. Jones, Ph.D.

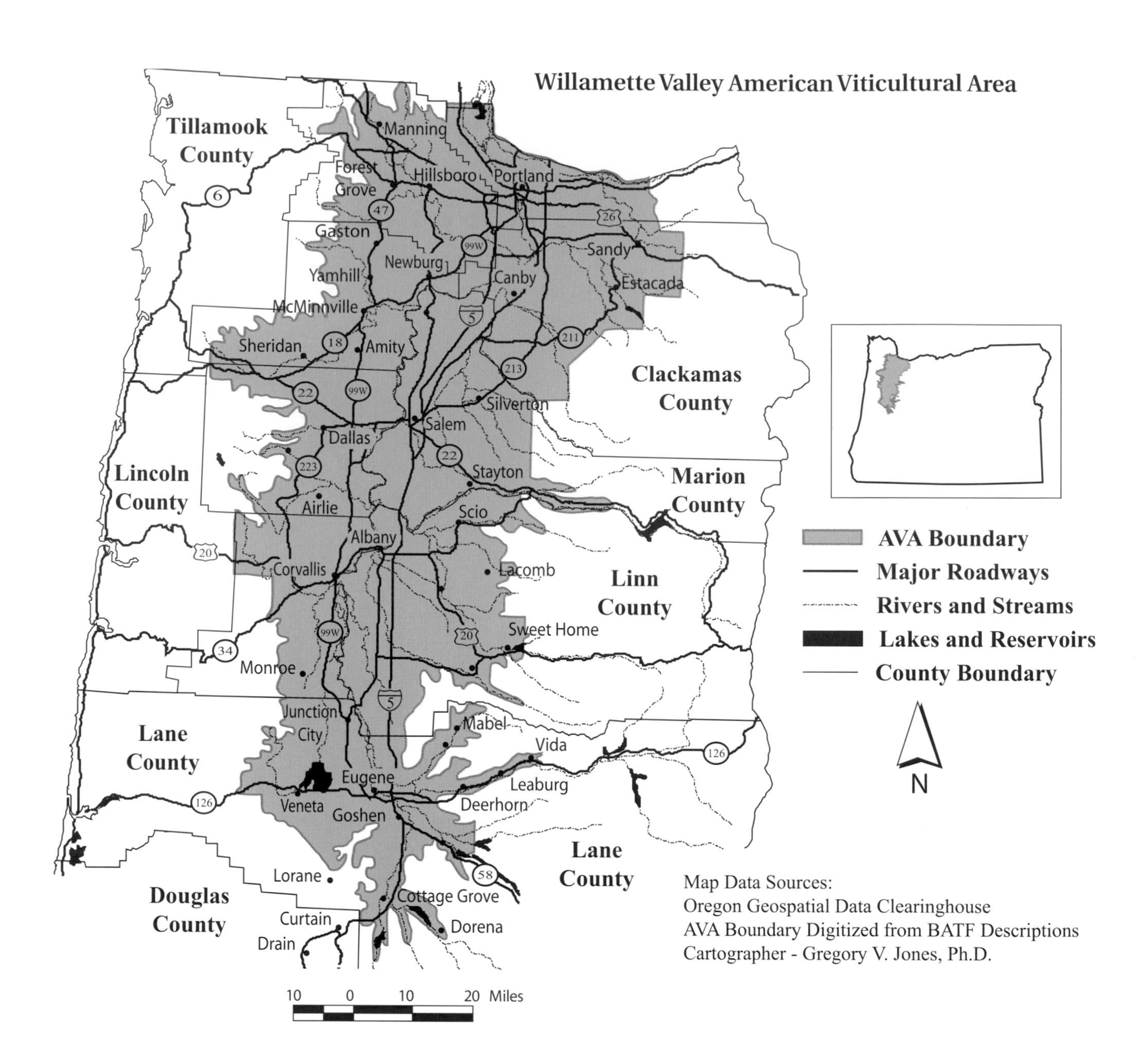
Willamette Valley American Viticultural Area
Tillamook County
Clackamas County
Lincoln County
Marion County
Linn County
Lane County
Lane County
Douglas County
Manning
Forest Grove
Hillsboro
Portland
Gaston
Newburg
Sandy
Yamhill
Canby
Estacada
McMinnville
Sheridan
Amity
Silverton
Salem
Dallas
Stayton
Airlie
Scio
Albany
Corvallis
Lacomb
Sweet Home
Monroe
Junction City
Mabel
Vida
Eugene
Leaburg
Veneta
Goshen
Deerhorn
Lorane
Cottage Grove
Curtain
Dorena
Drain
AVA Boundary
Major Roadways
Rivers and Streams
Lakes and Reservoirs
County Boundary
N
Map Data Sources:
Oregon Geospatial Data Clearinghouse
AVA Boundary Digitized from BATF Descriptions
Cartographer - Gregory V. Jones, Ph.D.
10 0 10 20 Miles

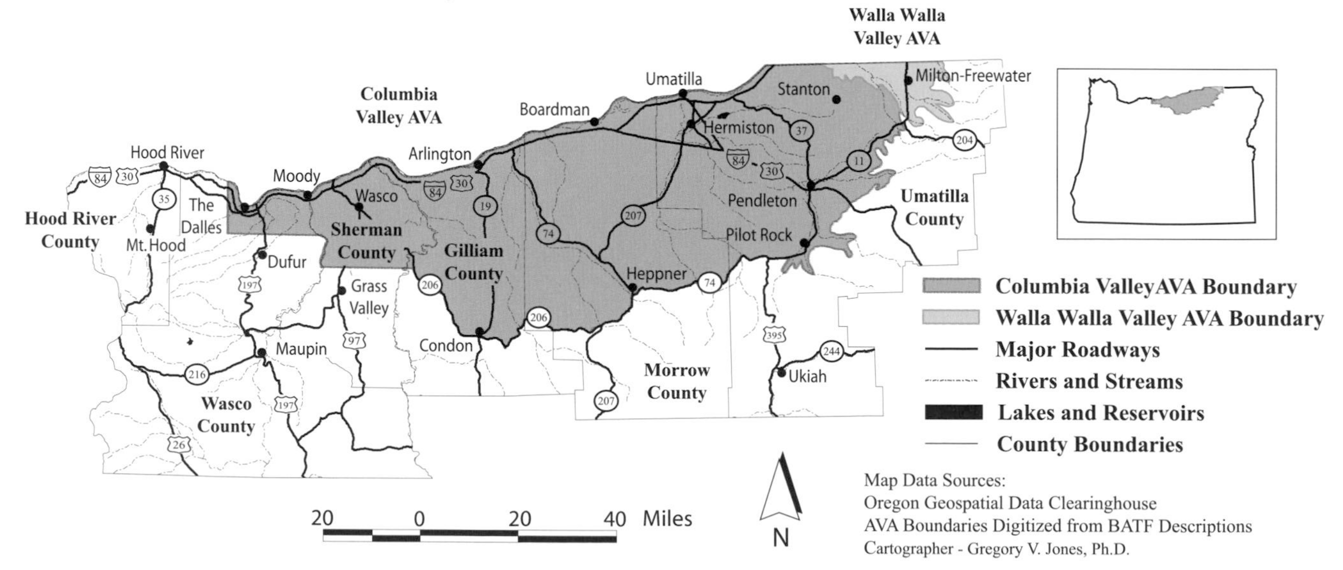
Columbia and Walla Walla Valley American Viticultural Areas
Walla Walla Valley AVA
Columbia Valley AVA
Umatilla
Milton-Freewater
Stanton
Boardman
Hermiston
Hood River
Arlington
Moody
Wasco
The Dalles
Pendleton
Hood River County
Mt. Hood
Sherman County
Gilliam County
Pilot Rock
Umatilla County
Dufur
Grass Valley
Heppner
Condon
Maupin
Morrow County
Ukiah
Wasco County
Columbia Valley AVA Boundary
Walla Walla Valley AVA Boundary
Major Roadways
Rivers and Streams
Lakes and Reservoirs
County Boundaries
N
Map Data Sources:
Oregon Geospatial Data Clearinghouse
AVA Boundaries Digitized from BATF Descriptions
Cartographer - Gregory V. Jones, Ph.D.
20 0 20 40 Miles

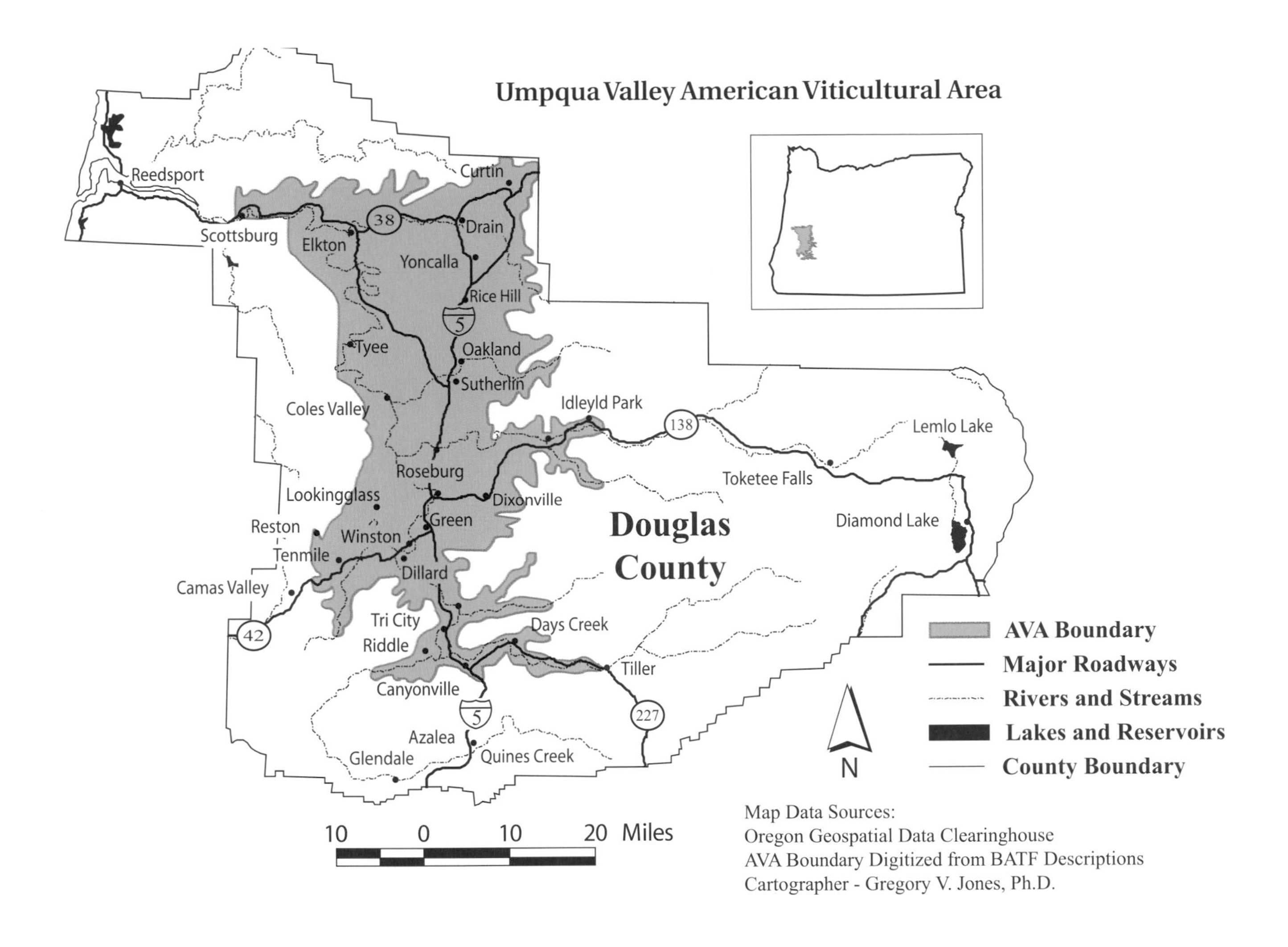
Umpqua Valley American Viticultural Area
Reedsport
Curtin
Scottsburg
38
Drain
Elkton
Yoncalla
Rice Hill
5
Tyee
Oakland
Sutherlin
Coles Valley
Idleyld Park
138
Lemlo Lake
Roseburg
Toketee Falls
Lookingglass
Dixonville
Diamond Lake
Reston
Green
Douglas
County
Winston
Tenmile
Dillard
Camas Valley
Tri City
Days Creek
42
Riddle
Tiller
Canyonville
5
227
Azalea
Glendale
Quines Creek
AVA Boundary
Major Roadways
Rivers and Streams
Lakes and Reservoirs
County Boundary
N
10 0 10 20 Miles
Map Data Sources:
Oregon Geospatial Data Clearinghouse
AVA Boundary Digitized from BATF Descriptions
Cartographer - Gregory V. Jones, Ph.D.

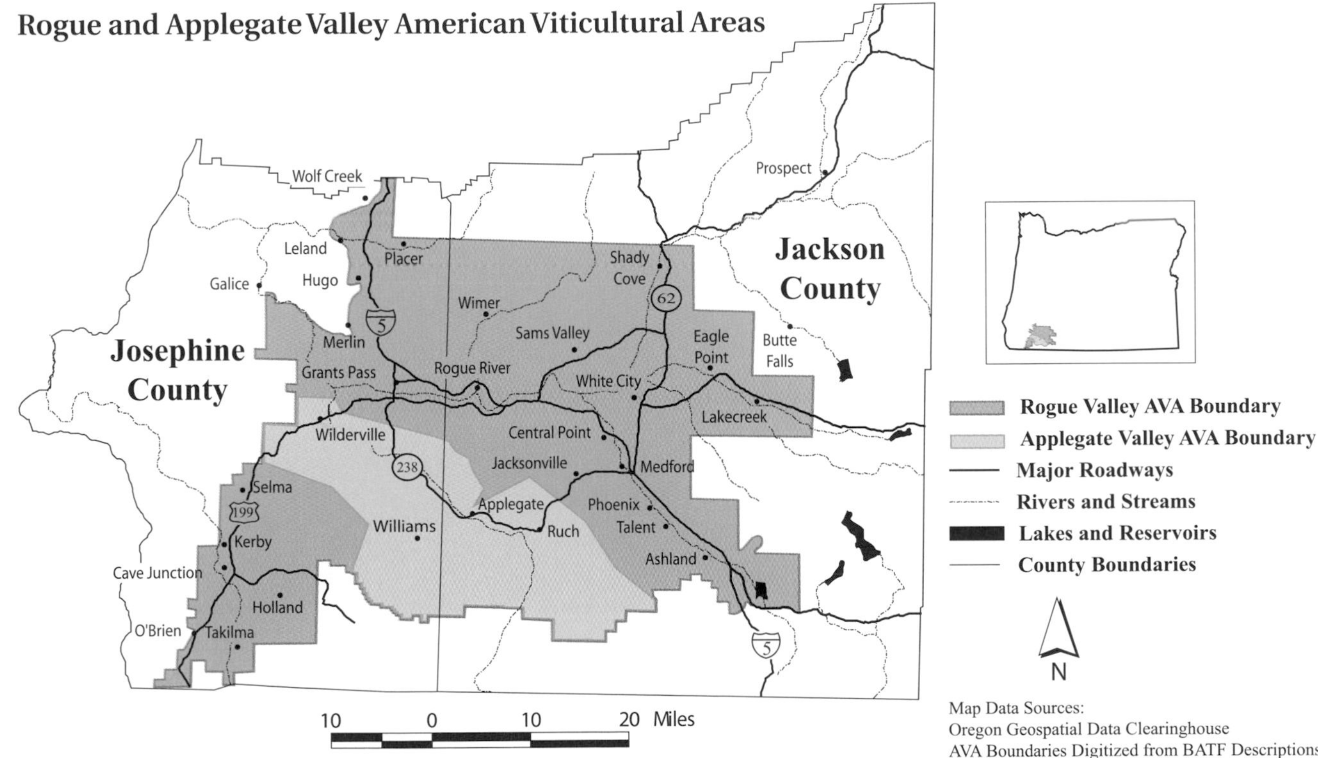
Rogue and Applegate Valley American Viticultural Areas
Wolf Creek
Prospect
Leland
Placer
Shady Cove
Jackson
County
Galice
Hugo
62
Wimer
Josephine
County
Merlin
5
Sams Valley
Eagle Point
Butte Falls
Grants Pass
Rogue River
White City
Lakecreek
Wilderville
Central Point
238
Jacksonville
Medford
Selma
199
Applegate
Phoenix
Williams
Ruch
Talent
Kerby
Ashland
Cave Junction
Holland
O'Brien
Takilma
5
Rogue Valley AVA Boundary
Applegate Valley AVA Boundary
Major Roadways
Rivers and Streams
Lakes and Reservoirs
County Boundaries
N
10 0 10 20 Miles
Map Data Sources:
Oregon Geospatial Data Clearinghouse
AVA Boundaries Digitized from BATF Descriptions
Cartographer - Gregory V. Jones, Ph.D.

Early spring in a vineyard on a high hillside in the Willamette Valley. (Photo by Dai Crisp)

Post-harvest fall colors in a Willamette Valley vineyard looking east. (Photo by Dai Crisp)

Head-trained, cane-pruned, vertical shoot positioned Pinot noir. (Photo by Dai Crisp)

Close view of the fruit zone of a Pinot noir vine. Clean morning sunlight exposure due to leaf removal in the east-facing fruit zone. (Photo by Dai Crisp)

Pinot noir nearing harvest, covered with morning dew. (Photo by Dai Crisp)

Pinot noir trained to the Scott Henry system. (Photo by Dai Crisp)

Pommard clone of Pinot noir close to harvest. (Photo by Dai Crisp)

Tractor pulling harvest bins between two vineyard blocks. (Photo by Dai Crisp)

Workers tying down canes on a cool spring day. (Photo by Dai Crisp)

Workers pulling brush out of the trellis in a shoot positioned vertical block of Pinot gris. (Photo by Dai Crisp)

Early spring management, showing both in-row weeds burned with a propane burner and weeds mowed close. (Photo by Dai Crisp)

Harvester picking Gewürztraminer in a single high wire hanging trellis. (Photo by Dai Crisp)

Workers pulling up catchwires and tucking in the shoots in a vertical shoot positioned training system. (Photo by Dai Crisp)

Winkler suggests that the proper "environmental-variety interrelationship" gives us a breadth and depth of flavors that are the best expression of varietal integrity in the resulting wine. He continues:

> *Its [climate] effect is largely an influence on the rates of change in the constituents during development and the composition of the grapes at maturity. Relatively cool conditions in which these changes, especially those of ripening, proceed at a slow rate have been found to be the most favorable for the production of dry wines of quality. These conditions foster the retention of a high degree of acidity and bring the aroma and flavoring constituents of the grapes to their highest degree of perfection in the mature fruit.... Under warmer climatic conditions the aromatic qualities of the grapes lose some of their delicacy and richness and the other constituents of the fruit are not so well balanced, hence the dry wines—even of the best varieties—cannot compare in quality with those of the cooler regions.... Yet this basic interrelationship of type with region and quality with variety is fundamental not only to the progress of our industry as a whole but also to the rational development of each section of our state.* (1938, pp. 14-15)

As previously mentioned, the vineyards of the Willamette Valley in the mid-1960s were the first in the United States to reflect the concepts of variety/environment relationship so well known through-out Europe for centuries. Sadly, however, the twentieth-century studies of Winkler and Amerine seem to have been largely ignored in the development of newer wine regions (even within established viticultural areas) as the "wine boom" has taken place in the United States since the mid-1980s. Indeed, we often see Syrah, Mourvèdre, and Marsanne of the Rhone growing alongside Pinot noir, Pinot gris, and Chardonnay of Burgundy next to Cabernet Sauvignon, Merlot, Cabernet Franc, and Sauvignon blanc of Bordeaux—in the same climate. Obviously, this is quite a departure from the European model.

The Willamette Valley has the coolest growing season of any agricultural area of its size in the United States, with average degree-day accumulations at favorable sites generally ranging from 2,000 to 2,200. Therefore, using the European experience, the choice of varieties to fit this marginal (for the maturity of grapes) climate became as easy as ABC...Alsace, Burgundy, and Champagne.

Figure 1 illustrates the similarity in all the important climatic factors between the Willamette Valley and Burgundy. The choice of Burgundian varieties (and varieties from relatively similar cool regions of Alsace and Champagne) was theoretically obvious to those who first planted vines in the Willamette Valley in the mid-1960s. All of the varieties chosen were early-maturing Period I (Pulliat) varieties.

The risk of growing even Period I varieties in the Willamette Valley is great, although it can be mitigated by proper site selection. Nonetheless, there will be vintage variations from year to year, and our cool northern clime tends to cause vines to produce less abundant crops than warmer regions. Simply put, the *only* reason for growing grapes in the Willamette Valley is flavor not high tonnage per acre. The complex varietal characteristics of the Pinot family are fragile and ephemeral. In warmer climates, these characteristics do not develop because the grapes are rushed to maturity too rapidly by excess heat. Although "correct sugar" content may be achieved, the more subtle varietal characteristics are often lost. In the case of Pinot noir in particular, the transition from perfect ripeness (and thus the varietal integrity) to just another "Big Red" wine from overripe grapes can be a matter of only a day or two.

In the Willamette Valley, generally speaking, Pinot noir achieves fruit maturity about six weeks later than the cooler areas of essentially Mediterranean climates, in early to mid-October, just at the end of the growing season. The grapes have ripened slowly in the cool and even temperatures that Pinot noir requires to develop its fullest potentials of flavor. Although these extra six weeks are nail-biters for the grower, they are essential in achieving the finest wines possible from this variety. Indeed, it is often the cooler, damper and more difficult growing seasons that produce the most flavorful and balanced Pinots in the Willamette Valley.

As stated previously, the climate constitutes both the risk and the reason for fine wine production in Oregon. And though climate is the single most important influence on ultimate wine quality, other environmental factors certainly come into play (particularly in a marginal climate) to create the terroir of a region. Although other chapters in this volume discuss soil, site, and other factors, some aspects of terroir must be at least briefly mentioned in this discussion of variety adaptation.

Soil

Much has been written and said over the years on the subject of soil and its influence on winegrape quality. A good deal of this is mythology. We have heard, for example, that it is the flint in the soils of Chablis that gives the wine a "hardness" or flintiness of taste on the palate. Flint is about as inert a geologic form as exists on the planet; it has no flavor and imparts none to a vine. Likewise, we have heard and read that Pinot noir must grow in calcareous (limestone-based) soil for it to achieve all the flavors that are possible in the variety. At the first International Symposium on Cool Climate Viticulture and Enology, held in Eugene, Oregon, in 1984, Raymond Bernard of the University of Dijon was asked to comment on the subject of *calcaire* (calcareous or limestone-based soil) and Pinot noir. His response was engagingly honest: "Yes," he said (paraphrased), "We in Burgundy feel that Pinot noir must grow in *calcaire* soil to achieve all of the nuance and subtlety of the characteristics inherent in the varietal...and the reason we feel so strongly this way is because that's the kind of soil we have." During a class at the University of California, the first author asked Maynard Amerine why the Burgundian concept of terroir was so imbedded in soil. He answered the question with another question: "Mr. Lett, if soil was the predominant factor in wine quality in Burgundy, how do you account for the incredible variation of the wines from vintage to vintage?" The answer was fairly obvious: the soil does not change dramatically from year to year, but the weather certainly does.

Comparison of Average Monthly Temperature

Average Monthly Temperature (ºC)
30
25
20
15
10
5
0
J F M A M J J A S O N D
Burgundy
Willamette Valley

Comparison of Average Monthly Sunlight

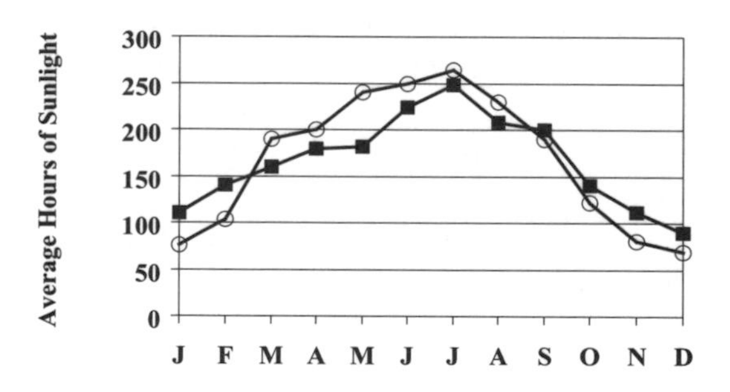

Comparison of Average Monthly Rainfall

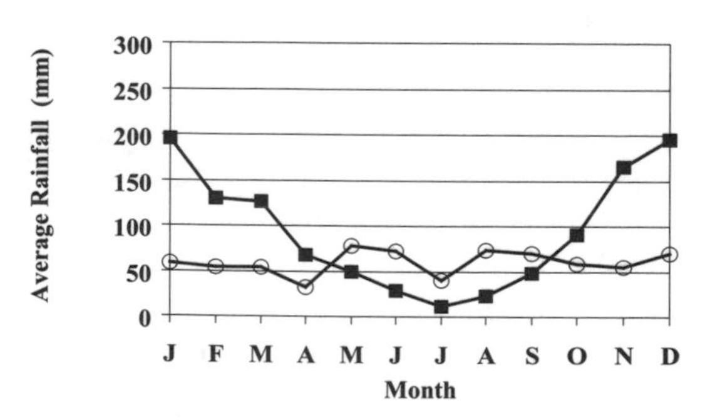

Figure 1. Climatic comparison of the Willamette Valley and Burgundy, France. Adapted from Bernard (1990).

It becomes clear that in a marginal climate like that of Burgundy or the Willamette Valley, the climate is obviously the most influential factor in wine quality. The late Louis Martini was quoted as joking that, "soil is something to hold the vine up." Louis was quite aware, however, that soil cannot be totally ignored in the development of fine wines. As an example, soil texture, chemistry, and depth have a profound influence on moisture and nutrient availability for vine health. Heavy alluvial soils are often poorly drained and too rich in nutrient value, thus promoting vegetative growth instead of fruit development and delaying maturity. On the other hand, many sedimentary soils have poor moisture retention and nutrient availability, which can lead to ripening anomalies in the grape. The cooler and more marginal the climate, the more important soil factors such as these become.

Elevation

The choice of varieties for the Willamette Valley can be profoundly influenced by vineyard elevation because of its effect on temperature. In the vineyards at Eyrie, all within a mile of each other but at different elevations, a difference of as much as two weeks in maturation time is seen between vineyards at 300 feet and those at 800 feet elevation. In lower to midrange elevations (300–700 feet), most of the proven varieties usually mature properly. At the higher elevations (700–1,000 feet), caution is advised in variety selection.

Varieties in the Willamette Valley

In the fourth edition of the *Oregon Winegrape Grower's Guide*, varieties were classified not only on their interaction with the climate but also on their marketability (Lett, 1992). With the rapid expansion of vineyard acreage in recent years and a great increase in the number of wineries, an established viticultural pattern has emerged in the Willamette Valley.

Pinot noir is far and away the most widely planted variety in the Willamette Valley, with Pinot gris having recently overtaken Chardonnay for second place (Eklund et al., 2002). Riesling acreage has declined steadily since 1990. This all suggests that, though a market plan still should be decided before ever putting a grapevine in the ground, what was a theory almost four decades ago seems to have worked. The Willamette Valley wine picture is clearly dominated by three varieties: Pinot noir, Pinot gris, and Chardonnay.

Pinot noir. It has been more than twenty years since the 1979 Gault Millau tasting in Paris and the subsequent Drouhin tasting in Beaune. These tastings demonstrated to the world for the first time that a Pinot noir from Oregon could not only hold its own but best some of the best wines of Burgundy. The rush to plant Pinot noir in Oregon has been non-stop ever since. These tastings also led to the initiation or expansion of Pinot noir production in other regions such as California, New Zealand, Australia, South Africa, and elsewhere.

Oregon's Willamette Valley is one of the few grape-growing regions in the world with those factors of climate, soils, rainfall, daylength, and myriad others that interact to capture the ephemeral essence of this elegant variety. Bigger, heavier, more alcoholic Pinot noir wines can and are being made in the warmer climates (and even in the Willamette Valley's warmer years). But Pinot noir, unlike varieties such as Cabernet Sauvignon or Merlot, has a fragile and elusive varietal character that tends to disappear quickly when the grapes become overripe. As mentioned before, the reason for growing Pinot noir in Oregon is the ability to capture, in most years, including and perhaps especially the more cool and difficult vintages, the true meaning and beauty of Pinot noir as a variety.

Almost every winery in the Willamette Valley produces a Pinot noir, though there continues to be a certain amount of market resistance (or indifference) to Pinot noir in the United States. Most consumers simply do not have a broad tasting knowledge of fine Pinot noirs. Still, Oregon's reputation for producing some exceptional Pinot noir at relatively reasonable prices continues to grow. I believe this decade will continue to be a dynamic market for these wines (as were the 1990s) as consumers are exposed to better and better examples of the variety.

The first Oregon Pinot noir vineyards were established with the Wädenswil (FPMS 2A) and Pommard (FPMS 4) clones. These clones produced the wines that put Oregon on the map, and they still dominate production in the region. Many recent plantings have included the French clones 113, 114, 115, 667, 777, and a few others. Most wineries intend to blend clones to enhance wine complexity, but pure Wädenswil and Pommard wines were common in the past and a few single-clone wines are being produced from the newer clones. Additional clones have been imported and tested by Oregon State University. Evaluations of twenty Pinot noir clones are reported in the Oregon Wine Advisory Board (1993-2001) reports. These clones were loosely classified into four groups: (1) Pinot fin, typically characterized by small clusters and a prostrate growth habit; (2) Mariafeld, loose-clustered types; (3) Upright, referring to their erect growth habit; and (4) Fertile, characterized by large clusters and prostrate growth habit. Wädenswil, Pommard, and the Dijon clones cited above are all classified as Pinot fin types.

Pinot gris. Pinot gris has been grown in cooler areas of northern Europe for centuries and under different names. Its origin is thought by some to be in Burgundy, where it is known as Pinot Beurot. It is more commonly grown in Alsace, where it is known as Pinot gris or Tokay d'Alsace. It is Pinot grigio in Italy, where it grows in cooler, higher-elevation vineyards in the north. In Germany it is known as Ruländer or Grauburgunder, and in Hungary it is Szürkebarát. In North America, Pinot gris had its beginnings at the Eyrie Vineyards in the Willamette Valley in 1966.

Pinot gris is now being planted in many areas throughout the New World, but alas, often where growing conditions are too warm to yield wines of varietal integrity. Pinot gris is considered to be a mutation of Pinot noir, so it is also an early-maturing variety and shares with Pinot noir an elusive and fragile varietal character. It too requires a cool climate to achieve its finest expression.

In the fourth edition of the *Oregon Winegrape Grower's Guide*, I (first author) wrote, "I have, somewhat reluctantly, moved this variety from the category of 'Viticulturally Established, Market Potential' to 'Market Established' status" (Lett, 1992). As of this writing and considering the rapid increase in acreage in the Willamette Valley, we can safely eliminate the term "reluctant." Indeed, in the year 2000, Pinot gris tonnage passed that of Chardonnay for the first time to become the second-largest winegrape variety produced in Oregon.

The Eyrie Vineyards' Pinot gris was obtained from the University of California as clone 1A, which originated in Alsace. Other clones, including the Colmar clones 146 and 152, are also grown in Oregon. Systematic comparative evaluations of these and other Pinot gris clones have not been done in the state.

Chardonnay. Since Pinot noir does so well in the Willamette Valley, it stands to reason (and fits the variety/climate adaptation profile) that the major white variety of Burgundy should also do well in the Willamette Valley's Burgundy-like climate. Unfortunately, all but the oldest and the most recent plantings of Chardonnay in the Willamette Valley utilized a clone developed in California for warmer climates. Clone 108 (FPMS 4 and 5[1]), as it is commonly known in Oregon, is designed for high tonnage, late ripening, and high acid—all characteristics that are contrary to high quality in a cool climate.

The first planting of Chardonnay in the Willamette Valley was with a group of French clones collectively known in the United States as the "Draper Selection," which were imported from Burgundy in the 1930s. The Draper Selection clones have produced excellent wines, and now wines from the more recently introduced Dijon and Espiguette clones are beginning to produce high-quality wines as well. Reports of Chardonnay clonal evaluations by Oregon State University are found in the research progress reports of the Oregon Wine Advisory Board (1993–2001). The French clones from Dijon, particularly 95, 96, 76, and ESP 352, have received the most attention at this time. The principal advantages of the newer clones compared to Clone 108 are earlier ripening (at least one week), smaller clusters, higher sugar content and better flavors.

Marketing of Oregon Chardonnay continues to be a challenge. Indeed, some Oregon grape growers are pulling Chardonnay out all together, just at a time when the market seems to be growing bored with warm-climate Chardonnays.

Although the "Big Three" varieties dominate production in the Willamette Valley, the climate is also conducive to producing wonderful wines from lesser-known varieties. Oregon has produced some high-quality Gewürztraminer and Riesling wines. The same is true for Pinot blanc, Gamay, and several other early-maturing varieties. Other varieties with potential in the Willamette Valley include Pinot Meunier, Dolcetto, Early Muscat, Muscat Ottonel, Müller-Thurgau, and Sylvaner.

Variety Selection in Southern Oregon

Initially, all of southern Oregon, like the Willamette Valley, was considered a cool climate for grape growing. Consequently, the first vineyards were planted with early-maturing varieties: Pinot noir, Chardonnay, Gewürztraminer, and Riesling. Considerable success has been achieved in many areas of southern Oregon with these varieties, as well as with Pinot gris, Early Muscat, and others. But it was soon recognized that the region's mesoclimates are both warmer and much more diverse than the Willamette Valley, suggesting that later-maturing varieties with higher heat requirements could also be successfully grown in many of these regions.

Umpqua Valley

The Umpqua Valley is somewhat warmer and drier than the Willamette Valley. Depending on elevation, the growing season averages 180 to over 215 days, and degree-days range from 2,200 to over 2,500 in most of the lowlands. The many subvalleys within the Umpqua region provide considerable mesoclimate diversity from site to site, so site assessment is critical to proper selection of varieties. Cooler sites, located principally in the northern and western portions of the AVA, have traditionally grown early-maturing varieties, whereas the area south of Roseburg provides warmer sites that have done well with somewhat later-maturing varieties.

The principal varieties of the region are, at present, the same cool-climate varieties as in the Willamette Valley: Pinot noir, Chardonnay, Riesling, and Pinot gris (Eklund et al., 2002). As in the Willamette Valley, a shift in Pinot noir clones has been occurring, with the original vineyards in the area planted to Pommard (FPMS 4) and Wädenswil (FPMS 2A) but more recent vineyards planting the French clones 113, 114, and 115. There is, however, a trend away from planting cool-climate varieties. The present emphasis is on warmer-climate varieties from Spain and the Bordeaux and Rhone regions of France.

The most exciting recent development in the region has been the success achieved at warmer sites with Tempranillo, Syrah, Dolcetto, and Merlot. Acreage has been increasing for all four varieties, and the recognition brought from some recent award-winning wines may stimulate continued expansion. Smaller acreages of Cabernet Franc, Malbec, Grenache, and Viognier have created interest with some high-quality wines. Sangiovese and Fresia have also shown good potential, but there is very little of either variety planted in the region.

Rogue Valley

Oregon's southernmost winegrowing region, the Rogue Valley, has the broadest range of climatic conditions in the western part of the state. Average heat unit accumulations generally range between 2,200 to 2,900 degree-days. The length of the growing season is generally shorter than in the Willamette and Umpqua Valleys, ranging from 155 to 185 days, so the risk of damage from spring and fall frosts is much greater here and can be a limiting factor at some sites.

Three tributaries of the Rogue River create the principal growing areas of the Rogue Valley: the Illinois Valley, Applegate Valley, and Bear Creek Valley. Considerable mesoclimate variation among these growing regions accounts for the broad range of conditions within the overall region. The Illinois Valley is the coolest area, with average heat accumulation similar to the Umpqua Valley, ranging between 2,200 to 2,600 degree-days. The Applegate Valley is warmer; heat accumulations range from 2,500–2,700 degree-days. The relatively large Bear Creek Valley is the warmest region, generally accumulating 2,400 to 2,900 degree-days.

It is difficult to generalize about variety adaptability in the Rogue Valley because of the high degree of climatic diversity there, even among sites within the same subregion. A thorough knowledge of a site's characteristics is critical to proper selection of varieties, and comparisons with neighboring vineyards must account for significant climatic differences. The performance history of established varieties at a vineyard site should be an indicator of what other varieties might be suited for the site's climate. For example, if cool-climate varieties are routinely harvested well before the end of the growing season, later-maturing varieties could be better adapted to the site. Degree-day accumulation data for the site will help with this decision, and experimentation is always recommended.

Recent production statistics (Eklund et al., 2002) indicate that the principal varieties in the Rogue Valley, in decreasing order of acreage, are Merlot, Pinot noir, Cabernet Sauvignon, Chardonnay, Pinot gris, and Syrah. Pinot noir is predominantly grown at cooler sites in the Illinois Valley. Pinot gris and Chardonnay are grown in all regions of the Rogue Valley. The Illinois Valley also grows other early-maturing varieties, including Gewürztraminer, Riesling, and Early Muscat.

Merlot and Cabernet Sauvignon have been successful in the Bear Creek Valley. These varieties have also done well on southern slopes of the Applegate Valley. A small amount of Merlot is grown in the Illinois Valley, but it and other later-maturing varieties are limited to sites with higher heat accumulation and a longer growing season. Syrah has only recently been planted to significant acreage in the Rogue Valley, and it shows good promise for the region. Lesser amounts of other warm-climate varieties have been successful in the Applegate and Bear Creek Valleys, including Cabernet Franc, Sémillon, Sauvignon blanc, and Zinfandel. Several other later-maturing varieties, including Tempranillo, Sangiovese, Viognier, Petit Verdot, and Malbec, have been planted for evaluation in various areas of the Rogue Valley region.

Merlot. Merlot has produced some good varietal and Bordeaux-style wines. Viticulturally, the variety is more difficult than Cabernet Sauvignon and other Bordeaux varieties. Vines are less vigorous than Cabernet Sauvignon and more difficult to train. Its earlier bud break can lead to spring frost injury, and poor fruit set can result if bloom occurs during cool weather. Merlot is also more susceptible to winter cold damage. Recently, young vines that had not yet hardened off were seriously damaged from low fall temperatures. The most common clones are FPMS 3 and the French clone 181. Other Merlot clones planted are FPMS 1 and 6 and French clone 314. A trial of Merlot clones is underway at the Southern Oregon Research and Extension Center and has begun to provide evaluative data on clonal performance in the vineyard and winery (Sugar and Vasconcelos, 2002).

Cabernet Sauvignon. The first later-maturing variety to have success in southern Oregon was Cabernet Sauvignon. It has been used to produce high-quality varietal and Bordeaux-style blends. Performance of Cabernet Sauvignon, however, can be very site specific. It tends to do better on well-drained soils, and the best fruit quality often comes from low crop levels at warmer sites. Generally, it is easy to grow, has an upright growth habit, and good fruit set. Cabernet Sauvignon matures about mid to late October in the warmer areas, or late October to early November in somewhat cooler sites or years. Because it ripens just prior to the first fall frost, varieties that mature after Cabernet Sauvignon seldom reach proper maturity in the Rogue Valley. A beneficial aspect of the variety is that it has more frost tolerance than Merlot and many other varieties. The predominate clones are FPMS 15 and the French clone 337, and some FPMS 7 has also been planted.

Variety Selection in Eastern Oregon

The winegrowing regions of eastern Oregon are the Columbia River region, the Hood River area, and the Walla Walla region. The mesoclimates within these regions are more highly variable than any other region in the state, and heat accumulation and length of growing season are very site specific (see Chapter 4). Overall, the region is generally warmer than western Oregon AVAs, with growing degree-days peaking at over 3,300 for some locations, which is the highest heat accumulation in the state. The more common vineyard sites in the Columbia River region accumulate 2,500–2,900 degree-days. The Hood River area has growing season temperatures and degree-days somewhat lower than the Columbia Valley AVA. Heat accumulation ranges from about 2,600 to just over 3,000 degree-days for the Walla Walla Region.

The principal varieties grown in eastern Oregon are, in descending order, Cabernet Sauvignon, Riesling, Merlot, and Syrah (Eklund et al., 2002). Riesling, which is also a prominent and successful variety in eastern Washington, is grown primarily in the Columbia River region. Cabernet Sauvignon and Merlot have been the most successful warmer climate varieties so far, but Syrah has only recently been planted in the area. Other warmer-climate varieties being grown include Viognier and Sémillon. Systematic evaluations of varieties and clones in eastern Oregon are needed to identify additional opportunities for this area.

Final Comments

This chapter has emphasized the importance of matching variety and clone to climate and site characteristics. The development of the Willamette Valley viticulture region is a model for variety–climate matchmaking. Pinot noir wines being produced in the region are considered to be among the finest in the world. Outstanding Pinot gris and other classic early-maturity varieties are also produced, but the marginal climate precludes consistent ripening of later-maturing varieties. Such classic winegrapes as Cabernet Sauvignon, Merlot, Syrah, Tempranillo, and others, however, are well suited to the warmer Umpqua, Rogue, Columbia, and Walla Walla regions. It is critical for Oregon growers in every production region to match varieties to their climate where, with good winemaking, the best expression of those varieties' flavors can be translated into wines that reflect a "sense of place" unlike any other on the planet.

Acknowledgments

The authors gratefully acknowledge the valuable comments on varieties and clones provided by Randy Gold, Scott Hendricks, Earl Jones, Porter Lombard, Laura Lotspeich, Kurt Lotspeich, and Philip VanBuskirk.

Note

1. Different selections of the same clone that have undergone different virus elimination treatments.

References and Additional Resources

Amerine, M. A., and A. J. Winkler. 1944. Composition and quality of musts and wines of California grapes. *Hilgardia* 15:493-675.

Bernard, R. 1990. Le Vignoble D'Oregon. *Progrès Agricole et Viticole,* 107(7):165-166.

Caldwell, J. 1998. *A Concise Guide to Wine Grape Clones for Professional,* 2d ed. John Caldwell Viticultural Services, Napa, California.

Eklund, B., L. Burgess, and R. F. Kriesel. 2002. *2001 Oregon Vineyard and Winery Report.* Oregon Agricultural Statistics Service, Portland.

ENTAV-INRA. 1995. *Catalogue of Selected Wine Grape Varieties and Clones Cultivated in France.* Ministry of Agriculture, Fisheries and Food. CTPS.

Galet, P. 1998. *Grape Varieties and Rootstock Varieties.* Oenoplurimédia, Chaintré, France.

Gladstones, J. 1992. *Viticulture and Environment.* Winetitles, Adelaide, Australia.

Hartmann, H. T., D. E. Kester, and F. T. Davies, Jr. 1990. *Plant Propagation Principles and Practices* ,5th ed. Prentice Hall. Englewood Cliffs, N. J.

Hyams, Edward. 1965. *Dinoysus: A Social History of the Wine Vine.* Macmillan, New York.

Jackson, D. I., and N. J. Cherry. 1988. Prediction of a district's grape-ripening capacity using a latitude-temperature index (LTI). *American Journal of Enology and Viticulture* 39:19-26.

Lett, D. R. 1992. Grape variety and site selection with an emphasis on the Willamette Valley. In T. Casteel (ed.), *Oregon Winegrape Grower's Guide,* 4th ed. Oregon Winegrowers' Association, Portland.

McIntyre, G. N., W. M. Kliewer, and L. A. Lider. 1987. Some limitations of the degree day system as used in viticulture in California. *American Journal of Enology and Viticulture* 38:128-132.

Oregon Wine Advisory Board. 1993/94–2000/01 (issued annually). *OSU Winegrape Research Progress Reports.* Oregon State University Agricultural Experiment Station, Corvallis.

Price, S. F., and B. T. Watson. 1995. Preliminary results from an Oregon Pinot noir clonal trial. In J. M. Rantz (ed.), *Proceedings of the International Symposium on Clonal Selection.* American Society for Enology and Viticulture, Davis, Calif.

Robinson, J. 1996. *Jancis Robinson's Guide to Wine Grapes.* Oxford University Press. New York.

Smart, R. E., and P. R. Dry. 1980. A climatic classification for Australian viticultural regions. *Australian Grapegrower & Winemaker* 17(196):8, 10, 16.

Sugar, D., and C. Vasconcelos. 2002. Evaluation of clonal selections of Merlot, and performance of Sangiovese, Syrah, and Viognier in southern Oregon. Oregon Wine Advisory Board, *OSU Winegrape Research Progress Reports 2001–2002.* Oregon State University Agricultural Experiment Station, Corvallis.

Wilson, J. E. 1998. *Terroir: The Role of Geology, Climate, and Culture in the Making of French Wines.* University of California Press, Berkeley.

Winkler, A. J. 1938. The effect of climatic regions. *Wine Review* 6(6):14-16, 32.

6

Rootstocks

Edward W. Hellman and Joel Myers

Rootstocks play an important role in the design of a successful vineyard by contributing to pest resistance, providing adaptation to different soil conditions, and influencing vine size and fruit ripening. Although rootstocks are used in Oregon primarily for phylloxera control, they also can be used as a tool to influence other aspects of vine behavior. The rootstock provides the root system of the grafted vine and therefore determines water and nutrient uptake, and it may influence, directly or indirectly to varying degrees, vine vigor, fruitfulness, ripening date, and fruit quality.

Important Rootstock Characteristics

Oregon viticulture has a short experience with growing grapevines grafted onto phylloxera-resistant rootstocks. A few vineyards began the transition from self-rooted vines to grafted vines in the late 1980s, but interest in rootstocks accelerated dramatically when phylloxera was discovered in several Oregon vineyards in 1990. Phylloxera has subsequently been found in every grape production region and is commonplace in areas of the Willamette Valley. Resistant rootstocks are the best method to control phylloxera and have become an integral part of new or replanted vineyard design decisions. A survey conducted in 2000 indicated that approximately 36% of Oregon's grape acreage was planted with rootstock (Rowley et al., 2001). The most popular rootstocks reported in the survey were 3309 C (17%), 101-14 (8%), and Riparia Gloire (3%). More than six other rootstocks were being utilized to a lesser degree, including SO 4, 5 C, 44-53 M, and 420 A.

Phylloxera Resistance. First and foremost the rootstock must have a high level of resistance to phylloxera. Avoid rootstocks with a *Vitis vinifera* variety in their parentage, because vinifera has insufficient resistance to phylloxera. California experienced a breakdown in the phylloxera resistance of the rootstock AXR#1, which has 50% vinifera parentage. All of the rootstocks discussed later in this chapter have high levels of resistance.

There is no single best rootstock for all sites and situations, so rootstock selection must be customized for each block of a vineyard. The characteristics currently considered most important for rootstocks in Oregon are phylloxera resistance, soil adaptation, drought tolerance, and influence on the fruiting variety's vigor and growth cycle. Rootstocks also must propagate easily, have a high rate of grafting success, and be compatible with the desired fruiting variety. Nematode or disease resistance, nutrient uptake efficiency, or other additional factors may become more important in the future. In summary, Oregon growers usually prefer rootstocks with high phylloxera resistance that make small vines and advance the maturity of the fruiting variety.

Soil Adaptation. Rootstocks adapt to a given soil poorly or well depending on such soil factors as texture, depth, drainage, pH, lime content, and nutrient availability. A few rootstocks seem to be adapted to a variety of soil situations, but many are better suited to specific conditions. Water availability in soils is generally a function of soil depth and texture. Deep soils and soils with higher clay content have greater water-holding capacity. Shallow soils or any soil with low water-holding capacity periodically experience drought conditions. An important aspect of a rootstock's adaptation to soil is the ability to withstand drought.

Drought Tolerance. Drought tolerance is highly desirable if irrigation is unavailable, but one must recognize that the rootstocks with the greatest drought tolerance are highly vigorous. Conversely, most of the devigorating rootstocks favored for high-quality wine production tend to have low drought tolerance. The exception may be 44-53, which shows good promise in Oregon but is still relatively new to the region. Careful consideration of soil water-holding capacity, vine spacing, cover crop competition, and other water management strategies should be made when selecting a drought-intolerant rootstock. Consider using a high-vigor rootstock with good drought resistance for a dry, nonirrigated site. It is not advisable to plant a rootstock with poor drought tolerance on a low water-holding capacity soil unless irrigation is present.

Irrigation is a valuable tool that makes available more options for vineyard design and management, including a greater variety of rootstocks.

Modified Vigor. Some rootstocks can modify the inherent vigor of the fruiting (scion) variety grafted upon it. Generally, Oregon vineyard managers prefer a relatively low-vigor vine. Smaller vines make canopy management easier, and good canopy management is critical for the production of high-quality fruit and wine. Larger vines are appropriate in locations that are suitable for divided canopy production systems, and for wine varieties or styles that can produce higher yields while maintaining wine quality.

One should keep in mind that potential vine vigor is determined by the combination of site characteristics such as soil type and depth, soil fertility, and water-holding capacity and the inherent vigor of the scion and rootstock varieties. Vigor, or ultimate vine size, can also be manipulated to some extent by vine density (spacing) and vineyard management practices such as irrigation, fertilization, cover-cropping, and hedging. The influence of rootstock on vine vigor may be much less significant than the effect of site and management practices. A rootstock evaluation trial by Oregon State University reported that site differences had more of an influence than rootstocks on vine vigor, yield, and fruit quality (Candolfi-Vasconcelos and Castagnoli, 1996). It is logical, however, to select a rootstock that is compatible with the objectives of the vineyard production system. For example, use a devigorating rootstock to produce smaller vines for a high-density (close-spaced) vineyard. If the rootstock cannot devigorate vines enough, perhaps because of a deep fertile soil, then canopy management tools must be employed to maintain appropriate vine size.

Advanced Maturity. Some rootstocks have a shorter or longer growth cycle compared to fruiting varieties. The growth cycle of the rootstock influences the entire vine, so a short cycle rootstock can advance the maturity of the fruiting variety. Earlier fruit maturity in Oregon's cool climate is a highly desirable characteristic for a rootstock. A short growth cycle may also enable the vine to become acclimated to cold temperatures earlier in the autumn.

Rootstock Descriptions

Potential rootstock performance on a specific site can be best predicted by observing performance on a similar site. The rootstocks described herein have performed well in western Oregon. The descriptions of their attributes are based on published reports (Candolfi-Vasconcelos, 1995; Howell, 1987), research trials conducted by Oregon State University (Candolfi-Vasconcelos and Castagnoli, 1996; McAuley and Vasconcelos, 2000; Taylor et al., 2000), and field evaluations in commercial vineyards. These rootstocks all have high phylloxera resistance, advance fruit maturity, and have produced wine of exceptional quality. They vary in their soil adaptation, drought tolerance, and influence on vigor. Following this list, several promising rootstocks are described with which there is less experience in Oregon, and a few more are suggested for trial evaluations.

Proven Performers

101-14

Phylloxera Resistance: High
Soil Adaptation: Requires deep soils with good water-holding capacity; tolerates clay and poor drainage
Drought Tolerance: Low, not recommended for dry nonirrigated sites
Modified Vigor: Low to moderate, smaller than 3309; performs well in closer vine spacings of 3.5–5 feet
Advanced Maturity: Advances maturity slightly more than 3309

3309 C

Phylloxera Resistance: High
Soil Adaptation: Prefers deep, well-drained soil with good water-holding capacity, but has performed well on a variety of Oregon soils
Drought Tolerance: Low, not recommended for dry nonirrigated sites
Modified Vigor: Low to moderate, slightly larger than 101-14; performs well in closer vine spacings of 3.5–5 feet
Advanced Maturity: Advances maturity, slightly behind 101-14

Riparia gloire

Phylloxera Resistance: Very high
Soil Adaptation: Requires deep, fertile, well-drained soil with good water-holding capacity
Drought Tolerance: Low, not recommended for dry nonirrigated sites
Modified Vigor: Low to moderate, perhaps slightly larger than 101-14; performs well in closer vine spacings of 3.5–5 feet
Advanced Maturity: Advances maturity, probably earliest of this group

1616 C

Phylloxera Resistance: Moderate to high
Soil Adaptation: Requires deep fertile soil; does well in poorly drained soils
Drought Tolerance: Low, not recommended for dry nonirrigated sites
Modified Vigor: Low to moderate. Performs well in closer vine spacings of 3.5–5 feet
Advanced Maturity: Advances maturity

5 C

Phylloxera Resistance: High
Soil Adaptation: Deep, fertile, well-drained clay soils
Drought Tolerance: Low, not recommended for dry nonirrigated sites
Modified Vigor: Moderate to high, more vigorous than 101-14 and 3309; performs well in wider vine row spacing of 5–6 feet
Advanced Maturity: Advances maturity

SO 4

Phylloxera Resistance: High
Soil Adaptation: Best in light, well-drained soils, but has performed well on a variety of soils
Drought Tolerance: Low, not recommended for dry nonirrigated sites
Modified Vigor: Moderate to high, more vigorous than 101-14 and 3309; performs well in wider vine row spacing of 5–6 feet
Advanced Maturity: Advances maturity slightly

Promising Rootstocks

44-53 M. This rootstock has looked very good in trials. It should perform similar to 3309 with respect to reduced vigor and advanced ripening. 44-53 has more drought tolerance than any of the devigorating rootstocks listed above. It should be suitable for many western Oregon vineyard situations.

Schwarzmann. This rootstock is moderately vigorous and has performed similar to 101-14. It should work well in deeper soils for premium wine production. Schwarzman has high resistance to phylloxera, but there are conflicting reports on drought tolerance.

Rootstocks to Test

110 R. A very successful rootstock in other growing regions; there are some plantings of 110 R in western Oregon. Vines are performing similar to self-rooted vines, but the growth cycle may be lengthened and harvest delayed. This rootstock could have a place in extremely poor soil conditions where increased vigor is needed. Very good drought tolerance

5 BB. This rootstock has performed well in other regions and could be considered for planting in poor soils where higher vigor is needed. It has performed with mixed results in western Oregon. It is not recommended for dry, nonirrigated sites, and it delays fruit maturity.

Gravesac. This is a new rootstock from Bordeaux that was bred specifically for tolerance to acidic soils. It is reported to be moderately vigorous, with high resistance to phylloxera.

Final Thought

Although the task of selecting a rootstock for wine growing in western Oregon may seem difficult, there are many examples of successful commercial vineyards to use as reference. Careful consideration must be given to soil conditions, the production system, and ultimately the kind of wine you wish to produce. Much has been learned in Oregon in the past thirty years regarding the correct varieties for premium wine production. We are now in the second decade of that same process with rootstocks and can make informed decisions. But we must continue to study and test rootstocks to refine our abilities to match rootstocks to a range of sites and situations as Oregon viticulture continues to both expand and redevelop our vineyards.

References and Additional Resources

Candolfi-Vasconcelos, M. C. 1995. Phylloxera-resistant rootstocks for grapevines. *In Phylloxera: Strategies for Management in Oregon's Vineyards.* EC 1463, Oregon State University Extension Service, Corvallis.

Candolfi-Vasconcelos, M. C., and S. Castagnoli. 1996. Effect of rootstock on Pinot noir performance at different sites in Oregon. *Proceedings of Oregon Horticultural Society* 87:192-198.

Galet, P. 1998. *Grape Varieties and Rootstock Varieties.* Oenoplurimédia, Chaintré, France.

Howell, G. S. 1987. Vitis rootstocks. In R. C. Rom and R. B. Carlson (eds.), *Rootstocks for Fruit Crops.* John Wiley and Sons, New York.

May, P. 1994. *Using Grapevine Rootstocks:The Australian Perspective.* Winetitles, Adelaide, Australia.

McAuley, M., and M. C. Vasconcelos. 2000. Evaluation of phylloxera-resistant rootstocks for the cultivars Pinot noir, Chardonnay, Pinot gris, and Merlot. Oregon Wine Advisory Board, *OSU Winegrape Research Progress Reports 2001–2002.* Oregon State University Agricultural Experiment Station, Corvallis.

Pongrácz, D. P. 1983. *Rootstocks for Grape-vines.* David Philip, Cape Town, South Africa.

Rowley, H., B. Eklund, L. Burgess, and R. F. Kriesel. 2001. *2000 Oregon Vineyard Report.* Oregon Agricultural Statistics Service, Portland. http://www.nass.usda.gov/or/vinewine.htm

Strik, B. C., and V. S. Freeman (eds.). 1992. Grape Rootstock Meeting Proceedings. Department of Horticulture, Oregon State University, Corvallis.

Taylor, P., M. C. Vasconcelos, and S. Castagnoli. 2000. Evaluation of the performance of Pinot noir grafted to five rootstocks. Oregon Wine Advisory Board, *OSU Winegrape Research Progress Reports.* Oregon State University Agricultural Experiment Station, Corvallis.

Wolpert, J. A., M. A. Walker, and E. Weber (eds.). 1992. *Rootstock Seminar: A Worldwide Perspective.* American Society for Enology and Viticulture, Davis, Calif.

7

Vineyard Design

Diane Kenworthy and Edward W. Hellman

The most important decisions in vineyard development, once the site and varieties are selected, are choices of the vine spacing and the training and trellis system. These design factors combined with the characteristics of the site (soil, elevation, aspect, and mesoclimate) and grape variety determine the entire management system for the vineyard. Collectively they determine the potential yield, fruit ripeness, ability to mechanize operations, and canopy and disease management practices. There is no single "right" choice, rather a whole range of "good" choice. A quick trip around the world of grape growing shows that vines are incredibly adaptive to different systems and growing conditions. Grape growers have evolved systems that work for their conditions and desired outcomes. Unlike some Old World wine regions, we in the New World can do anything we want, with no appellation law or tradition to restrict our choices.

One of the challenges of making the spacing and training decisions is that it is not a simple, linear process. It is more a gestalt, with each part of the whole interacting with all the other parts to make the whole. Truly, when all goes well, the sum is greater than the parts. Unfortunately, when not all goes well, it requires a lot of continual inputs to alleviate the problems.

Design Factors

The following factors are important to consider when designing your vineyard:

Resources. A vineyard is an expensive investment (see Chapter 1), and the spacing and trellising systems chosen have a big impact on both the capital investment and the ongoing operating costs of the vineyard. The major costs of vineyard development, beyond labor, are plants and trellis materials. In general, if there are more plants per acre, the trellis can be simpler and will be less expensive to build. If there are fewer plants per acre, the trellis needs to be more complicated to produce a comparable yield, and it will be more expensive to build. But money is not the only resource necessary to the successful vineyard installation; there is also time and expertise. Whether you choose to hire management and labor or provide that yourself, there is an associated cost.

Site and Vine Capacity. A good vineyard plan matches the design components to the expected size of the grapevines and their crop, making the operation efficient to manage and appropriate for the economic goals of the vineyard. Vine size is primarily determined by the inherent capacities of the site and the vine, but it can be modified to some degree by the addition of vineyard inputs and management practices. By *site capacity,* we mean the ability of a vineyard site to support vine growth and fruit production. We make a distinction here between the terms *capacity* and *vigor.* Many people refer to a location as a high-vigor site, for example, by which is actually meant a high-capacity site. Vigor is the rate of growth, and capacity is the potential total amount of growth supported by a given area. High-capacity sites tend to promote high growth rates, hence the commonplace, but sometimes confusing, equation of vigor and capacity. However, a rapid growth rate—that is, "high vigor"— can occur for varying time periods in sites of different capacity.

Site characteristics are the most influential factor determining a site's capacity. A site with deep soil and good water availability would be considered a high-capacity site because it has the potential to produce relatively large vines and crops. *Vine capacity* is the total amount of vegetative growth and ripe fruit that a specific vine is capable of producing. Vine capacity is highly dependent on site capacity, but it is also influenced by variety and rootstock characteristics, and it can be modified by vineyard inputs such as irrigation and fertilization and by management practices related to cover crops, pruning, and the canopy.

Ultimately, vineyard design must focus on the expected vine size as determined by the combination of site capacity, vine capacity, vineyard inputs, and management practices. A vineyard design that is poorly matched to the vine size will be more difficult and costly to manage and less likely to achieve good economic returns and high-quality fruit.

Vine Balance. The best way to optimize a vineyard is with balanced vines. A balanced vine has medium diameter shoots with few laterals and moderate-length internodes. The shoots should have about 15 leaves and be spaced about 3 inches apart on the fruit wire. In a balanced vineyard, each vine has enough room to express its capacity with an adequate number of medium-sized shoots and enough room on the trellis to display most of the leaves to sun. A balanced vine produces enough, but not too many, shoots and leaves to properly ripen a crop of the desired quantity and quality.

A high-capacity vine grows many vigorous (fast-growing) shoots. If pruning or training does not allow enough shoots to develop, the out-of-balance vine will have excessively vigorous shoots that produce many lateral shoots. The result is a large, shaded canopy with low yield and poor fruit quality. Conversely, a low-capacity vine can produce only a few vigorous shoots, so to be balanced it should be planted at a close spacing on a simple trellis. Smaller vines that are allowed to produce too many shoots will be out of balance; most shoots will be small and spindly, with little chance of fully ripening the clusters.

Marketing. It is important to remember that a successful vineyard requires the grapes to be transformed into wine that can be profitably sold. Regardless of whether the grape grower plans to make the wine or sell the grapes, to be in business, something has to be sold. There are always "hot" ideas about growing grapes, and it may be tempting to ignore the basic principles of vineyard planning to follow the current fad. This can be an expensive error if the considerations listed above are not also included in the decision. Grape growing and wine fads will certainly come and go during the thirty-year or more lifespan of the vineyard. Balanced, well-managed vines have the best chance of producing high-quality, balanced wines, and thus give the best chance of profitably selling the grapes or wine in the long term.

Determining Site Capacity

There are many aspects to understanding a site (see Chapter 3), but for the purposes of trellising and spacing decisions the most important factor is capacity. The capacity of a site is a function of soil type, depth, and fertility. The type and depth control the amount of water available to the plant, and the fertility determines the nutrition available to the plant. Determine the soil types and characteristics within your site, then visit several different vineyards in the area that have the same soil types to observe the vigor of their vines. Pay particular attention to how well the vines fill the trellis system and whether they require hedging during the season to maintain a well-exposed canopy.

Although grapevines grow well in a wide range of soil conditions, there are four common situations in the Willamette Valley. Comparable situations are present in Oregon's other growing regions.

1. Moderately shallow, low water-holding capacity, low-fertility soils such as the Willakenzie series. Low site capacity.
2. Moderately shallow, medium water-holding capacity, medium-fertility soils such as the Nekia series. Low to moderate site capacity.
3. Moderately deep, high water-holding capacity, medium-fertility soils such as the Jory series. Moderate site capacity.
4. Deep alluvial, high water-holding capacity, fertile soils such as the Woodburn series. High site capacity.

As you move down this list from 1 to 4, the vines tend to grow a larger number of vigorous shoots. The spacing and trellising must allow all the shoots to be well displayed for light interception, and for the fruit to be well exposed to light and air to prevent fungal disease and enable good ripening.

It is also important to consider whether the capacity of a site is going to be modified by soil amendments, fertilization, or irrigation before deciding the spacing and trellising. Irrigation has the greatest chance of significantly increasing the capacity for vine growth in a low- or moderate-capacity site.

Variety and rootstock also influence the amount of vine growth and capacity. For instance, Cabernet Sauvignon has greater vine capacity than Pinot noir. Most rootstocks have moderate capacity, about the same as own-rooted vines, except Rupestris St. George, which has very high capacity, and Riparia gloire, which has very low capacity. The more widely used hybrids, such as 3309C or 101-14, tend to be more intermediate in capacity.

Although factors such as irrigation, soil amendments, variety, and rootstock must be considered when designing the vineyard, the inherent capacity of the site remains the most important determining factor for vineyard design.

Common Spacing and Training Systems

Grapes have been cultivated for thousands of years. There have been many refinements on spacing and trellising over the course of this long history, most driven by changes in technology. For instance, grapes were originally planted at very high density, by layering, with no support system at all. A vineyard resembled a cornfield, with many single trunks randomly placed. All access to the vineyard was on foot, probably only at pruning and harvest. When people started using plow animals, rows were created to allow horse or mule access. If the soil was heavy, the rows were wider to fit two horses or an ox, since one horse could not pull the plow in a heavy clay soil. Because each plant got a little bigger when it had more room to grow, trellis systems were added to provide a support structure. Different parts of the world adapted different ideas about spacing and training, depending on local conditions and the traditions of the people who started the local industry. New World wine regions have not adopted a single point of view about how to plant vineyards. The most common systems currently used in Oregon, listed by typical row width, are these:

- **Vertical Shoot Position (VSP), narrow rows** (3–6 feet). One fruit wire, all shoots trained upward with the use of multiple catch wires.
- **Vertical Shoot Position (VSP), moderate rows** (6–8 feet). One fruit wire, all shoots trained upward with the use of multiple catch wires.
- **Scott Henry, moderate to relatively wide rows** (8–10 feet). Two fruit wires separated vertically by 10–12 inches, the upper tier trained upward with the use of multiple catch wires, the lower tier trained downward.
- **Single High Curtain System (High Hanging), relatively wide rows** (about 10 feet). Simple trellis of a single high fruit wire, shoots allowed to fall in a single curtain.
- **Geneva Double Curtain (GDC), wide rows** (10–12 feet). Trellis with two high fruit wires, separated horizontally by at least 4 feet, shoots allowed to fall in a double curtain.

Any of these systems can be adapted to most moderate-capacity sites. High-capacity sites produce large vines that can take advantage of elaborate trellis systems such as the Geneva Double Curtain or the Scott Henry system, which allow more shoots per vine to be well displayed to sunlight. Low-capacity sites are best suited to simple trellis systems to support the smaller vines.

Deciding Row Spacing

Row spacing is largely determined by the kind of equipment and trellis to be used in the vineyard and, increasingly, the cost of land. All trellis designs require room for the tractor to pass, as well as some room for the foliage. Vine-to-vine competition between rows is much less important than within-row competition, unless rows are 3 feet apart or less.

Row spacing is extremely important in determining yield potential for a site. If the fruit wire of each row is fully used, an increase in the number of rows gives a proportional increase in potential yield. If the row spacing is less than 6 feet, plans should be made for over-the-row equipment, otherwise all mechanical operations such as mowing, weed control, and spraying are difficult and costly. In moderate-capacity sites such as Jory soils, there are successful examples of wide-spaced elaborate trellis vineyards, moderate-spaced VSP vineyards, and close-spaced (3 feet by 3 feet) VSP vineyards. These examples demonstrate that management can be adjusted to make different systems work in moderate-capacity situations. In low-capacity sites like the Willakenzie series, it is difficult to make each vine produce enough grapes to justify an elaborate trellis, wide-spaced vineyard. In high-capacity sites, wide spacing is more appropriate; very close row spacing requires many trips through the vineyard to trim vines, and fruit quality usually suffers.

Deciding Vine Spacing

Vine spacing should be matched to site and vine capacity. The goal should be to have the fruit wire completely filled after pruning with well-spaced fruitful buds, neither too close nor too far apart. If a site has low capacity, vines should be planted closer together; they should be farther apart if a site has high capacity. If the vineyard is going to be cane pruned, as is traditional in the Willamette Valley, the vines should not be more than about 6 feet apart. Canes that are longer than 3 feet tend to have poor bud break at mid-cane positions, leading to inefficient use of the fruit wire. Inappropriate vine spacing undermines management efforts to achieve vine balance.

If possible, visit vineyards in sites of similar capacity at pruning time. If there are well-spaced, moderately vigorous canes with few laterals, the vine spacing is successful. Observe whether there is enough room on the wire to leave as many buds as there are good canes from the previous year. For example, if 24 medium-sized shoots grew last year, there should be room on the wire for two 10-bud canes and two 2-bud spurs (24 buds). Excessively close spacing does not provide enough room on the wire to leave the appropriate number of buds. Too wide spacing results in unfilled trellis wire, which reduces potential yield below the site capacity.

Considerable debate is generated whenever a discussion turns to the effect of training system or spacing on wine quality. Many opinions have emerged in the past twenty years about the relative merits of very close spacing versus wider spacing. In our experience, the best wines come from good sites with appropriate management that achieves good vine balance between canopy size and crop size, regardless of vine spacing.

Once again, there is no single “right” choice; trellising and spacing that are appropriate for the site capacity produces a balanced vine, which has the best potential for consistently producing high-quality fruit.

Successful Strategies in Oregon

The design and layout of the vineyard are most critical in extreme situations. Moderate-capacity situations are more flexible, and the design can reflect personal preference without getting into serious trouble. Although it is possible to vary all of the factors independently, allowing many different combinations, it is helpful to consider some examples of successful strategies that have been applied to sites of differing capacity. The following examples are not all-inclusive, but they encompass the common variations of spacing, training and trellis systems applied to appropriate situations.

Site Capacity: Low
Trellis/Training System: VSP Close Spacing
Trellis Expense: Low
Row Spacing: Close (3 feet)
Vine Spacing: Close (3 feet)
Special Labor: Shoot positioning required.
Special Mechanization: Over-the-row tractor or hand work required. Mechanical pruning possible, but not if cane pruned. Mechanical harvest possible, but equipment may not be available locally.
Special Considerations: Very capital-intensive to establish. If site capacity higher than anticipated, requires many trimming operations. Over-the-row tractor requires a low trained vine, which makes hand operations onerous. They are also expensive, may not be available locally, and may be dangerous to operate on steeper slopes.

Site Capacity: Moderate
Trellis/Training System: VSP Standard Spacing
Trellis Expense: Low
Row Spacing: Moderate (6–8 feet)
Vine Spacing: Depending on site, 3–6 feet
Special Labor: Shoot positioning and leaf removal required.
Special Mechanization: Standard or narrow size between-row tractor. Mechanical pruning possible, but not if cane pruned. Mechanical harvest possible.
Special Considerations: Local standard for much of Oregon. Makes locating management, equipment, or labor easy. Possible to shift to Scott Henry if site is higher capacity than anticipated.

Site Capacity: Moderate
Trellis/Training System: Single High Wire (High Hanging)
Trellis Expense: Low
Row Spacing: Moderate to wide (8–10 feet)
Vine Spacing: Depending on site, 5–6 feet
Special Labor: No shoot positioning required. May require leaf removal to assure good fruit exposure
Special Mechanization: Standard size between-row tractor. Spray equipment needs to assure good coverage, orchard-type spray equipment. Mechanical pruning possible, but not if cane pruned. Mechanical harvest possible, but may not be locally available
Special Considerations: Fruit exposure varies between very high and very low due to the falling nature of the canopy. Bud fruitfulness high since basal buds very exposed to light unless in too high capacity site. Pruning may be fatiguing due to high fruit wire. Harder to train if variety is naturally upright (e.g., Cabernet Sauvignon, Sauvignon blanc, or the Gamay Beaujolais clone of Pinot noir).

Site Capacity: Moderate to high
Trellis/Training System: Scott Henry
Trellis Expense: Low
Row Spacing: Moderate to wide (8–10 feet)
Vine Spacing: Depending on site capacity, 5–6 feet
Special Labor: Shoot positioning required
Special Mechanization: Standard size between-row tractor. Mechanical pruning possible if spur pruned. Mechanical harvest possible.
Special Considerations: Fruit exposure tends to be good because growth is separated. Bud fruitfulness high since basal buds very exposed to light. Timing of separating the two tiers is critical to success. On narrower rows, the lower tier can impede tractor access in the spring before downward shoot positioning is possible, since the early growth is right at tractor wheel height. Easier to train if variety is naturally drooping (like many clones of Pinot noir, Merlot).

Site Capacity: High
Trellis/Training System: Geneva Double Curtain
Trellis Expense: Moderately high
Row Spacing: Must be wide, at least 10 feet
Vine Spacing: Depending on site, 5–6 feet
Special Labor: Pruning requires 2 people/row or 2 passes. Shoots fall naturally, but centers must be kept clear.
Special Mechanization: Standard size between-row tractor. Spray equipment needs to assure good coverage, orchard-type spray equipment. Mechanical pruning possible, but not if cane pruned. Mechanical harvest possible, but may not be locally available.
Special Considerations: Fruit exposure varies between very high and very low due to the falling nature of the canopy. Bud fruitfulness high since basal buds very exposed to light. Pruning may be fatiguing due to high fruit wire. Easier to train if variety is naturally drooping (like many clones of Pinot noir, Merlot).

Field Layout

Vineyards larger than a few acres are usually divided into blocks to create reasonably uniform management units appropriate for the unique characteristics of the site (i.e., topography, soils, mesoclimates). Smaller blocks typically consist of a single variety/rootstock combination and occupy a somewhat uniform area. Generally the elevation is within a relatively small range and, preferably, the soil type is fairly uniform within a block. Larger blocks may contain several variety/rootstock combinations, but the overall objective is to create a block that can be managed uniformly and efficiently.

Within these constraints, blocks should be designed to make the most effective use of the available land. The size and shape of blocks are highly variable on hillside sites. Rows longer than about 400 feet should be divided midway with an alleyway to facilitate picking access to the entire block, or plans should be made to provide mechanical support to picking crews. Leave at least 25 feet of room at the end of rows for equipment to turn around. Plan the location of equipment travel lanes and loading and staging areas. Measure and mark off row and vine locations carefully to ensure that spacing is uniform and rows are straight. A little extra effort in layout and planting avoids later headaches created by crooked rows and unevenly spaced vines.

Row Orientation

The topography of the site is often the determining factor for row orientation. Many Oregon vineyard sites are on slopes, and there is a strong preference to orient the rows down, rather than across, the slope. There is a much greater risk of machinery tipping over when driving across slopes. Down-slope rows may also facilitate cold air drainage during spring or fall frost events, hence reducing frost damage. Terracing slopes is generally not recommended unless the slopes are severe (greater than 30%); it is questionable whether such sites are economically feasible to farm. Soil erosion can be reduced in down-slope rows by planting winter cover crops under the grape row and either a permanent or winter cover crop between rows.

When slope is not a factor, the preferred compass orientation of vineyard rows is north–south, which provides the most even exposure to sunlight for both sides of the canopy and more net solar radiation capture. Care should be taken on the west side of the canopy to prevent excessive fruit exposure to the hot afternoon sun, especially in warmer sites. An east-west row orientation provides direct sunlight on the south side of the canopy but only diffused and reflected light on the north side. This orientation is sometimes preferred in warmer sites to prevent excess fruit exposure on the west side.

Conclusion

Designing a vineyard is a challenging but rewarding task. It is essential to think of the vineyard as a whole system, and to reconsider the entire system when changing one variable. Site characteristics and economics should drive your development decision-making. Allowing preconceived ideas to override objective observations results in a less successful vineyard. On the other hand, there is room for personal preference. Many approaches to vineyard development have been equally successful. Choose a vineyard system that makes sense for both you and your site.

8

Training Systems

Edward W. Hellman and Scott Henry III

The natural climbing growth habit of grapevines requires a supporting structure, so commercial vineyards utilize some form of trellis system consisting of posts and wires. The trellis system must be constructed with sufficient strength to support the weight of fully developed vine canopies and their fruit. Vines are trained onto the trellis in a specific, predefined shape and size known as the *training system;* the number and arrangement of wires on the trellis is defined by the training system. Innumerable training systems have been devised for growing grapes under different climates and circumstances, but the overall objectives of a training system are to optimize fruit yield and quality and to facilitate management operations.

Selection of a training system is a critical aspect of vineyard planning, as discussed in Chapter 7, and is integrated into decisions on vine and row spacing, grape variety, rootstocks, and other cultural considerations. Many of these decisions initially should be based on the capacity of the site; in other words, a training system should be selected that is appropriate for the site. High-capacity sites that encourage growth of large vines are best served by training systems designed to accommodate abundant shoot growth. Likewise, a lower-capacity site promotes smaller vines that are best suited for simple training systems with small canopies. Other considerations commonly influence the selection of a training system as well, including local familiarity, labor requirements, and potential influence on fruit quality attributes.

This chapter describes several training systems utilized in Oregon. Chapter 9 is devoted to a thorough discussion of the Scott Henry system, which was developed in Oregon. Additional training systems have been described elsewhere (Adelsheim, 1992; Freeman, et al., 1992; Jackson, 1997; Jackson and Schuster, 1997; Smart and Robinson, 1991). It should be noted that the dimensions of training and trellis systems can be variable and should be customized for the vineyard's needs. Trellis height is probably the most critical factor; it should not exceed row spacing, so that shade is not cast on adjacent rows.

Vertical Shoot Positioned System

The most common training system in cool-climate viticulture regions of the world is the vertically shoot positioned (VSP) system. VSP also is by far the most common system used in the cool climate regions of Oregon. There are several variants of the system in use, including the Single and Double Guyot and the Arc-Cane (Pendelbogen) systems. Two fruiting canes are commonly used, except for high-density vineyards that use a single cane. Cane pruning is far more common in Oregon than cordon-trained/spur-pruned systems.

The objective of VSP is to develop a single narrow, upright, vertical curtain of foliage that provides excellent sunlight exposure to the entire canopy. This is achieved with a relatively short trunk containing one or two arms near the head. The arm is the structure from which fruiting canes and renewal spurs arise. Fruiting canes are tied to a lower trellis wire, often referred to as the *fruiting wire* or the *training wire,* and the subsequent shoot growth is directed in a vertically upright orientation (Figure 1). The fruiting wire is often located about waist-height for worker convenience, but it can be positioned lower. Close row spacing requires a shorter trellis height, so the fruiting wires can be less than 24 inches. The height of the trellis must provide for adequate length of supported shoot growth above the fruiting wires.

Support for the growing shoots is provided by various arrangements of permanent wires and pairs of movable "catch" wires. Support wire positions are more or less evenly distributed vertically, with the first wire or pair of wires 8–10 inches above the fruiting wire. Early shoot growth is upright and the shoots themselves provide some self-support by coiling tendrils around permanent support wires. Paired catch wires provide the necessary additional support to create the desired narrow, upright canopy. These movable wires are positioned below or just above the fruiting wire at the beginning of the season. As shoots grow and require additional support, the catch wires are lifted up and secured by slipping them over a permanently installed nail, hook, or notch on the trellis post. Lifting the wires, when conducted at the proper growth stage, "catches" most of the shoots and

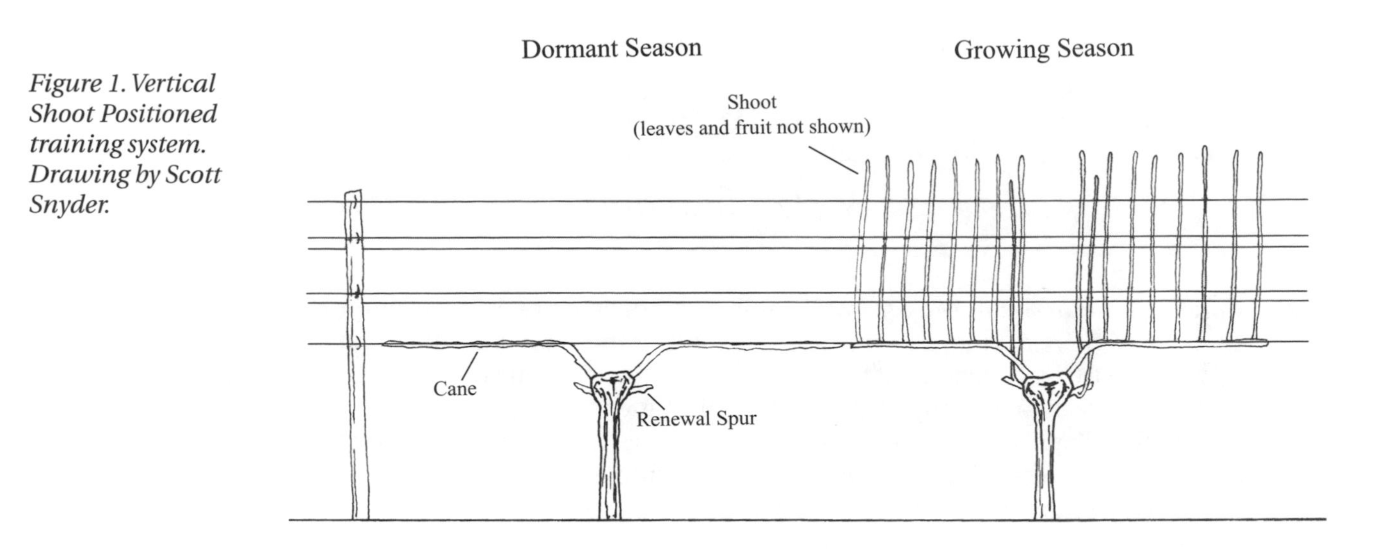

Figure 1. Vertical Shoot Positioned training system. Drawing by Scott Snyder.

pulls them together within the confines of the two wires. Often, some shoots are missed by the catch wires and require subsequent "tucking'" to position them between the wires.

The paired catch wires, at their point of attachment, are separated only by the diameter of the trellis post, so the confined shoots form a narrow canopy. Wire clips are commonly used to keep the wires close together at distant positions from the trellis posts. Two pairs of movable catch wires and one or more permanent catch wires are commonly used. The movable wires are raised to successively higher positions as shoot growth progresses, until each pair reaches its final position. This provides an even distribution of wires to support fully developed shoots with their ripening fruit.

The VSP system produces an efficient canopy with excellent sunlight exposure for low- to moderate-capacity vines. The favorable canopy microclimate promotes optimal ripening of fruit, particularly in a cool climate. It is common in Oregon to further enhance sunlight exposure of Pinot noir fruit by leaf removal in the fruit zone, which improves ripening, increasing wine color and phenolic content (Price et al., 1995). The VSP system is well suited to the production of very high quality fruit, particularly in cool climates where yields typically must be restricted to obtain fully ripe fruit.

The inherently low shoot density of a cane-pruned, single canopy limits the cropping potential of vines trained to the VSP system. Therefore, higher yields on a per-acre basis can be obtained only by using closer row spacing (more vines per acre), which necessitates a shorter trellis to avoid adjacent-row shading. Very close row spacing also requires specialized equipment for floor management and pesticide applications, as well as additional considerations.

The upright orientation of growth in the VSP system encourages greater shoot vigor, so some tipping may be necessary if shoots grow significantly beyond the top trellis wire. Shoot tipping tends to stimulate greater development of lateral shoots, which if excessive can compromise the favorable sunlight exposure of the VSP canopy. High-capacity vines have high shoot vigor that may require tipping several times per season, and additional canopy management practices may be needed such as hedging to remove laterals. Therefore, VSP is not suited for higher-capacity vines. It is often possible to convert a VSP system to a higher-capacity training system such as the Scott Henry, if the row spacing is wide enough to accommodate the taller trellis.

Single High Wire System

A simple, single hanging curtain system has been used in Oregon, variously referred to as the Single High Wire system (Adelsheim, 1992), the High Hanging system, or the Single Curtain system. This system most commonly uses cane pruning, but cordon training/spur pruning is used where vine spacing is greater than 5 feet. The fruiting canes and renewal spurs arise from arms at the head of a tall trunk and are tied to a single trellis wire near the top of the post, commonly 64–68 inches (Figure 2). Shoots tend to grow downward, or to bend over and hang down as they increase in weight with growth. The downward growth orientation reduces shoot vigor and development of lateral shoots.

The Single High Wire system produces a single hanging canopy with generally good sunlight exposure at the base of shoots. Exposure varies, however, depending on the natural growth habit of the variety. Fruit exposure is higher, and can result in sunburned fruit, on varieties with a drooping rather than an upright growth habit. The extent of fruit exposure can

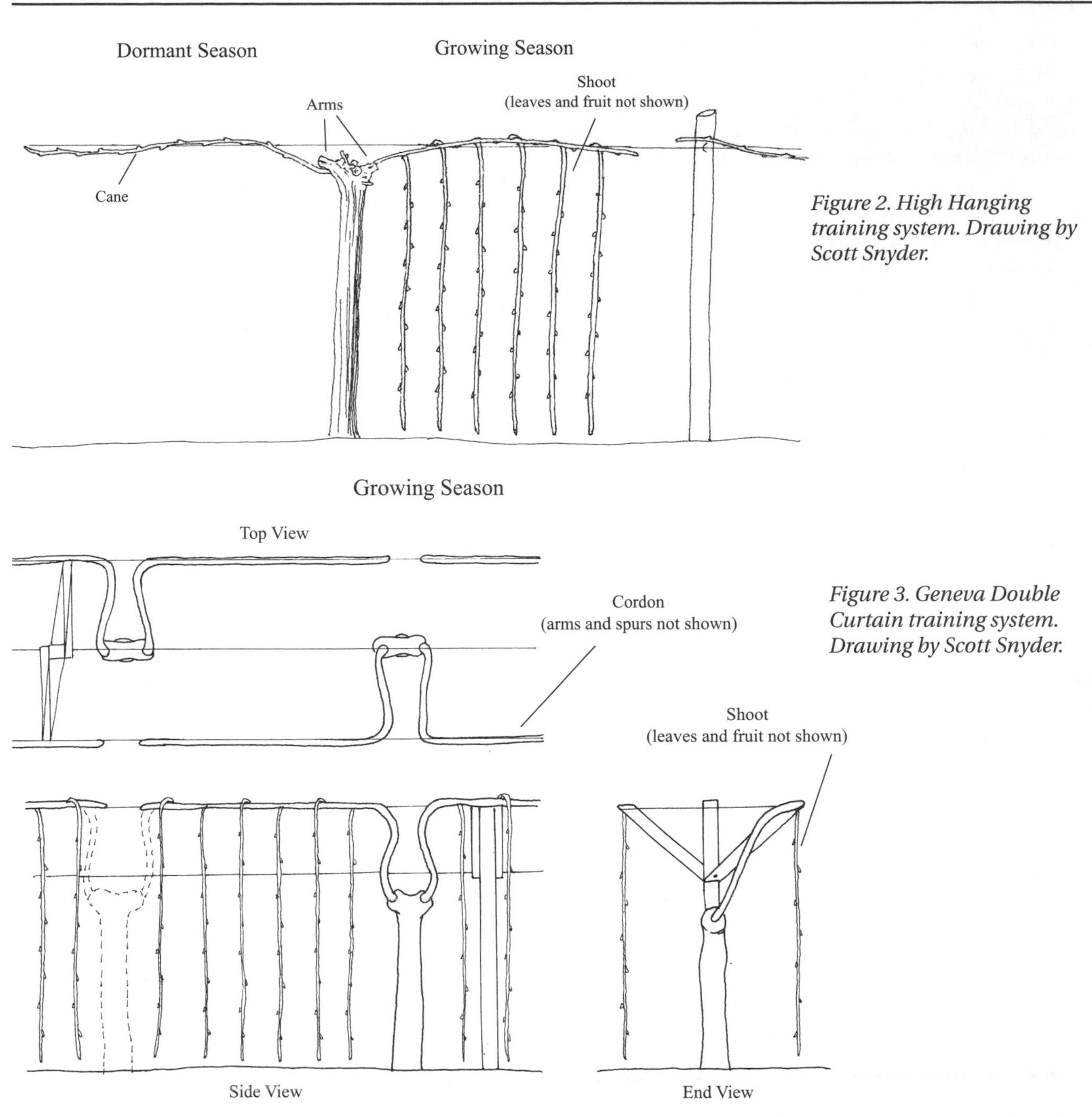

Figure 2. High Hanging training system. Drawing by Scott Snyder.

Figure 3. Geneva Double Curtain training system. Drawing by Scott Snyder.

influence the phenolic structure of the grapes and ultimately the wine. The major advantages of this system are its low cost and simplicity. It utilizes the most simple trellis system and does not require shoot positioning.

Geneva Double Curtain

High-capacity sites have the potential to produce large vines that develop an undesirably dense, shaded canopy. Such sites require a training system that can accommodate their greater shoot growth and take advantage of their higher cropping potential. The Geneva Double Curtain (GDC) system was developed in New York to enable production of a large canopy that has good sunlight exposure. It achieves this by dividing the canopy horizontally into two parallel hanging curtains, 4 feet apart (Figure 3). Thus it is similar to the Single High Wire system, but with double the canopy size. The GDC system has been utilized to a small extent in Oregon.

The GDC system is most commonly cordon-trained/spur-pruned to enable a greater bud count on high-capacity vines. The trunk is trained to a stake or a trellis support wire positioned 12-14 inches below a pair of fruiting wires. The fruiting wires are attached to the ends of a 4-foot cross-arm at or near the top of the trellis post, commonly 5.5–6 feet above the ground. Various configurations are possible with one or two trunks and two or four cordons per vine. One common method (see Figure 3) is to either extend the trunk or develop a cordon that is trained over to a fruiting wire. Close to the fruiting wire, the cordon is branched to form two cordons that are trained along one of the fruiting wires, extending in opposite directions until they reach the trunks of the adjacent vines. The next vine is trained in a similar manner on the opposite fruiting wire. Arms with fruiting and renewal spurs are developed along the underside of the cordons; they should be evenly spaced to provide the desired shoot density and oriented downward.

The key to managing the GDC is that shoots must be positioned downward to prevent the curtains from growing into each other. The space between curtains must remain open to enable thorough sunlight exposure of the canopy. Failure to keep the curtains divided negates the advantages of the system, causing a dense, shaded canopy with all of its accompanying problems. Do not make the mistake of building the trellis with fruiting wires closer than 4 feet apart. Shorter cross-arms may be less expensive and enable closer row spacing, but it is nearly impossible to keep the two curtains separate if they are closer than 4 feet.

The GDC system is well adapted to mechanization, and equipment is commercially available to perform many vineyard operations, including harvesting, pruning, and shoot positioning. One obvious advantage of the GDC is that larger canopies produce greater yields. The GDC system does double the canopy size, but it does not double the yields of a comparable single curtain system (Jackson, 1997).

The GDC has some disadvantages, including a potential for excessive sunlight exposure effects on fruit phenolics. It is also more challenging to achieve thorough spray coverage of pesticides on both sides of the two curtains.

Trellis Construction

At first glance a trellis system seems to have many similarities to a simple wire fence, but a trellis is subject to considerable forces and must support significant loads. Furthermore, the lifespan of the vineyard should be long, so it is highly desirable for the trellis system to remain functional as long as possible. Careful planning and selection of materials are important aspects of building a strong, durable trellis. The important structural factors are extensively covered in Freeman et al. (1992) and Smart and Robinson (1991). Additional practical information on installation of a trellis is available in Adelsheim (1992).

References

Adelsheim, D. 1992. Spacing, training, and trellising vinifera grapes in western Oregon. In T. Casteel (ed.), *Oregon Winegrape Grower's Guide,* 4th ed. Oregon Winegrowers' Association, Portland.

Freeman, B. M., E. Tassie, and M. D. Rebbechi. 1992. Training and trellising. In B. G. Coombe and P. R. Dry. (eds.), *Viticulture,* Vol 2: *Practice.* Winetitles, Adelaide, Australia.

Jackson, D. 1997. *Monographs in Cool Climate Viticulture: 1. Pruning and Training.* Lincoln University Press, Canterbury, New Zealand.

Jackson, D., and D. Schuster. 1997. *The Production of Grapes and Wines in Cool Climates.* Lincoln University Press, Canterbury, New Zealand.

Price, S. F., P. J. Breen, M. Valladao, and B. T. Watson. 1995. Cluster sun exposure and quercetin in Pinot noir grapes and wine. *American Journal of Enology and Viticulture* 46:187-194.

Smart, R., and M. Robinson. 1991. *Sunlight into Wine.* Winetitles, Adelaide, Australia.

9

Scott Henry Training System

Scott Henry III

The Scott Henry training system was designed to improve grape yields and quality in a high-capacity, cool-climate site. It has proved subsequently to be well adapted for growing high-capacity vines in other climatic situations. Large vines are trained on a unique trellis that divides the canopy vertically and provides good sunlight exposure to foliage and fruit. The system enables a large leaf area to be displayed with a relatively low leaf density. The resulting leaf exposure to sunlight promotes fruit ripening and a high degree of bud fruitfulness. The well-exposed clusters are less susceptible to fungal diseases, which contributes to high fruit quality. Significant improvements in both wine quality and grape yield have been obtained at the Henry Estate Vineyard by the development of the Scott Henry training system.

Site Description

The Henry Estate vineyard is situated in Coles Valley (one of the "100 valleys" of the Umpqua), which is the last valley the Umpqua River passes through before reaching the Coast Range. The vineyard is on the valley floor, which is slightly undulated in profile, at about 400 feet above sea level. It is a high-capacity site with fertile, mostly clay loam soils varying in depth between 2 and 20 feet. The soil is underlain with gravel from the remains of the old riverbed. Normal summer water level is about 13 feet. The vineyard is situated toward the center of the valley, which allows for early morning and late afternoon sun. West winds come upriver from the Pacific Ocean almost every afternoon, influencing the mesoclimate of the vineyard. The wind carries moisture in the spring, which protects the vineyard from frosts. Later in the year, these moisture-laden winds tend to increase powdery mildew and bunch rot problems. Degree-day accumulation in most years is in the 2,200–2,300 range.

History of Henry Estate

The first planting of 12 acres was made in 1972 and consisted mostly of Pinot noir (UCD-4, Pommard clone) and Chardonnay (UCD-5, commonly referred to in Oregon as Davis 108 clone). A couple of rows of Gewürztraminer were also planted at this time to provide propagation material. All vines were from heat-treated, mist-propagated nursery stock that originated from the mother block at the University of California, Davis. The vineyard has subsequently been expanded to about 40 acres, including 3 acres of Riesling and 3 acres of Pinot gris. Fruit was first harvested in 1975, the third leaf, and wines were made by other wineries for several years before the Henry Estate Winery was built for the 1978 crush.

The initial training and trellising was a standard three-wire vertical shoot positioned (VSP) system on a 7-foot plant spacing and 12-foot row spacing. The plants were headed at the first wire and cane pruned to two canes. All vine growth was trained upward.

As the vineyard matured, the high capacity of the site made control of vine growth an increasing problem. Wines from the early years of the vineyard, such as 1978, 1979, and 1980, won many medals, but with time a smaller portion of the crop was utilized for the top grade wines. By 1980, when vines were eight years old, it became obvious that the dense canopy and shading was reducing Pinot noir fruit intensity and color and increasing the pH. In addition, more and more fruit was being lost to powdery mildew and bunch rot. The degradation in fruit quality reached a low in 1982, when a very light colored, low-intensity Pinot noir was produced and the Chardonnay was totally lost to bunch rot.

Canopy Management Trials

Many of the standard modifications to improve grape quality such as shoot thinning, hedging, leaf removal and crop reduction were tested in vineyard trials. Wines produced from these trials indicated that several of the management practices had positive results. First, the best wines were produced when the crop load was in balance with vine growth. Balance is evident when laterals are slow to develop and only a very small second crop is set on the vine; shoot tips stop growing at about veraison and the canes begin to harden. Crop levels that were either too low or too high resulted in less desirable wines.

Second, wine quality, especially for Pinot noir, increased significantly when the fruit was exposed to the sun. Increased fruit exposure also reduced the amount of bunch rot, which was especially significant for Chardonnay.

One of the field trials revealed that vine vigor could be reduced by increasing the number of canes from two to four per vine. More canes produced a higher yield by increasing the number of clusters per vine. But the large crop and canopy created by four canes had inadequate sunlight exposure, resulting in delayed fruit ripening. A trellis and training system was needed that would increase leaf area while reducing canopy density.

Birth of the Scott Henry System

The Scott Henry system was created with the objective of achieving balance in relatively large vines while maintaining good sunlight exposure in the canopy and on the fruit. To do this, it was necessary to develop a training system with two fruiting levels instead of one. The system created a lower fruiting level and an upper fruiting level, about 10 inches apart. Shoots from the upper level were trained upward while shoots from the lower level were trained downward, thereby preventing the upper fruiting level from being shaded by growth from below. Separation of the shoot growth from the two levels opened up a window in the center of the canopy.

The window is a critical component of the system, contributing to higher quality fruit and better vineyard economics. Fruit quality was improved by the increased exposure to sunlight. Air circulation through the two fruiting levels also improved, resulting in more rapid cluster drying after rain or heavy dew. This provided an environment less favorable to the development of powdery mildew and bunch rot. The window also improved spray coverage for disease control, resulting in fewer fungicide applications. Better sunlight exposure of the buds increased their fruitfulness by about 30%. Consequently, fewer shoots were needed to achieve the target crop load, which also opened up the canopy for better sun exposure to the fruit and leaves.

The addition of the downward-trained lower fruiting level increased the leaf area by about 40% over a standard VSP system. Greater leaf area enabled complete ripening of the larger crop loads that were necessary for proper balance in our bigger vines.

In summary, the Scott Henry training system achieved everything we were looking for and a bit more; balanced vines, exposed fruit with open canopies, top quality wines, less incidence of powdery mildew, reduced bunch rot, and fewer number of fungicide applications in the vineyard.

System Description

Basic Scott Henry System. A schematic view of the basic Scott Henry system is shown in Figure 1. Four canes, two on the lower wire and two on the upper fruiting wire, are used to provide the fruit. Four replacement spurs, one with each cane, are used for renewal growth. Shoot growth from the top canes is trained upward and shoot growth from the lower canes is trained downward. The growth between the two levels of fruit is separated shortly before bloom. The upward shoot growth is held in place by two fixed-position foliage wires and two movable catch wires,

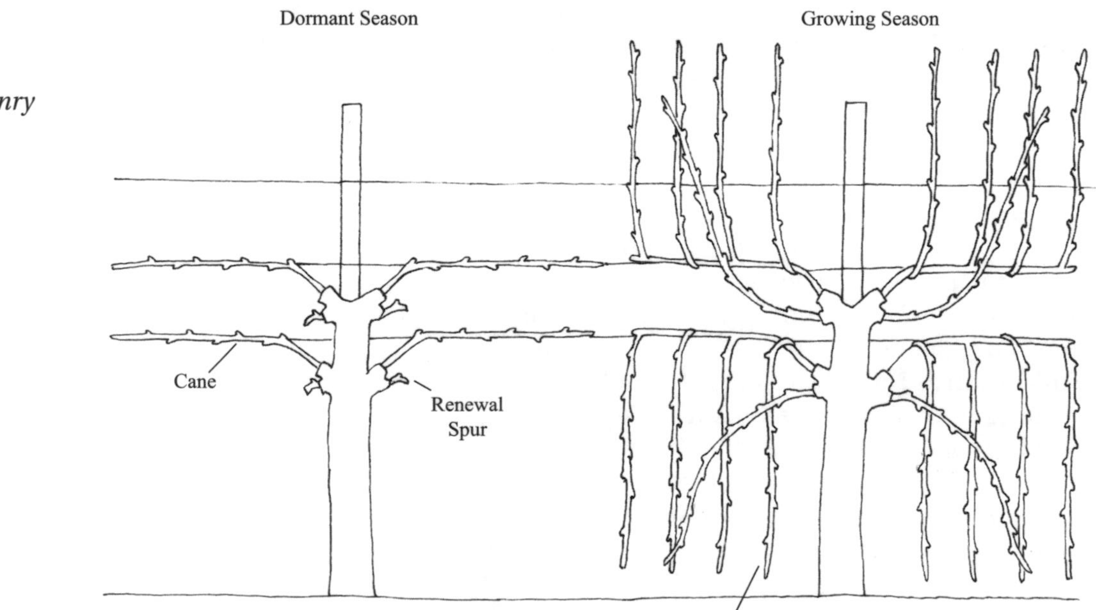

Figure 1. Scott Henry training system. Drawing by Scott Snyder.

one on each side of the row. The movable catch wires are placed midway between the two fruiting levels at the beginning of the season; later, these are moved up to the fixed-wire position when shoot growth is long enough to require additional support. As the two catch wires are moved up, the growth from the lower wire is wiped downward by hand. The upward shoot growth is hedged when it is about 10 inches above the trellis to prevent development of an umbrella-like canopy. When the vines are pruned in the dormant season, the lower canes are loosely tied to the wire so that it is easy for the entire cane to roll. Sometimes this occurs naturally in the direction of the wind. The canes on the upper fruiting wire can be wrapped and tied as normal. The idea is to divide the growth on each side of a vertical curtain so that a window is opened between the two levels of fruit. The lower growth is allowed to grow downward and along the ground out into the row middles.

Each vineyard site has its own growth characteristics, so adjustments to the system's basic parameters—number and height of wires, plant and row spacing, and crop load—are necessary to optimize the system at any given site.

Number of Wires. The basic system uses six wires: two wires are used to tie the upper and lower fruiting canes, two fixed wires are used to support shoots of the upper fruiting level, and two moveable catch wires, one on each side of the trellis post, are used to train shoot growth vertically. Catch wires can be moved as necessary to contain shoot growth of the upper canopy. The last movement of the catch wires usually positions the wires at the top of the trellis post. Catch wires do not have to be moved to specific positions at scheduled times. It is best to move them independently to whatever height is needed to support shoot growth.

Wire Height. The first dimension to select is the distance between the two fruiting wires. I prefer 10–12 inches; trials in New Zealand utilize only 6–8 inches. The wider spacing ensures that the window between the two fruiting levels stays open later in the season, allowing air to move through the window and quickly dry the fruit after a rain or heavy dew. With equal amounts of fruit on each level, the canopy leaf area between upward and downward growth should be about the same. Therefore, if the distance from the ground to the top of the stake is 78 inches, the lower fruiting wire should be at 38 inches, the upper fruiting wire at 50 inches, the third wire at 64 inches, and the top wire at 78 inches. The upward growth has about 38 inches of vertical growth (the shoots extend about 10 inches above the top of the stake before hedging is necessary) and the downward growth has about 38 inches before reaching the ground.

Plant Spacing. The basic Scott Henry system was developed for a plant spacing of about 7 feet. Wider spacing can be accommodated if strong, fruitful canes can be developed to fill the distance between plants. Keep in mind that bud break can be erratic on longer canes, and that canes must be healthy and strong, indicated by the diameter, firmness, and color throughout their entire length. Spacing at 7 feet requires a cane length of about 39 inches: half of the plant spacing minus 3 inches for the gap between cane tips. The cane length requirement increases to 45 inches for an 8-foot spacing. Some cultivars such as Pinot noir do not have a reputation for generating long, strong canes. To overcome this shortcoming, cane pruning may be replaced with a cordon system (see "System Modification" below). Vine spacing closer than 7 feet provides the vineyard manager another option, which I call the Scott Henry High/Low system (see below). I presently use a 6-foot vine spacing at Henry Estate.

Row Spacing. The 12-foot row spacing used at the Henry Estate Vineyard was initially selected to utilize our existing tractors. For our high-capacity site, it appears the optimum should be narrower, about 10 feet. If the row spacing is not sufficient to prevent the tractor from tugging on the shoots as it moves down the row, movable catch wires can be used to position lower fruiting shoots downward to keep the foliage growth out of the way.

I believe the ratio of canopy height to row width should be about 0.75 for the Scott Henry system. As row spacing increases, so should the height of the trellis. The maximum practical height of the trellis from the ground is about 78 inches, to facilitate moving the catch wires to the top of the post or stake. If the trellis is too high for the row spacing, fruit ripening is delayed in the lower canopy because of early afternoon shading.

Basic System Management

Pruning. A strong cane, preferably from last year's replacement spur, is selected for each fruiting cane. It should be cut to length so that a 4–6-inch gap is left between the tips of adjacent canes when they are positioned on the wire. In addition to the canes, a two-bud replacement spur is retained to provide the growth for next year's cane and replacement spur. The lower canes are hung loosely on the wire with ties, similar to fastening a drip irrigation line. This loose approach is necessary to minimize the effort and damage later when the growth is turned downward by hand. The top canes are wrapped around the wire several times and tied tightly. The replacement spurs should be selected about 4–6 inches below the wire and in a direction that allows the growth to be in line with the vertical curtain.

Crop Level. Each grower must experiment with crop level until the desired compromise is reached between quantity and quality. We have varied crop level in our trials from 2 tons/acre to 12 tons/acre. Surprisingly, the best wines did not come from the lowest crop levels. At low yields, vines responded with an increase in the number of lateral shoots, resulting in increased growing tips, second crop production, and shading. Excessively high yields lead to an undesirable delay in ripening. The bud count may be varied between the upper and lower canes if a difference in ripening is experienced between the two fruiting levels. Fewer buds are retained on the level that is experiencing delayed ripening and correspondingly more buds are kept on the other fruiting level. Trials in our vineyard have not found a significant difference in wine quality between the two fruiting levels as long as the total bud count maintains vine balance.

Fruit ripening can also be delayed on lower canes when row spacing is too narrow for the trellis height, creating shade in the lower canopy. One solution to this problem is to shift the height of the fruiting wires higher, thereby increasing the canopy area of the lower wire and correspondingly decreasing the upper canopy.

Canopy Division. Division of the canopy takes place about two weeks before bloom and extends through bloom. It is important to wait until sufficient growth has occurred to make the movement of the catch wires effective, but it is also important not to wait until the tendrils are so tough that shoot damage occurs during separation. Early separation allows more light into the bloom area, which results in better fruit set and higher bud fruitfulness next year.

Separation is done by hand with two people, one on each side of the row. The lower growth is separated from the upper growth and wiped downward. Often the whole cane can be rolled over, positioning the shoots downward. The movable catch wires can be used now if there are enough upper shoots that have reached adequate length. The catch wire should not be moved at this time if too many shoots are missed. We position the catch wires about 12 inches above the upper fruiting wire by hooking them over a nail in wooden stakes or in a slot for a steel stake. With a little practice, two people should be able to separate about 1.5–2 acres per day.

If the trellis is high enough, the catch wires need to be moved again when sufficient growth has occurred to make the movement effective. For our vineyard, this moves the wire to the top of the stake. This is also a good time to wipe downward the growth from the lower cane that has persisted to grow upright.

Hedging. As the shoot growth from the upper fruiting canes continues to grow above the top wire, the shoots begin to bend over. At this time, the growth needs to be hedged with a cutter bar. It is important that the open window between the two canes be left open for good air movement. Sometimes lateral growth tends to close off this window. Therefore, this window is also hedged when the tops of the vine are hedged.

Shoot Thinning. In our vineyard, both canes and trunks are shoot-thinned by hand. The canes and spurs are thinned to obtain a desired shoot count. The trunks are thinned to allow light penetration into the spur areas. This allows for an even distribution of foliage over the trellis system while maintaining a desired crop level.

Weed Control. We utilize a cover crop between the rows and spray a post-emergence herbicide in a 2-foot wide section under the vines. Standard weed management approaches can be used with the Scott Henry Trellis system, but weed growth under the vine tends to be restricted naturally because the lower canopy shades this area.

Disease Control. A standard fungicide spray program is more than sufficient for vines trained to the Scott Henry Trellis system. Exposed fruit enables excellent spray coverage, which optimizes fungicide efficacy. This has enabled us to use lower fungicide application rates and reduce the number of spray applications. The double level of fruit does, however, require a slightly wider spray pattern than a VSP system requires.

Harvest. Our harvest is done by hand. Pickers are happy with the Scott Henry system since the fruit is exposed and easy to locate. This results in little fruit being missed during the harvest process. Mechanical harvesting is also effective with the system.

System Modification

The discussion to this point has been about a basic Scott Henry system utilizing four fruiting canes. Because the system is so adaptable, however, modifications abound. There are at least six versions of cane-pruned and three versions of cordon-trained/spur-pruned Scott Henry system used commercially. Every time I visit wine-growing regions that utilize our system, I find a variation that is working for a grower or winery.

Currently, I prefer modifications that utilize only two fruiting canes. Pruning, tying, and thinning costs are lower for two canes compared to four. One modification that we now frequently use is the High/Low.

High/Low. Using the same trellis configuration, the High/Low modification (Figure 2) uses only two fruiting canes or cordons per vine. Each cane or cordon extends completely across to within 2–3 inches of the trunk of the next vine. Adjacent vines in the row are trained alternately to either the upper fruiting wire or the lower wire. Again, shoot growth is trained upward from the upper wire and downward from the lower wire. The alternating training pattern enables each vine to have all of its fruit at one level. Another option for cane-pruning (not shown in Figure 2) is to develop all of the trunk heads just below the upper fruiting wire with two arms (renewal zones) at each wire. Renewal spurs are retained on the level that does not have a fruiting cane. Thus, for individual vines, the fruiting zone alternates from high to low from one season to the next.

Cordons are preferred for the High/Low modification if vine spacing is greater than 6 feet or if the grape variety tends to have poor bud break on long canes. Cordon training has the further advantage of being adapted to mechanical pruning.

S Modification. The modification currently preferred for cane pruning is the S modification (Figure 3), because it seems to make it easier at pruning to find good fruiting canes in proper positions. Two canes are used for each plant, but one cane is on the upper fruiting wire and the other is on the lower fruiting wire. A two-year study in our vineyard by Oregon State University showed that fruit ripened evenly on the two fruiting levels independent of variations in crop loads between fruiting levels (Ratliff-Peacock et al., 1998).

Scott Henry System Advantages

Balanced Vines. The two levels of fruit allow crop loads to be varied until a balance between fruit and vine growth is achieved. This enables the shoot growth of moderate- to high-capacity sites to be channeled into the production of high-quality fruit.

Sun Exposure. The fruit on both fruiting levels has good exposure to the sun. This is an important wine quality factor. We have found no advantage in wine quality by leaf pulling, which is quite a cost savings. The balanced vine produces a narrow canopy with good light penetration, thus avoiding yellowing of interior leaves.

Crop Losses. The open window between the two levels of fruit allows good air movement through the vine, which facilitates rapid drying after a rain or heavy dew. The downward shoot position exposes the fruit similar to a Geneva Double Curtain system, and since the growth is away from the upper fruit level, the upper level fruit has fewer leaves to produce shade. This enables an effective spray program. Prior to the Scott Henry system in our vineyard, we could expect about 15% of the Pinot noir and 25% of the Chardonnay to be lost to bunch rot. Since we introduced the Scott Henry system, these types of losses have been essentially eliminated.

Ripening. The Scott Henry system has resulted in fruit ripening 4–6 days ahead of a trellis that was hedged but not separated. In addition, the pH levels of the juice have been reduced, which enhanced wine quality. Fruit intensity has been improved and a significant increase in color has been achieved in Pinot noir because of better light penetration into the fruiting zones of the vine and the additional leaf area (about 40%) from the downward canopy.

Crop Level. Utilization of the Scott Henry system results in a significant increase in crop level (about 30%) for the same total bud count on the vine; greater bud fruitfulness results from better sunlight exposure of the buds on the fruiting canes. Therefore, to meet crop level objectives, fewer shoots are required, which helps open the canopy for better light penetration.

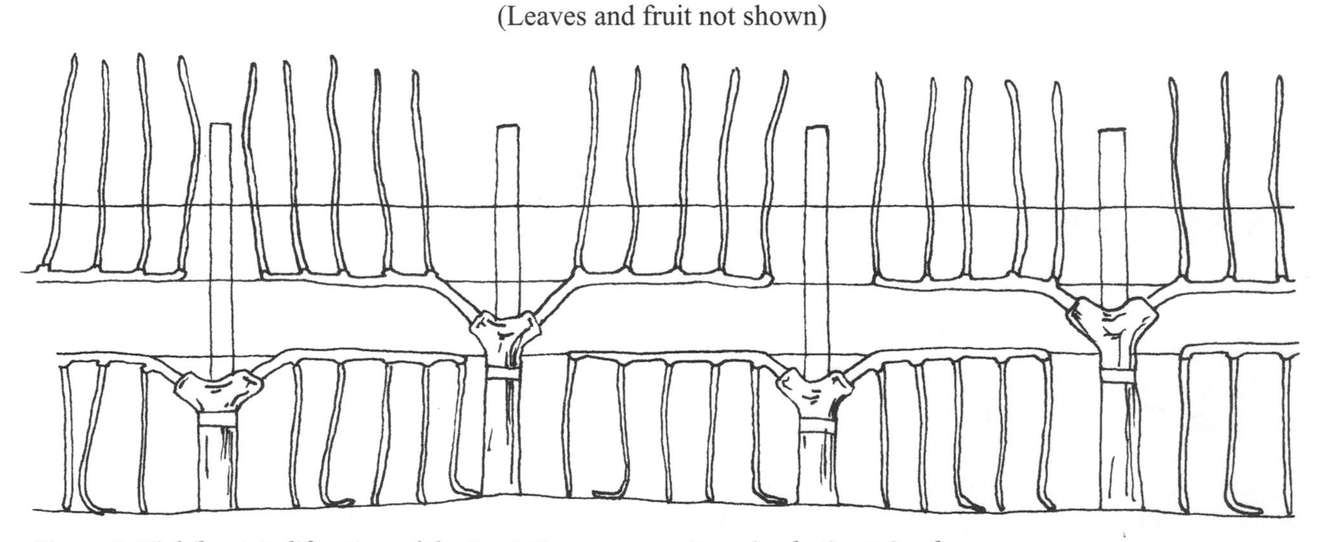

Figure 2. High/low Modification of the Scott Henry system. Drawing by Scott Snyder.

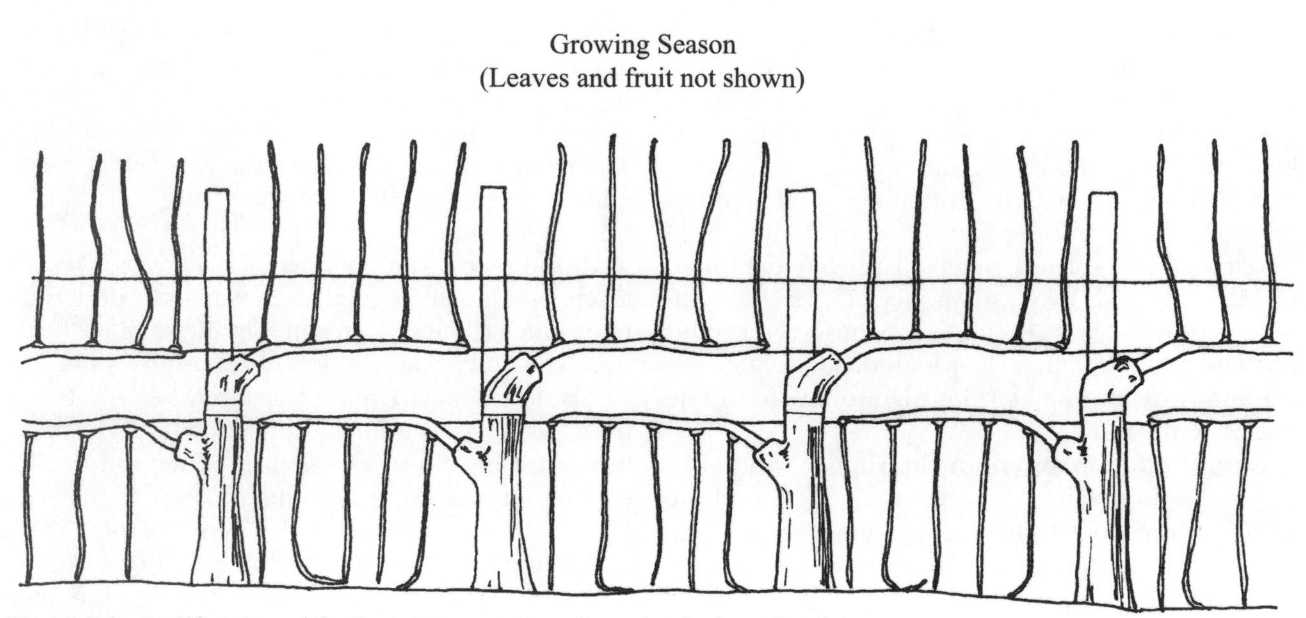

Figure 3. S-modification of the Scott Henry system. Drawing by Scott Snyder.

Economics. This system not only is able to achieve most of the results of a complicated double canopy system but is also practical. This is especially true for vineyards that have already been trained to a vertical canopy. Modification time and costs to adapt to a Scott Henry system are minimal. Some additional costs are associated with the separation of the growth between the two fruiting levels. The ability to harvest mechanically with existing harvesters is retained by the Scott Henry system.

System Concerns

Uneven ripening between fruiting levels is a concern often mentioned in discussions of the Scott Henry training system. The Australians have a great saying that applies: "No worries." One of the strengths of the Scott Henry system is its adaptability. If a particular configuration does not give satisfactory results, minor adjustments can often correct the problem. Even ripening between the fruiting levels is important and should be monitored. Any significant differences in ripening between fruiting levels can be eliminated by one or more of the following adjustments: (1) increase the shoot length of the level with delayed ripening by shifting the fruiting wires up or down; (2) at pruning, retain fewer buds at the level with delayed ripening, leave correspondingly more buds on the other fruiting level; or (3) thin shoots more aggressively on the level with delayed ripening.

Summary

The Scott Henry training system was originated in the vineyard by a grower trying to solve a canopy problem without incurring overbearing costs. This system has been largely responsible for bringing our winery back to the medal-winning road. It is not, however, the answer for every situation. If it is being considered, I suggest conducting trials in the vineyard before committing significant amounts of acreage to this system. This experience also enables the grower to try some modifications that may be more suitable for a particular site.

Acknowledgments

I would like to thank Richard Smart for his important contribution to the enhancement of this training and trellis system throughout the world. If this system does help some growers, my reward is that better wine always makes for a better world.

References

Ratliff-Peacock, J., M. C. Candolfi-Vasconcelos, and S. Henry. 1998. Development of viticultural practices to improve winegrape performance. Experiment II: effect of canopy location on yield components and fruit composition in Pinot noir grapevines trained to the Scott Henry trellis system. Oregon Wine Advisory Board, *Oregon State University Winegrape Research Progress Reports,* Oregon State University Agricultural Experiment Station, Corvallis.

II
Vineyard Development

Vineyard development begins well in advance of planting the vines; proper site preparation is necessary for establishment of a successful vineyard. The considerations and methods for preparing a site are described in Chapter 10. The period before planting is your best, and sometimes only, opportunity to make significant modifications or improvements to the vineyard site. Preparation of the site prior to planting often involves numerous steps and requires time; do not attempt to rush the project. Remember that the vines can be productive for thirty years or more, so you will have to live with any mistakes for a long time. Take the time to prepare the site properly, addressing soil drainage, mineral nutrients, soil pH, and other concerns.

Similarly, do not rush into purchasing plants. In Chapter 11 we discuss the different types of plant materials that can be used to establish a vineyard. High-quality, disease-free plants are the cornerstone for establishment of a successful vineyard. Place your nursery order at least one year in advance of planting. If given a short lead-time, nurseries may be out of the particular varieties, clones, or rootstocks called for by your plan. Accepting substitutes negates your careful planning efforts. Work with a reputable nursery and purchase certified planting stock.

Be prepared at planting time with all of the necessary supplies and materials for planting and caring for young vines. The important considerations and methods for planting, managing, and training young vines are discussed in Chapters 12 and 13. Considerable effort and attention to detail go into the planning and preparation of a vineyard site, and equal care must be exercised at planting time to ensure that vines are properly planted. Young vines have small root systems and do not compete well with other vegetation, so weed control is critical and vines must be provided with adequate water and nutrients. Intensive management is required in the first few years of a vineyard; vines require frequent individual attention during the training process that develops their ultimate structure. Common wisdom holds that vines should produce their first (partial) fruit crop in their third year, but it is more important to the long-term health and productivity of the vineyard to base fruiting age and cropping level on the size and capacity of each vine.

If irrigation is to be utilized, the complete system should be installed and operational in time to water newly planted vines. Chapter 14 provides a thorough overview of the common components of a vineyard drip irrigation system and the necessary factors to consider when designing a system.

Thoughtful vineyard planning creates the blueprint for a successful vineyard, but careful execution of this plan is necessary to achieve the desired goals. Most of the activities and practices of vineyard development are conducted only once, so they must be done properly to establish a uniform and efficient vineyard. Correcting mistakes later is never as good as doing it right the first time, and it is always more expensive. Shortcuts and use of lower-quality materials to save costs are counter-productive.

10

Site Preparation

Andy Humphrey and Edward W. Hellman

Remember those Tonka Toys you played with as a kid? This is the real deal, and it's fun! Think of it—bulldozers, scrapers, trucks, graders, backhoes and excavators. What a blast, and it's very therapeutic for stress relief. All of these machines are really noisy, so if you want to shout at the top of your lungs questioning your own sanity, no one will really be able to hear or understand you. But before you jump on a bulldozer and start moving earth, consider that this is your best opportunity to make significant changes or improvements to the site. A thorough site assessment should have identified the particular needs and opportunities of the site. Every site is different and requires varying practices and equipment to prepare for a vineyard. We outline the most common preparation activities, but in most cases your site will not require all of these.

The development of a vineyard often is considered to be spread over a period of two or three years, with site preparation happening in the same year as planting. Although it is possible to accomplish this, given a relatively dry spring, it is more realistic to plan for two years to complete preparation and planting. A cleaner, less expensive, better prepared, and easier installation can be done if year zero is added to the front of the development time line.

Previously Cultivated Sites

Sites that were previously cultivated cropland or mostly clear pasture need relatively little site preparation compared to sites covered with trees and brush. Preparation practices for the former include the same final steps as preparing an uncultivated site; so if you do not need to clear the land, skip down to "Deep Ripping."

Previously Uncultivated Sites

Land Clearing. Fast, cheap, and good. It would be nice to have the best of all worlds, but you only get to choose two out of three. If you want the work done fast and good, it won't be cheap. If you want it done fast and cheap, it won't be very good. It is possible to get cheap and good, but it won't get done very fast.

Many new landowners make the choice of hiring a small, one-person outfit with old equipment. Although this may seem the least expensive approach, it usually is not very fast. Old equipment got that way through thousands of hours of pushing dirt, stumps, brush, and rocks. Old equipment breaks down frequently and parts are expensive and hard to find. For a one-person excavation company with limited cash flow, a breakdown can shut down the operation for weeks or months. On the other hand, a well-equipped company with a larger pool of experienced operators can bring the appropriate equipment and get the job done quickly. But it won't be cheap.

Logging. Don't forget your logging permit. If you are taking down marketable trees and plan to sell them, you must obtain a logging permit from the Oregon Department of Forestry. The permit is free, but you must pay taxes on the value of the trees if they are sold.

Traditional logging leaves the tree stumps in the ground. The most efficient way to deal with stumps is to use an excavator with a "thumb." Once the trees have been cut down and removed and all of the slash (branches, small trees, underbrush, etc.) has been cleaned up, the stumps are dug with the excavator. Stumps are added to the slash piles and burned along with everything else. If the area around the stumps is cleaned up before they are dug out, clean material is then pushed into the void left by the stump. The alternative is that the stumps are removed during the clean up of the slash, allowing branches, brush, and debris to fall or get pushed into the stump hole.

Burning Slash Piles. The cleaner and more compacted the pile is, the faster and more completely it burns. In bone-dry conditions, a bulldozer with a brush rake attached to the front blade is efficient at pushing piles together cleanly. The rakes are 12–18 inches apart and extend past the bottom of the blade. As the slash is pushed and rolled with the bulldozer, most of the dirt falls away, leaving a fairly clean pile. In moist or wet conditions the dirt sticks to the slash, which makes it nearly impossible to burn up completely. In these conditions, an excavator with a thumb can be used in conjunction with the bulldozer to clean up the pile. The excavator picks apart the brush and literally shakes the dirt or mud out of it, then consolidates it into a compact clean pile. It is usually cleaner and better to move the debris with the excavator in the first place, thereby avoiding any dirt in the pile. This gives you a hotter, cleaner and more complete burn.

Check with your local fire department before burning any brush. In general, the Oregon Department of Environmental Quality allows agricultural burning on a day-to-day basis depending on current conditions. Your local fire department will let you know if burning is allowed. If you burn when not allowed, you can be fined.

If your pile is clean, you will have a large, hot bonfire. If your pile is dirty and goes out before it burns up, it will have to be picked apart, reconsolidated, and relit. In the worst conditions, when your slash pile is muddy, green, and it's the middle of winter and raining but you have to get the burning done, it is possible but not much fun.

Bring your chainsaw, splitting maul, and a "brush fan." The fan is a large, gas-powered wind machine that not only adds oxygen to fire but pushes the fire deeper into the pile. Start with a campfire on the downhill edge of the pile. Cut and split wood and continue to add to the fire until it is approximately the size of a large pickup truck. Place the fan downhill from the fire in a spot where it won't get covered if the pile falls over. The fan will push the fire into the pile and help create and maintain a hot spot. Like I said, it's not much fun, but it works.

Once the pile has burned up, a pile of ash remains. The ash contains high concentrations of potassium. This is not necessarily a bad thing to add to your soil, but you don't want it all in one spot. Spread it out as much as is feasible.

Cleanup. When the trees are gone and the slash piles are burned, your property will look like a battlefield that was hit by a tornado. Go over the field with the rippers on the back of the bulldozer. The rippers themselves don't pick up more brush, but they pop up roots, sticks, and rocks that were missed or covered. Now it's time for the fun part—picking sticks.

You will need some personpower for this task. The idea is to create a human broom. The stick pickers sweep across the field picking up sticks and rocks and throwing them into a dump-bed truck or a trailer pulled by a farm tractor. Consolidate the roots, sticks, and rocks into piles and burn them. After the burn, the rocks can be picked up as a pile with an excavator and carted off. After the initial pass by the human broom, your field will start looking like a field, but there are a lot more sticks and roots that you can't see. They will be in your way later if you don't take care of them now.

A spring tooth harrow or chisel plow is effective for popping and floating sticks, roots, and rocks after the initial cleanup with the rippers. Pass over the entire field and pick sticks again. On a heavily wooded property, six or eight times through with the harrow and human broom should clean up the field quite well.

Deep Ripping. Soils with a shallow hardpan or subterranean outcroppings of shale benefit greatly by deep ripping to fracture the soil for better root penetration and to allow water to drain more freely. In some soils, the effects of deep ripping are short term and seem to make little difference in long-term vine vigor or overall vine health. Deep ripping the soil down to as much as 4 feet requires power. Use a large bulldozer to pull multiple or a single subsoiler rip shank. A single shank can be pulled deeper than multiple shanks. A more thorough breakup of the subsoil is accomplished by ripping the field first in one direction, then ripping again perpendicularly. Deep ripping should be done when the soil is relatively dry and prior to land grading, before the topsoil is replaced in the leveled area.

Land Grading. Before you start the final steps of getting your field to seedbed condition, take a look at the overall topography. Often a little extra time and effort spent on mitigating slight or moderate irregularities on your property pays you back with greater ease of farming, better conditions for the vines, or increased plantable acreage. Abrupt changes such as cliffs, high spots and low spots can all be changed if the grading equipment is big enough. Just remember to save the topsoil somewhere and put it back on top when you are finished. Steep slopes can be terraced to gain acreage. D-9 bulldozers can work wonders.

Surface Water Drainage. Now is the time to consider managing water runoff from heavy rainfalls. Assess the natural surface drainage pathways on the site. Do they run through areas planned for the vineyard? Could alternative routes be constructed that would redirect surface water flow to areas outside of planned vineyard blocks? If it is not feasible to alter the natural drainage, vineyard blocks should be located outside the path of water. Grassed waterways can be used to channel runoff from the vineyard site. Consider capturing and holding the runoff in a pond for an irrigation water source. Contact your local office of the USDA Natural Resources Conservation Service for assistance in assessing your site's surface drainage needs and developing a plan for managing runoff.

Leveling. The rough work is done so now it is time for a little fine tuning. Humps and ditches, caterpillar tracks and dirt clods can be leveled out in a variety of ways. Multiple passes with a disc followed by a drag on the final passes levels out most small inconsistencies in the field. In some cases, you may want to use a road grader to level the slope of a field before smoothing with the disc.

Drain Tile. If you walk through portions of your field in May and sink up to your ankles in mud, you may want to consider installing drain tile in those areas to carry away the excess subsurface water. Wet areas are usually easy to spot. They usually appear in swales or low spots and have a greater concentration of vegetation. Many qualified excavation companies can help you design a drain tile system for your situation. In some cases it may be more appropriate to engage the services of a qualified civil engineer to better ensure that the tile system is adequate for the given area.

Other Site Improvements. In addition to preparing the footprint of the vineyard to be planted, future needs should be considered while you have the equipment on site. Are access roads needed? Is a home site, barn site, or loading area planned? If these things are considered and dealt with before planting while you have equipment on site and lots of room to work, you can avoid headaches and greater expenses later.

Soil Amendments. Pre-plant is the best time logistically to make any additions or modifications to the soil, including lime and other mineral nutrients or organic matter. A thorough soil analysis is recommended for a new site to provide baseline information on soil pH and to identify any severe nutrient deficiency or toxicity problems. It is important, however, to consider a soil analysis report as merely a rough indicator of soil condition. *Do not* use the results as a recipe for ordering a custom fertilizer mix or lime application. Unless severe nutrient deficiency or toxicity is indicated, it may be best to take a "wait and see" approach to fertilization.

There is, at present, no consensus on whether Oregon vineyard soils should be limed, and if so, how much lime should be applied. If you decide to lime your soil prior to planting, this may be the one situation where you actually can attain Fast, Cheap, and Good all in one. Most custom liming companies do lime applications with semi trucks to get the product to the field and large spreaders to apply it. The newer spreaders can use global positioning systems and computer technology to change the rate of material applied to different areas of your field. Use of this system requires that your fields first be mapped with complete soil analysis results so the applicator can differentially apply the desired amount to each portion of the field.

The alternative to a pre-planting application of lime is to use smaller equipment that fits between the rows in a planted vineyard. Spin spreaders or custom-built spreaders can do the job adequately, but the process is slow and expensive compared to a pre-planting application done by a custom liming company.

Plow/Disc/Harrow. The final step in preparing the site is to cultivate the soil by going over the field in succession with plow, disc, and finally harrow. The soil is now in good condition to accept seed for a winter cover crop.

Winter Cover Crop. The newly prepared site is highly susceptible to soil erosion during the winter if it is not protected by a cover crop. Cereal rye or winter oats grow quickly and provide a good root system for holding topsoil during winter rains. If you are trying to improve the available nitrogen content of your soil, you may want to consider an appropriate proportion of legumes with your seed mix.

On the valley floor, rye grain farmers plant at a rate of around 40 pounds per acre. On the poorer soils and higher elevations of hillsides, germination of the seed is generally lower. To ensure a good stand of cover over the winter, 100 pounds per acre is necessary. The seed should be planted with a drill if possible. Broadcasting the seed, then rolling or shallow cultivating at 2 inches also works well. The idea is to cover the seed with soil. More than 2 inches or less than 1 inch of depth can lead to a lower germination of seed, resulting in a thinner cover crop. One obvious consequence of a thinner cover is less prevention of soil erosion over the winter.

The timing of planting the seed is important; it should be done in late August or early September. If the seed is planted later, during colder conditions, germination can be poor. If the seed is planted earlier, before the rest of the tractor work is finished in the vineyard, there is the risk of disturbing the seed to the point where germination is reduced. In late August through early September, the conditions are usually dry and warm. The cover crop begins to grow when the first rains of fall moisten the soil. There is always a slight risk that a late summer rain will initiate germination and subsequent dry weather cause the young seedlings to fail. If this happens, reseed immediately. The winter cover crop is disced under in the following spring, prior to final soil preparations for planting.

11

Planting Stock

Steve Castagnoli and John Miller

Planting high-quality nursery stock is an important key to both short- and long-term success of a vineyard. Healthy, well-grown nursery stock in good condition contributes to quick establishment of the vineyard and an early return on investment. Planting nursery stock that is known to be free of serious viruses or other harmful diseases and pests contributes to the long-term productivity of the vineyard.

Clean nursery stock is important to the health of individual vineyards and to the entire Oregon wine industry. Diseases and pests can be unintentionally introduced into a region by careless importation of plants that have not been produced under a certification program. State regulations have been established to protect the Oregon wine industry from diseases and pests by imposing a quarantine on importation of grapevines and establishing a grapevine certification program.

Quarantines

State and federal quarantines restrict the importation of grapevine propagative materials and finished nursery stock into Oregon. The list of potentially devastating grapevine pests and diseases currently *not* found in Oregon is long. Quarantines are an important safeguard in maintaining the relatively soft approach to pest and disease management currently available to Oregon growers, and they aid in preventing the ravages of serious viral diseases such as leafroll and fanleaf.

The federal quarantine prohibits the importation of all grapevine material from outside the United States except under permit from the USDA Animal and Plant Health Inspection Service (APHIS). Post-entry quarantine conditions require a specific protocol for disease testing and elimination prior to release from quarantine. This includes extensive virus indexing and treatment of infected plants. Currently there are three programs permitted by USDA-APHIS to import grapevine materials into the United States—through Cornell University, Southwest Missouri State University, and the University of California, Davis. Contact information for the three importation programs is provided at the end of this chapter. A less restrictive permit may be obtained for importing some Canadian material.

The Oregon Grape Quarantine, which is administered by the Oregon Department of Agriculture (ODA), requires that all grapevine material from outside of Oregon be certified by an official state agency that it is free of fanleaf and leafroll virus diseases and grape phylloxera. Additionally, all rooted plants, whether potted or bare-rooted at time of delivery, must be grown in soilless sterile media. This precludes the importation of field-grown rooted plants. Several states, including California, have certification programs recognized by the ODA. Canadian material meeting the above criteria may be imported by obtaining a federal import permit from the USDA-APHIS.

The future status of the Oregon and U.S. quarantines is unclear. Under the auspices of the Food and Agriculture Organization of the United Nations, the North American Plant Protection Organization (NAPPO), a regional organization of the International Plant Protection Convention, is currently working to develop guidelines for the movement of grapevine propagative materials within North America (Canada, Mexico, United States). NAPPO's mission statement indicates the dual priorities of promoting both trade and plant protection: "NAPPO ...coordinates the efforts among Canada, the United States and Mexico to protect their plant resources from the entry, establishment and spread of regulated plant pests, while facilitating intra/interregional trade."

Adoption of a NAPPO standard for grapevines is controversial. Critics argue that a single North American standard may actually increase the potential for the importation of grapevine pests and diseases if those currently on regional quarantine lists are reclassified as "non-quarantine" by the NAPPO standard.

Certified versus Non-Certified Planting Stock

Vineyards that are established with Oregon-grown nursery plants have the choice between certified or non-certified plants. Established in 1970, the Oregon Grapevine Certification Program was created in conjunction with the Oregon Grape Quarantine to provide Oregon growers with a readily accessible source of planting stock that is free of serious viral diseases and phylloxera. The Oregon Grapevine Certification Program is administered by the ODA, which oversees the establishment and regular inspection of certified nursery increase blocks.

In 1999, the program rules were revised to create a two-tiered certification system. Under the current rules, an "Elite" class of nursery stock is produced from an increase block with no detectable viral infections of any kind. A "Registered" class is produced from an increase block that is free of serious viral diseases. Oregon's two-tiered certification program is responsive to the changing technology in virus detection. As more sensitive tests are developed, plant material that was clean today may not be considered so tomorrow, even though the new virus detected (such as stem pitting or fleck) is not considered a serious threat to vine health. A current list of grapevine nurseries participating in the Oregon Grapevine Certification Program can be obtained from the ODA Plant Division.

Some grapevine materials brought into the state illegally (so-called "suitcase material") have been infected with serious viral diseases or carry damaging pests. It is widely believed that phylloxera was brought into Oregon on non-certified grape plants, circumventing the established quarantine designed to protect vineyards from the pest. Growers are strongly advised against the illegal importation of non-certified grapevines or propagation wood.

Whether you buy plant material or receive cuttings from a friend, always ask for the parentage and source of both rootstock and scion. All certified nurseries must have documentation showing both. Without expensive tests, it is impossible to determine which selection or clone of a given variety was used for scion or rootstock.

Sources of Planting Stock

It is recommended that new vineyards be established with vines grafted onto phylloxera-resistant rootstocks. Growers can obtain planting stock from commercial nurseries, or some growers may choose to produce their own grafted plants. If you decide to do your own grafting, it may be tempting to collect scionwood from an established vineyard block. However, unless the source vineyard was established with certified vines, there is a significant risk of unknowingly propagating virus-infected plants. Similarly, rootstock materials should also originate from certified planting stock. Rootstock vines can be grown in an increase block to provide an adequate supply of cuttings for grafting. Alternatively, cuttings or rootings may be purchased for a fraction of the cost of a finished plant.

Successful production of grafted vines may be achieved with careful attention to a few principles: use clean, healthy scionwood and rootstock, carefully match the stock and scion pieces, provide the proper environmental conditions, and use good sanitation practices. Control of environmental conditions necessary for callusing and rooting of bench-grafted plants is critical to success. It is recommended that one start on a small scale until enough experience is gained to produce strong grafted vines consistently. Producing grafted plants may, however, be beyond the scope of many vineyard operations. Therefore, many growers choose to purchase planting stock from a nursery. Most purchased, grafted vines in Oregon vineyards originated in Oregon or California nurseries.

Types of Grafted Vines

Field-grown *dormant bench grafts* are bare-rooted vines that were bench grafted using dormant cuttings, grown in the field, and dug approximately one year after they were grafted. They may be held in cold storage prior to shipment and planting. When grafting onto difficult-to-root rootstocks, the nursery may graft onto a dormant rootstock rooting that was grown in the field for a year prior to grafting. These are generally not grown in the field for an additional year.

Dormant potted vines are usually bench grafted using dormant rootstock and scion cuttings. Rather than being field grown, they are grown in nurseries in containers, usually plastic pots or paper sleeves.

Green bench grafts are potted vines that are grafted using dormant rootstock and scionwood and planted in the same year. Vines are potted after callusing, then growth is initiated in a greenhouse. Acclimation to outdoor environmental conditions is achieved by moving the plants out of the greenhouse and gradually exposing them to full sunlight. Vines are planted in the vineyard after the danger of frost has passed. Because production costs for green bench grafts are generally lower than for dormant vines, the cost to the grape grower is usually somewhat lower. Some growers believe that the lower cost, compared to dormant bench grafts, is not a true savings when the greater fragility and slower initial growth are taken into account.

Green grafted vines are potted vines that are grafted using actively growing rootstock and scionwood. The grafting material may be collected from field-grown or greenhouse-grown vines. Because the rootstock and scion material may be produced in the greenhouse, this method is advantageous for rapidly increasing selections that may be available only in limited supply, such as winegrape clones or rootstock varieties that are recently released from quarantine. Green grafted vines must be handled with care during transport, delivery, and pre-planting because they are often the most fragile type of grafts.

Ordering Plants from Nurseries

Be prepared to place your order for grafted plants twelve to eighteen months prior to the desired delivery date to ensure availability of the desired rootstock and scion combinations. Dormant vines are usually dug at the nursery and planted in the vineyard during March and April, so orders should be placed before the previous January, when actual grafting begins. Nurseries have varying requirements for deposits and payment: some require half down at the time of the order and the balance at delivery, others divide payment schedules into thirds. Prices may be better if the order does not contain small amounts of several different scion/rootstock combinations. Larger quantities may result in discounts.

Receiving, Handling, and Caring for Grafted Plants

Upon delivery, inspect all vines to make sure there is the proper number of plants, they are labeled as the correct variety and rootstock, and they appear healthy. Contact the nursery about any deficiencies within the "warranty" period, if any.

Although mandatory standards for grape plants are not in effect in Oregon or California, here are some commonly accepted guidelines: Dormant bench grafts should be approximately 14 inches long (including 1–2 inches of scion), and rootstock and scion calipers should be 5/16 inch and closely matched. The graft union should be healed around the vine and should be able to withstand the "thumb test," a gentle but firm lateral pressure on the scion, without breaking. There should be at least 8 inches of top growth from the scion (although the top growth may have been trimmed to two buds) and three or more roots of 5/32" diameter radially pointing several directions. The rootstock should be reasonably straight and round, and the base should be well callused with a solid pith. The graft union and disbudding wounds should be free of bulbous growths that may indicate crown gall.

Dormant vines may be held in cold storage for a considerable length of time prior to planting. It is generally recommended, however, that dormant bare-rooted or potted vines be planted in early spring so that the vines can benefit from mild temperatures and favorable soil moisture while they are becoming established.

Green growing bench grafts are more fragile than dormant bench grafts and require gentle handling during both transport and planting. Common guidelines for green growing bench grafts include an 8–10-inch total plant length (including a 1-inch scion), and rootstock caliper should be at least 1/4 inch. At least 6 inches of healthy new top growth and healthy (mostly white) root growth should be evident when vines are removed from their pots. As with dormant bench grafts, the graft union should be healed around the vine and be free of large, bulbous calluses indicative of crown gall. As with dormant bench grafts, rootstocks should be generally straight and round.

Regardless of the type of nursery stock, successful establishment of vines in the vineyard requires careful attention to weed, pest, and disease control as well as plant nutrition, staking, and irrigation. Green growing vines, whether they are green grafted or green benchgrafts, often require closer monitoring and more frequent irrigation than do dormant planted vines. For this reason, many nurseries require vineyards purchasing green growing vines to have the capability of irrigating on a regular basis.

Little or no published research has compared the performance of grafted grapevines produced with different propagation methods. It is difficult, therefore, to recommend one type of nursery stock over another. There may, however, be distinct advantages of one type over another depending on the specific situation in a given vineyard. A nursery representative should be able to advise you on the best type of planting stock for your situation.

Grapevine Importation Programs

Foundation Plant Materials Service
University of California
One Shields Avenue
Davis, CA 95616-8600
(530) 754-9294

Cornell University, Department of Plant Pathology
New York State Agricultural Experiment Station
Geneva, NY 14456
(315) 787-2334

Missouri Grape Importation and Certification Program
State Fruit Experiment Station
Southwest Missouri State University
9740 Red Spring Road
Mountain Grove, MO, 65711
(417) 926-4105

Resources

Oregon Department of Agriculture
Plant Division
635 Capitol St. NE
Salem, OR 97301-2532
(503) 986-4644

North American Plant Protection Organization http://www.nappo.org

12

Planting and Managing a Young Vineyard

Randy Gold and Porter Lombard

The success of a vineyard depends upon good planning and preparation, followed up with proper viticultural practices to establish and manage the grapevines. Planting vines looks easy, and it is, but preparation, precision, and care are required to avoid mistakes that can cause problems throughout the life of the vineyard. Be prepared prior to planting, either with the trellis posts already installed or the vine locations precisely marked. Have a vineyard map and planting plan, tools, and enough good labor to plant the vines quickly and properly. You should be ready to plant the vines soon after they are delivered. Most vineyards do not have facilities to store plants properly prior to planting. Extended holding periods under suboptimal conditions may stress or weaken the plants, slowing their establishment after planting.

Marking off the Vineyard
(adapted from Wolf and Poling, 1995)

The management efficiency of a vineyard is greatly enhanced by uniformly spaced vineyard blocks. Therefore, it is important to take care in setting up the blocks to have straight, parallel rows in a rectangular pattern with consistent spacing of rows and vines. Errors usually cannot be corrected after planting. The method described here uses a surveyor's transit to mark post or vine locations. Alternatively, rope, wire, or steel measuring tape can be used to mark straight lines, but a surveyor's transit is faster, easier, and more precise. There are many different ways to mark off the vineyard, and the methods described here are merely examples. Remember that the critical objective is to produce straight parallel rows that are evenly spaced, with uniformly spaced vines.

Vines can be planted either before or after installation of trellis posts, but planting after post installation avoids potential damage to tender vines caused by workers, equipment, and trellis materials. Planting after trellis installation usually must be done by digging individual planting holes. Planting before post installation can be done in holes or furrows and enables the use of planter equipment. New grape planters are laser-guided and can quickly and efficiently plant straight rows at precise vine spacings.

Depending on whether you are installing trellis or planting vines first, use the method below to first locate either the endpost or end vine positions. Start by defining the length and location of the first row. The first row is usually located to run parallel to a property boundary such as a fence, or in relation to a road, topographical feature, or other landmark. Row length is determined by the desired size and shape of the block. Measure a straight line that contains a whole number of vines at your selected vine spacing. Add to this length at each end, the space required for the endpost assembly, and at least 25 feet of headland for equipment turnaround.

Place a stake to mark the reference point (A) at one corner of the block; this represents either the endpost or the first vine in row 1 (Figure 1). From reference point A, measure and mark a straight line to reference point B, which locates either the opposite endpost or the last vine in row one.

The first row is used as a reference to develop a true right angle as the first step in designing the rectangular shape of the vineyard block. Right angles make a triangle with the length of sides in a ratio of 3:4:5, so a precise right angle can be created by marking off a triangle with side lengths in these proportions, for example, 60, 80, and 100 feet. Start by measuring a straight line from reference point A to the location (aa) 60 feet down row 1. Next, sight a perpendicular line through the endpost or first vine positions of every row to create the second side of the triangle. A transit makes this easier: from reference point A sight down row 1 to point B with the transit compass set at 0, then turn the transit to 90° and sight down the endpost or first vine positions and stake the corner location (C) in the last row. Measure and stake the position (cc) 80 feet from reference point A down this line. Check the accuracy of the 90° angle by measuring the third side of the triangle formed by a straight line between points aa and cc. The third side should measure 100 feet to fulfill the 3:4:5 ratio of a right triangle. Recheck your measurements and transit sightings if the third side measurement is incorrect, and move the stakes as necessary.

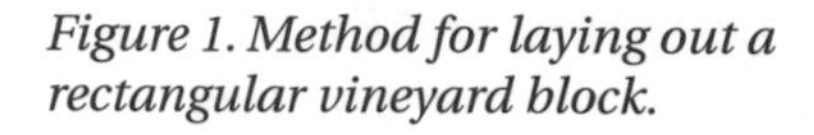

Figure 1. Method for laying out a rectangular vineyard block.

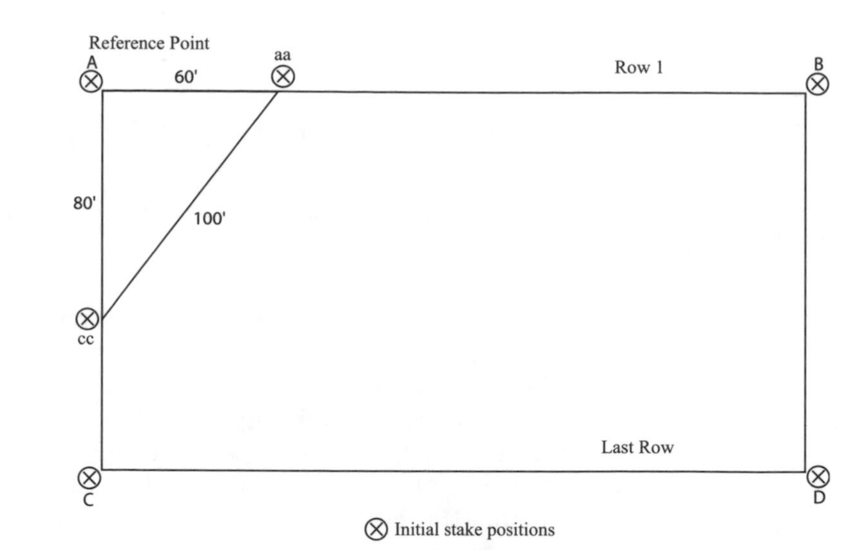

Complete the rectangle by making a second right angle at point B formed by reference row 1 and a perpendicular line through the endpost or last vine positions of each row. Position the transit at point B and sight point A with the transit compass set to zero. Rotate the transit to 90° and stake the location (D) at the corner position of the last row. Use the same procedure to check the accuracy of the 90° angle. Stake a point on reference row 1 that is 60 feet from point B, and a second point on line B-D that is 80 feet from Point B. The line between these two points should be 100 feet.

Once the precise rectangle has been established, row locations can be marked along the straight line A-C. Position the transit at point A and direct an assistant with a measuring tape to place stakes at the location of each row precisely. Move the transit to point B and repeat the process to mark the row locations along line B-D.

If trellis posts are to be installed prior to planting, endposts are installed first. Then, line-post positions can be located by tightly stretching a steel tape measure or pre-marked trellis wire between the endposts of a row. After the installation of all trellis posts, vine positions can be located by using a measuring tape or pre-marked wire as just described. The planting location of each vine can be marked in a variety of ways; a small amount of lime can be dropped on the soil, or a plastic drinking straw can be inserted on the mark. Straws can be quickly and accurately placed with the use of a filed-off screwdriver or punch tool that is slightly longer than the straw. Insert the tool into a straw and push it into the ground at the mark.

If vines are to be planted prior to trellis installation, individual holes can be dug or a furrow can be opened. Planting holes can be located by using the tape measure method described above; stretch the metal tape or pre-marked wire between temporary stakes positioned at the row ends. Planting into furrows can be facilitated by making a grid pattern of grooves in the prepared soil, marking the row spacing and location of individual vines. Grooves can be made with cultivator teeth that are precisely spaced on the toolbar. Rows are marked first by installing three cultivators on the toolbar at the proper row spacing. Mark the exact location of the full length of the second row by sighting down the row-end markers and periodically placing small stakes loosely in the ground. A tractor pulling the toolbar is carefully driven through the field so that the middle cultivator goes directly over the stakes marking the second row. The two outside cultivators simultaneously mark rows one and three. Subsequent rows are similarly marked in relation to previous rows and should align exactly with the previously staked row-end positions. Care should be taken to ensure that cultivators are firmly attached and do not shift, even a little, on the toolbar. Periodically check the consistency of the spacing and make the necessary adjustments.

Vine locations are marked in a similar fashion, running the toolbar, with cultivators adjusted to the proper vine spacing, through the field perpendicular to the row grooves. Furrows are dug by carefully following the row grooves with a plow or tree-planter blade, and vines are planted where the cross grooves intersect the furrow.

Planting the Vines

Nursery plants. Dormant grapevines should be planted as early as possible in late winter or spring, after the threat of low winter temperatures has passed. Depending on location, this could be anytime between mid-February and mid-April. Earlier planting enables vines to get established under less stressful environmental conditions, and usually with adequate soil moisture.

Fall planting of dormant vines is also possible if nursery plants are available. Make sure the plants were recently dug and completely dormant. Do not plant vines that were held in cold storage all summer. Cold-stored plants may begin to grow in the fall and will be severely injured when subjected to cold temperatures.

Green growing vines should be planted after the threat of late spring frost. However, nurseries can rarely deliver green growing vines before June 1. Avoid planting much later than early July or there will not be enough time for sufficient vine growth and maturation of the shoots for winter survival.

Make arrangements with the nursery to deliver the vines when you are ready to plant. When they arrive, inspect all vines to make sure they are the variety/rootstock combinations you ordered. Bareroot vines should be completely dormant and moist. Green growing plants should appear healthy and the potting mixture should be moist. See the section on receiving plants in Chapter 11 for a further discussion of evaluating nursery plant quality. Be prepared to store the plants in a shaded and cool, but not freezing, location if weather is not conducive to immediate planting. Keep the roots of dormant bareroot plants moist, and irrigate green growing vines until you are ready to plant.

Soil Preparation and Planting. Prior to planting, the soil should be tilled and have good moisture at planting time, but it should not be excessively wet. Irrigate the field immediately prior to planting if the soil is dry.

Vines can be planted in holes or in furrows. Planting holes are dug by hand or with an auger. Some growers have found it faster and easier to dig holes with a spade or hand trowel. Augers sometimes dig holes that are too large, requiring considerable effort to backfill with soil around the roots. Another drawback of augers is that they can glaze the sides of the hole when used in wet clay and loamy soils. The glazed sides of the planting hole act like a clay pot and restrict root growth. A shovel can be used to break up the glazing, but it is better to avoid using an auger under these conditions.

Opening a furrow with a V-shaped shovel plow or a modified tree planter can greatly facilitate planting. The tree planter opens a deeper, wider furrow to facilitate placement of large-rooted grape plants. Regardless of the method used, the objective is to make a hole that is big enough to accommodate the grape plant without trimming or compressing the root system.

Keep the roots of bareroot plants moist prior to planting. Bundles of vines can be transported in picking buckets partially filled with water. Bareroot bench-grafted plants should be placed in the hole deep enough to cover the top of the root system with several inches of soil, and the graft union should be about 6 inches above the soil surface. It is important for the scion to be well above the soil line so it does not produce roots, which would circumvent the phylloxera resistance of the rootstock.

Nursery vines in sleeves or pots should be watered 12–24 hours before planting. Potted or sleeved plants should be planted at the same depth at which they grew in the nursery, which should put the graft union at least 6 inches above the soil.

Handle all grafted vines carefully to prevent breakage of the graft union. The top growth from the graft union can be trimmed to leave two to four buds, or about 5–6 inches long. Do not trim the roots. It is better to make the hole bigger to accommodate the roots then to reduce the root mass. When planting into furrows, the walls sometimes need to be enlarged to fit the roots. Gently spread the root system out in all directions in the bottom of the planting hole or furrow. Simply shoving a vine in the hole and covering it with soil may create a "J-rooted" condition. Such vines have most of their roots on one side and the ends are turned up and close to the surface, giving a J-shaped appearance. J-rooted vines usually establish poorly and may never become strong plants.

Potted or sleeved plants have smaller root systems, so they require a smaller planting hole. Plant these vines at the same depth at which they were grown in the nursery. Paper sleeves disintegrate in the soil within a year, so they do not have to be removed prior to planting. Some growers cut one side of the paper or remove tubes from the roots because of concern about restricted root growth. Usually the side of the tube is cut with a knife to break the tube open without spilling the roots out.

Hold the plant straight while loose soil is filled in around the roots. Make sure the vines are in a straight row by sighting down the row and aligning plants with trellis posts or marking stakes. Firm, but do not pack, the soil with hands or feet to eliminate large air pockets around the roots. Water immediately after planting to help settle the soil around the roots.

Vine Support. The trellis wires can be installed after planting, but they are sometimes postponed until the second year. Vine training is begun in the second season, so it is imperative that the wires be installed prior to bud burst.

A 4–5-foot long stake, often made of bamboo, can be installed next to each vine to provide support for newly planted grapevines. Vines are loosely tied to the stakes as they grow, or grow tubes can be placed over the plants and attached to the stake. Several different sizes and colors of grow tubes are available. Although some growers believe that grow tubes increase vine growth in the first year, the more likely advantages are that vines do not require any tying the first season, vines are protected from deer and rabbits, and contact herbicides can be used safely in the row for weed control. Tubes must be supported by a stake or tied to a trellis wire. The tubes should be removed by late September to help harden off the vines for winter.

Care of Young Vines

It is important to maximize growth of young vines in the first three years to bring them into full production as soon as possible. Full production usually is achieved between the third and fifth leaf (season), but it can take as long as the eighth leaf under poor management. The critical factors for quick establishment are weed control and irrigation. Deer control is also important in areas where deer are numerous.

Preventing weed competition. Weed competition is probably the single greatest limiting factor for rapid establishment of grapevines. Grass or other weeds within 3 feet of young vines compete for moisture and nutrients, limiting growth of the vines. Therefore all vegetation within that distance should be controlled. Clean straw mulch or black plastic helps control weeds, but cultivation or manual weed control is still necessary. If no mulch is used, weeds should be cultivated on a weekly basis. Preemergence herbicides can be used to reduce weed growth, but only certain products can be used safely on young vines. Be sure to check the directions for use on the herbicide label carefully to make sure the product can be used on young vines. Follow label rates and directions.

Alternatively, contact herbicides can be used if they are applied in a manner that does not allow them to contact grapevines. Grow tubes can shield vines from herbicides if the spray pressure is low and nozzles are directed to the ground. If grow tubes are not used, cardboard, rolled newspaper, or sheet metal can be placed around the vine temporarily for protection from the spray. Row middles are usually clean-cultivated during the first two growing seasons to reduce weed competition. Winter cover crops are planted to control erosion. A permanent cover crop could be planted in row middles during the first or second year if the vines are irrigated and in-row weeds are controlled.

Irrigation. Young vines have limited root systems, so they must be watered regularly, either by hand or with an irrigation system. If an irrigation system is planned, it should be installed prior to planting so it can be used during the critical establishment period. The drip tube can be placed on the ground if mechanical weed control is not to be used, or attached to a low wire if the trellis is in place. Wire placement is preferred to keep the drip tube out of the way of cultivation and other equipment. Position drip emitters close enough to the plant to wet the root system but not to drip onto the trunk. Emitters should be slightly uphill of a vine when planted on a slope. Check the new irrigation system carefully and frequently to make sure the young vines are receiving water. See Chapters 14 and 19 for more information on irrigation.

The amount, frequency, and seasonal duration of irrigation vary depending on rainfall, temperature, and soil type. Irrigation, either once or twice a week, should start about early May and continue until mid-September. Recommendations range broadly for drip irrigation, from 2 to 8 gallons of water per vine weekly in the first leaf. Sprinklers should apply 2–3 inches per acre per week on newly planted vines. The lower water rate can be used early in the season and increased during the peak water requirement months of July and August. Generally, irrigation rates are reduced somewhat in the second year unless vine growth has been insufficient.

Organic and Plastic Mulch. Establishment of grapevines has been facilitated on nonirrigated sites by the use of straw mulch or black plastic in the vine row to conserve soil moisture and reduce competition from weeds. Straw mulch must be free of seeds or it may create a bigger weed problem. Mulch must be applied in a layer at least 3–4 inches thick to be effective in reducing weeds. Availability and high cost are often limiting factors for the use of organic mulch.

Generally, plastic is installed by a plastic mulch-laying machine shortly after planting, but before the new plants break bud. A 3–4-inch slit is cut in the plastic to allow the grape plant to protrude through the mulch. Later, weeds also grow up through the slit and should be removed by hand. Weeds that grow along the edges of the mulch can be pulled, mowed, or sprayed with herbicide. Prior to installation of the trellis system, holes are cut in the plastic to accommodate the posts.

Black plastic can be installed prior to planting if the plants were grown in paper tubes or sleeves. The root systems of these plants are compact enough to fit through a reasonable sized hole cut in the plastic. Holes can be punched at each vine location with a knife, but keep the flap to place around the trunk with some soil to hold it down. Planting can be done after covering the ends with soil to secure the plastic. All other planting procedures are the same as previously described. The plastic strip should be 5 feet wide to prevent weeds close to the plants and 2–4 mil thick to last one year, or 7 mil to last two or three years. Plan on removing the plastic just before it deteriorates to the point that pick-up is difficult.

Fertilization. Nitrogen fertilizer may be beneficial for good vine growth in the first season or two, particularly on a relatively infertile soil. Fertilizer does not, however, compensate for inadequate weed control or insufficient water. Dry fertilizer can be applied to the soil surface, or soluble fertilizer can be injected into a drip irrigation system. It is generally advised to fertilize young vines lightly, applying no more than 8–20 pounds of actual nitrogen per acre, depending on the existing soil fertility level. For more information on fertilizers and nutrition, see Chapter 18.

Training. The structure of the vine is created by selecting and tying a trunk and shoots in the proper positions to form the shape defined by the training system and trellis parameters. The time required to achieve the final vine size and shape can be highly variable, depending on growth rate and the vine spacing. Closely spaced vines have a much smaller trellis area to fill, so it should not require as much time to complete the training process. The process of training young vines is described in Chapter 13.

Disease Control. It is important to keep young vines healthy by controlling diseases, even though there is no fruit produced in the first year or two. The primary disease concern is powdery mildew, which can be controlled by canopy management practices in combination with fungicide sprays; see Chapter 23 for further information. When vines are producing fruit, botrytis bunch rot is an additional concern, and appropriate control measures should be applied.

Protecting the Graft Union. Some growers have used whitewash paint to protect first-leaf graft unions from freeze damage that can occur when sunlight reflection off snow warms trunks during the day, followed by sub-freezing temperatures at night. Whitewash, prepared by diluting white latex paint by 50% with water, is applied in the fall to the south side of the trunk, from the ground to 4 inches above the graft union.

References

Wolf, T. K., and E. B. Poling. 1995. *The Mid-Atlantic Winegrape Grower's Guide.* North Carolina Cooperative Extension Service, Raleigh.

13

Training Young Vines

Edward W. Hellman and Andy Humphrey

During vineyard planning, decisions are made regarding the training and trellis systems to be used in a new vineyard. Appropriate preparations are then made to obtain the required materials for training and trellis construction, and vines are planted at the selected row and vine spacing. Execution of the training system plan begins soon after the newly planted vines begin to grow.

General Considerations

Adequate resources must be available to provide proper care and training of all young vines. It is far better to provide good management to a smaller acreage than inadequate management to larger acreage. Some vineyard owners find it more appealing to invest in the purchase of additional plants, and thus a larger vineyard, than to invest in the necessary labor and materials to provide proper care for the vines. Making this mistake usually results in a poorly established vineyard that requires at least several additional years to reach the original fruit production goals, and one that may never achieve full expectations.

Training is the process of directing and controlling growth to form a vine with the desired shape and structure. The final outcome is a vine with specific dimensions, including head height and number, position, and length of arms, canes, or cordons. The goal is uniformity among all vines within the vineyard block. The reality is that all vines do not grow at the same rate, nor exactly in the same way. Thus, training young vines requires thoughtful judgment in the selection of which shoots to retain and which to remove, the extent of pruning, and the timing of training practices, all of which must take into consideration the overall health and vigor of the vine.

Training cannot be done according to a predetermined schedule. Implementation of training practices is dependent upon the extent of vine development. Often, growth in the first season is not sufficient to begin most training practices, even with good growing conditions and management. Remember that the typical objective in the first year is to establish a strong vine with a proportional root system. Under circumstances of significant weed competition or inadequate water, young vines may grow so slowly that they require two or more years to establish a strong enough vine to begin training. Attempting to train weak shoots into trunks, arms, or cordons can result in perpetuation of the low-vigor problem and decreased fruit quality in subsequent years, particularly if the vines are not properly defruited.

In contrast, under exceptional, ideal circumstances, vines producing healthy, balanced growth in their first year in the vineyard can sometimes be trained to begin a trunk by the end of the first season.

Remember that uniformity of the vineyard block is a major objective, so weaker vines must be treated differently than stronger vines, until they "catch up" in size. This frequently means more intensive care and crop removal while the weak vines are building up in size. For logistical and uniformity reasons, it may be desirable to treat an entire vineyard block relative to its weaker vines. For example, if 50–60% of the vines in a new block did not grow adequately in the first year, it may be simpler and ultimately more effective to instruct the vineyard crew during dormant pruning to "prune everything back to two buds," rather than attempt to have each vine custom-pruned according to its vigor.

Chronic weakness of newly planted vines despite apparently good growing conditions demands efforts to determine the true cause of the poor growth, and correction of the problem if possible. It is unreasonable to expect weak vines to somehow catch up to the more vigorous vines under the unchanged circumstances that are causing their poor growth. Solve the problem quickly, even if it means replacing vines that are still alive, but struggling. Areas of weak vines will always be less productive and require at least as much management input as healthy areas of the vineyard.

Limit fruiting. Young vines are quite capable of producing flowers and fruit, but they often lack enough leaf area to provide adequate photosynthates for fruit development in addition to shoot and root growth. All flower clusters must be removed in the first year and in all subsequent years in which vine size is inadequate. Overcropping is very easy to do on a small, young vine. The consequence is uneven and often stunted shoot growth, which weakens the vine and delays its full development. Repeated overcropping is likely to have a long-term debilitating effect on the vine as well as reduce fruit quality and ultimately wine quality.

Limited cropping can begin after the main structure of the vine has been established with strong, healthy wood. There are no rules stating that vines should produce a certain amount of fruit in year 2, 3, 4, or later years. The amount of fruit should be based on the size and vigor of each vine. Just as overcropping should be avoided, it is equally important to allow vines of adequate size to develop and ripen fruit. If relatively large young vines are excessively defruited, they produce overly vigorous shoot growth. Appropriate cropping levels help keep shoot growth in proper balance.

Removal of Suckers and Watersprouts. Suckers arise from belowground bud positions on young vines. If these are properly eliminated when the vines are young, less suckering is necessary in subsequent years. During the dry period of summer of the first two growing seasons, carefully dig around each vine and cut off the suckers without leaving any stubs. Like all canes, the basal region of suckers contains latent buds, and if these are not removed they subsequently produce more suckers.

Watersprouts commonly arise from latent buds on the upper part of the trunk and from the head and arms. These vigorous shoots do not usually arise in useful positions for training purposes, so those that are in poor positions should be removed promptly. Rubbing off watersprouts in a timely manner while they are still small completely removes the shoot without leaving a stub.

Vine Support. Regardless of the training system, it is advantageous to have straight, vertical trunks that are aligned in the row. Development of straight trunks requires that some means of support be provided, as well as guidance applied to the shoot selected to become the trunk. Bamboo, wooden, or metal stakes are commonly used to support shoot growth of young vines. The growing shoot is tied loosely or wrapped around the stake as it grows upward. Shoots can be tied to support structures with many different materials including twist ties, tying tape, and twine. Mechanical applicators are available for some types of ties to facilitate training. Some trellis systems include a permanent support stake at every vine. This is particularly advantageous for the use of in-row cultivation equipment that relies on a sensing arm to trigger retraction of the cultivator when it touches the stake.

A relatively recent development in vine training aids is translucent plastic "grow tubes," also known as "vine shelters." Grow tubes are used during the first season and typically removed by late summer or early autumn. They are installed over newly planted vines and anchored to the ground with a small stake or tied to a trellis wire. Commonly, about 3 inches of soil is mounded up around the base of the tube in spring to prevent airflow from beneath the tube, which helps retain heat within the tube. Later in the season, as air temperature increases, the tube is raised above soil level to facilitate airflow in the tube, preventing excessive heating within the tube. Grow tubes tend to accelerate shoot growth while they are within the confines of the tube. Once the shoot tip emerges from the top of the tube, growth rate becomes similar to vines without grow tubes. The major benefits of grow tubes appear to be simplified training in the early stages, easier weed control, and increased soil water retention. Shoots grow up straight within the supporting tube, without the need for tying until they outgrow the tube. Young vines are also somewhat protected from weed control implements (hand hoes) and herbicide sprays. An additional benefit may be some measure of protection against feeding damage by insects, small mammals, and deer while the shoots are within the tube.

Training Practices

The training process utilizes several viticultural practices to shape the vine:

- pruning: removal of all or portions of shoots or canes
- suckering: removal of trunk suckers
- disbudding: removal of dormant or lateral buds
- tipping: removal of shoot tips
- tying: attachment of vine parts to a supporting structure
- shoot positioning: placement of shoots in a specific location and orientation.

All of these operations must be conducted by hand, requiring frequent individual attention to each vine.

Training practices in the first year may be limited to simply tying shoots upright on a supporting structure. Most vine training practices commonly begin in the second growing season, starting with dormant pruning. Under exceptional circumstances, however, training can begin in the year of planting. One determining factor is how much growth can be reasonably anticipated from the vines. The other significant factor is whether sufficient finances are available in the first season to purchase the materials and labor required for training. Early planting with ideal growing conditions and intensive management can enable substantial growth in the first year. If adequate resources are available, it is reasonable to begin training practices to develop the shape and structure of such healthy, vigorous vines during the first season.

More commonly, first-year growth is not sufficient to enable much training, even with ideal growing conditions and vine management. There are still some advantages, however, to providing support for vines in the first season. Vines tied to a support make it easier to control weeds, and they are out of the way of vineyard tractors. It is also much easier to get good spray coverage on supported vines.

It is often preferable to allow the vines to grow the first season with little, if any pruning. A larger leaf area produces more photosynthates, enabling development of a larger root system and contributing to the buildup of food reserves for good growth in the following season. Under extreme circumstances, some vines may still be too weak after two years without training. These vines should be pruned back to the strongest cane, retaining only two to four buds to focus the vine's energy on a small number of shoots. Remember that size and health of the vine, not vine age, should be the determining factor for the initiation of training procedures.

The sequence of steps involved with training young grapevines is often described in association with time, such as year 1, year 2, and so on, which may refer to either the planting year or the years that training practices are applied. But, as previously described, vines often develop at varying rates. Therefore, training steps are described here with regards to the development of structures, trunk development or head establishment, for example. These may occur at varying times among vines of the same age, but the development process and sequence remain constant. Thus the steps described may take a single season or several seasons, depending on vine vigor, and the actions may be taken during the growing season or dormant season. When training spans over more than one growing season, canes are typically pruned back to healthy, balanced growth during dormant pruning, and the process is resumed with new shoot growth in the following season.

Similarly, the height and position of vine structures vary with the training system, but the development process is the same. For example, all cane-pruning systems use essentially the same training practices to develop the trunk, head, and arms. The training practices described here can be applied to the training systems commonly used in Oregon.

Developing the Trunk

Training commonly begins in the year after planting, and the first objective is to develop the trunk. The previous season's growth is thinned out at dormant pruning, retaining a single strong spur with two to four buds (the process is commonly called "2-budding"). Alternatively, if the previous year's growth was exceptionally good, one strong vertical-growing shoot can be selected and tied to the support structure to form the trunk. If the vine is "2-budded," some viticulturists prefer to direct all of the vine's energy into a single strong shoot by removing the second shoot in early spring. Frequently, young vines produce many small shoots from secondary, tertiary, or sometimes latent buds. These should be removed as early as possible, ideally when the shoots are no longer than 2–4 inches, by breaking or rubbing them off.

Others believe it better for the vine to have one or two additional shoots to provide more total leaf area for the production of photosynthates to support continued development of the entire vine, including the root system. The additional shoot number tends to reduce the vigor of shoots, resulting in shorter internodes. If not overdone, closer spaced internodes make it easier in the following year to select good positions for shoots to form the arms (cane pruning) or cordons. Retaining extra shoots also provides some compensation for wind damage, deer browsing, and other factors that can retard the development of young vines (Wolf and Poling, 1995). Ultimately, the extra shoots are removed at their base during dormant pruning.

The growing shoot selected for the trunk must be tied to the support structure about every 8–12 inches to ensure a straight trunk. This requires repeated visits to each vine, typically about once every one or two weeks. At the same time, lateral shoots arising from the lower portion of the shoot are pinched off with fingers or pruned off while still short. Lateral shoot growth is minimized if shoots are positioned upright and loosely fastened to the support structure (Wolf and Poling, 1995). As the shoot approaches the height at which the head of the trunk will be developed, the uppermost lateral shoots can be retained for potential development of arms (cane-pruning) or cordons.

On each visit, new shoots arising from older parts of the vine must be rubbed off. Suckers should be traced to their underground source and cut off as close to their base as possible. A good trunk shoot should have growth of sufficient size, about pencil diameter (1/4 inch), extending at least up to the fruiting (training) wire. If strong growth does not reach this far in a single season, it is usually pruned back to the first large bud and trunk development is continued the following season.

Establishing the Head

The top of the trunk is terminated at a specific height (head height) depending on the training system and trellis, varying from 8 to 56 inches. One method to establish the head is to allow the trunk shoot to grow 18–24 inches beyond head height, then cut it back to the head position. This "tipping" cut has the effect of stimulating lateral shoot growth in the leaf axils below the cut. The laterals are then used to develop arms or cordons, depending on the training system. An alternate method is to allow the trunk shoot to grow as long as it can be supported, then tipped back to strong growth, bent over at the head position, and tied to the fruiting wire. Pruning and tying can be done when the shoot is growing, or at dormant pruning time.

Developing the Arms and Fruiting Canes for Cane-pruned Systems

The most common cane-pruned training systems utilize one (Single Guyot), two (Double Guyot), or four (Scott Henry) arms. Below, we describe development of a double-arm system; similar procedures are used for the other training systems. Arms are developed from two lateral shoots arising from below the tipped head, 6–8 inches below the fruiting wire. The two shoots are selected on opposite sides of the trunk and growing in the same plane as the trellis wires. Ideally, internode length on the trunk is relatively short, enabling development of a fairly compact structure for the head and arms. If the extended trunk was used to create one arm, the opposite arm is created from a lateral shoot arising 6–8 inches below the fruiting wire. By either method, a trunk of specific height is developed, with two short arms at the top of the trunk, forming an overall "Y" shape.

At this point of vine development, assuming all retained growth is strong, the vine is capable of producing some fruit in the following season. The amount of fruit the vine is allowed to produce must be in balance with its size and capacity, as indicated by cane diameter (minimum of pencil size, 1/4 inch) and internode length. The arms (shoots) develop into canes in their first season and therefore have fruiting potential for the following year. Crop potential can be manipulated at dormant pruning by adjusting the length of arms (now canes), and thus the number of fruitful buds. Subsequent fruit thinning may be necessary to keep the vine in balance. Alternatively, full-length fruiting canes can be retained, and shoot thinning is later done to reduce the crop level and create a more open canopy environment. Overcropping at this point inhibits shoot and root development, weakening the vine and slowing the training process.

In the following dormant season, each arm (now two years old) is cut back to retain one strong basal cane arising from the arm, near the trunk. This will be the fruiting cane in the coming season; it is tipped, bent over, and tied to the fruiting wire. Fruitful shoots are produced from the dormant buds on the fruiting cane. In the next dormant season, the old fruiting cane (now two years old) is cut back to a strong basal cane, which becomes the new fruiting cane for the coming season. By this method, the fruiting cane serves the dual functions of current year fruit production and fruiting cane renewal. An alternate method is to retain two canes on each arm; one is selected to be the fruiting cane and the other becomes a renewal spur, pruned to one or two buds. The renewal spur produces a strong shoot in the coming season, which in the next dormant period is trained as the new fruiting cane.

Fruiting canes are shortened to extend halfway to the adjacent vine in a two-cane system, or all the way to the next vine in a single-cane system. If vine growth is inadequate to fill the allotted trellis space with good fruiting wood, canes are pruned to retain the desired number of fruitful buds for balance with vine capacity. The cane-pruned vine is now fully developed, with a permanent trunk and one or two semipermanent arms, with a fruiting cane and a renewal spur for each arm.

Developing the Cordons, Arms, and Spurs for Cordon Systems

Cordon training systems contain all of the same structures as cane-pruned systems, but their number, spacial arrangement, and orientation are different. Cordons are essentially horizontal trunks containing many arms, as opposed to a vertical trunk with two arms in many cane-pruning systems. As with cane-pruning, arms are the sites where fruiting canes arise, but most cordon systems short-prune the canes into "fruiting spurs." Similarly, the cane renewal function can be served by the fruiting spur or by a separate renewal spur elsewhere on the arm.

Cordon development follows a procedure similar to arm development for a cane-pruning system. Shoots arising from below the head of the trunk are trained along a trellis wire to create the permanent cordon structure. Lateral shoots growing from the cordon shoot are selected to develop short arms and eventually fruiting spurs.

Head height is an important factor for cordon systems. The cordon shoots must branch off the trunk about 6–8 inches below the cordon support wire. If head height is too close to the cordon wire, the branch angle will be too broad, which makes it weak and susceptible to splitting under a heavy fruit load.

Cordon shoots are gently bent onto the trellis wire and tied when they reach a length of about 18 inches. As each shoot grows, it is tied two or three more times, but never too close to the growing tip. The shoots may be tipped after they have grown more than 18 inches past the halfway point to the next vine. Final cordon length should reach to within about 6 inches of the cordon from the adjacent vine. Lateral shoots are retained during the growing season to provide greater leaf area for the vine.

During dormant pruning, all laterals are removed from the cordon canes. Each cordon is then straightened by weaving it around the trellis wire. They should be wound just enough to straightenand transfer the weight of the cordon to the wire—no more than one and a half turns, since the cordon must be unwound the next fall to prevent breaking the wire as the cordon thickens. Strong plastic ties are used to attach the cordon to the wire at several locations. The cordon cane produces fruitful shoots from dormant buds in the following season, and a limited amount of fruit can be allowed to develop. Remove flower clusters to control the amount of crop. Again, cropping levels must be appropriate for the size and capacity of the vine.

Where vine spacing is wide, requiring cordons up to 4 feet long, it is recommended that the cordon be developed over two seasons (Wolf and Poling, 1995). Long canes often have poor shoot growth from midcane buds because of apical dominance of the shoots near the distal (apex) end of the cane. To establish a long cordon, dormant prune the first season's cordon cane back to the halfway point of the final cordon length. During the next season, extend the cordon by training a shoot from a distal bud position. A similar procedure is used to extend any cordon cane that did not have strong growth out to the full, required length. At dormant pruning, the weaker tips of these cordon canes are cut back to the point where they are approximately 5/16 inch in diameter, to a bud positioned on the underside of the cane. Vigorous shoot growth from this bud is used to extend the cordon.

Once the cordon is developed to its full length, during the next growing season all extraneous shoots growing from the cordon should be rubbed off while they are small. Vertical shoot positioned cordons require shoot removal on the underside of the cordon; hanging cordon systems require removal of shoots from the upper side of the cordon. The disbudding process on the fruiting side of the cordon retains shoots spaced about 8–10 inches apart. Disbudding or shoot removal should also be performed on any watersprouts or shoots growing from the cordon below where it is bent onto the wire.

Cordon systems that use vertical shoot positioning require that the shoots be supported and positioned by doubled catch wires. Timing is critical to prevent wind breakage of shoots or twisting of the cordon from the weight of the shoots. Any shoots extending more than a foot above the top wire should be tipped.

In the subsequent dormant pruning period, the canes arising from the cordon are pruned to fruiting spurs containing one to three buds. The cordon is untied and unwound from the trellis wire, then retied straight alongside that wire. Cropping level is manipulated by adjusting the number of buds retained and subsequent fruit thinning. The pruning procedure is repeated in subsequent years, with new fruiting canes selected from strong basal canes that grew from the previous year's spur. Arms gradually grow larger from the annual addition of the basal node of old fruiting spurs retained every year in the renewal process.

References

Wolf, T. K., and E. B. Poling. 1995. *The Mid-Atltantic Winegrape Grower's Guide.* North Carolina Cooperative Extension Service, Raleigh.

14

Microirrigation Design Considerations

Robert Burney and Edward W. Hellman

Irrigation can be vital to the health and productivity of vineyards in many grape-growing areas of the world, including drier areas in Oregon. Many western Oregon sites receive adequate rainfall and have deep soils with good water-holding capacity. Irrigation is not essential to maintaining a vineyard on such favorable sites, but it can provide several benefits to perhaps justify the cost; see Chapter 19 for a discussion of the potential benefits of irrigation and economic considerations.

The purpose of this chapter is to provide information about the major components of vineyard irrigation systems and their use. The emphasis is more on design considerations than actual construction and maintenance. You will not be an irrigation expert after reading this chapter, but you should be a more aware consumer capable of asking good questions and making informed choices. Because of the complexity of irrigation systems and the number of variables involved, consultation with an irrigation design professional is recommended. Additional information can be found in Schwankl et al. (1998).

Microirrigation has become the standard water delivery method for vineyards, primarily utilizing drip emitters; microsprinklers may be appropriate for some situations. Microirrigation is the most water- and power-efficient irrigation method available. Recent developments extend this efficiency to sprinkler frost protection systems. Lower flows allow either smaller system sizes or more vine area to be watered with a given water supply. Still, the installed cost of the water distribution portion of an irrigation system can easily reach or exceed $2.00 per vine. Water source development costs are widely variable because they depend on the type of water source, power requirements and source, and pumping and filtration requirements. Sprinkler frost protection demands a larger water supply and more power and would be a major addition to what is needed for a drip irrigation system alone. The development cost of a simple, modest water supply for a drip-irrigated vineyard of several to a few dozen acres could be between $8,000 to $20,000. Costs for larger vineyards, or to add frost protection capability, can range several times higher. Determine whether the benefit to your vineyard is worth the initial and ongoing cost and labor. Keep in mind that costs continue beyond the installation; inspection, maintenance, and operation of the system will be a significant annual expense.

Preliminary Considerations

Water Supply

There are a few essential preliminary factors to investigate prior to developing an irrigation system. First, adequate water supplies may not be available in some areas. In Oregon, water belongs to the public, not the landowner. Contact the local watermaster to find out about local water rights, priorities, and conditions. Possessing a water right is not a guarantee of availability.

There may be little or no surface water legally available when it is needed most, even if it is flowing through the vineyard property. New agricultural irrigation is essentially prohibited in a few designated areas of Oregon because of low water availability. Collection and storage of seasonal surface runoff and subsurface drainage are, however, allowed. Damming a perennial stream is not permitted.

The water source should be evaluated for sufficient flow and volume during critical times of the growing season. Streams, ponds, and wells often decline during periods of drought. A brief well test is no guarantee of sustained irrigation capacity in late summer. Find out about the water source before spending thousands of dollars on an irrigation system.

Irrigation water requirements are highly variable and site dependent. Local climatic factors of importance are quantity and seasonal distribution of rainfall, temperature, humidity, and wind. Other site-specific factors include soil water-holding capacity, soil depth, vine planting density and age, and cover crop.

Block Layout and System Sizing

Managing your vines for uniformity and quality is an important goal. Most sites have some variation in soil characteristics, topography, and perhaps even climate. Thus all but the smallest sites will probably be subdivided into blocks to achieve both uniformity and practicality. The irrigation system should likewise serve and promote these twin goals.

There are two common mistakes in designing irrigation: not allowing for potential expansion, and not including a margin of extra capacity. A system that is designed to barely meet criteria under-performs in practice. There are always changes, and small errors are sometimes made in design or installation of a system. Additionally, wear and deterioration develop over time. "Little" vine additions here and there may significantly increase demand on the system. So it is always prudent to build in a margin of extra capacity beyond today's peak demand.

Ideally, irrigation system design proceeds concurrently with vineyard design. Coordinated planning enables blocks to be sized to match their irrigation demand with the capability of both the water source and the delivery and distribution systems. Demand primarily means flow (gallons per minute), and maximum flow is often a system's limiting factor. Blocks may have to be sized to match the well or pump output, rather than vice-versa. Developing additional water sources may be the only solution to providing the required water in a reasonable time frame.

System planning must calculate the water demands of each block in terms of both volume of water applied and the time required to distribute it. Relevant factors are the number of vines, emitter size and number per vine, volume applied per plant, and scheduling.

There is no fixed amount of water required for irrigation. Optimally, an irrigation system would be capable of watering one-third of the vineyard at a time. This capacity balances cost and size versus performance and convenience, responding reasonably quickly to extreme conditions. Individual circumstances may warrant a different capability. Make sure that the irrigation capability is sufficient in a heat wave or drought. Capacity less than one-fourth or one-fifth of the whole vineyard may compel some reduction of the area to be irrigated.

What does this mean in practice? Most commonly, vines each have one half-gallon emitter (for vine spacing of 5 feet or less) or two half-gallon emitters (for spacing of 5 feet or more). It must be noted that "gallon" emitters actually are sized in liters. Thus a half-gallon emitter delivers 2 liters, or 0.53 gallons, per hour, which is nearly 6% more water than the "half-gallon" nomenclature implies. Actual delivery rate should always be used in calculations. Convert liters to gallons by dividing by 3.78.

We use a hypothetical vineyard for several calculations in this chapter. Our example is an 18-acre vineyard divided into three 6-acre blocks for irrigation scheduling. Vine density is 1,245 vines per acre (5- by 7-foot spacing), with one "half-gallon" emitter per vine. Thus, one block requires flow of 3,960 gallons per hour, or 66 gallons per minute.

6 acres x 1,245 (vines/acre) x 0.53 (gallons per hour/vine)
x (1 hour/60 minutes) = 66 gallons per minute

If less flow than required by this design is available from our water source, there are a few modifications that can be made to reduce the flow needed. The vineyard could be broken up into more blocks of smaller size, which increases materials and labor expense. Our example vineyard is already using the smallest emitter size and only one emitter per vine, so no reduction of emitters is feasible. If available volume or prolonged scheduling is limiting, that may force a choice not to irrigate some portions of the vineyard. Exclude perhaps a less important block or one for which irrigation is less needed. Determine how long it would take to get through a complete irrigation cycle. Blocks with different characteristics or irrigation strategies may need different schedules. Work out schedules on paper to ensure real-time feasibility. Complicated irrigation schedules are much easier to manage with an automated controller.

Irrigation System Elements

The major elements of a drip irrigation system are illustrated in Figure 1. Note that a specific system may have a different sequence or not utilize all of the components shown in the diagram. We discuss these components individually, beginning with the water source and moving downstream.

Water Source

It is advantageous to have access to a perennial stream, spring, or lake. More often, however, one or more wells or ponds must be developed. Determine if multiple sources are truly independent or instead drawing from the same primary source. Water quality must be analyzed, regardless of the source and preferably during the season of use, to determine the required filtration treatment. The type and amount of contamination determines how elaborate the filtration and treatment must be.

Surface Water. The solid contaminant load of surface water is apparent. Algae and water plants are especially prevalent in slow or still water in summer. Ponds often require chemical treatment or pre-filtering to make final filtering manageable. If a pond has no outflow during the irrigation season, algae and aquatic plants can be controlled with chemicals. Most surface water warrants an automatic backflushing filter.

Pond construction and maintenance greatly influence aquatic plant growth. Making the sides steep, clearing plants off the banks, and keeping out

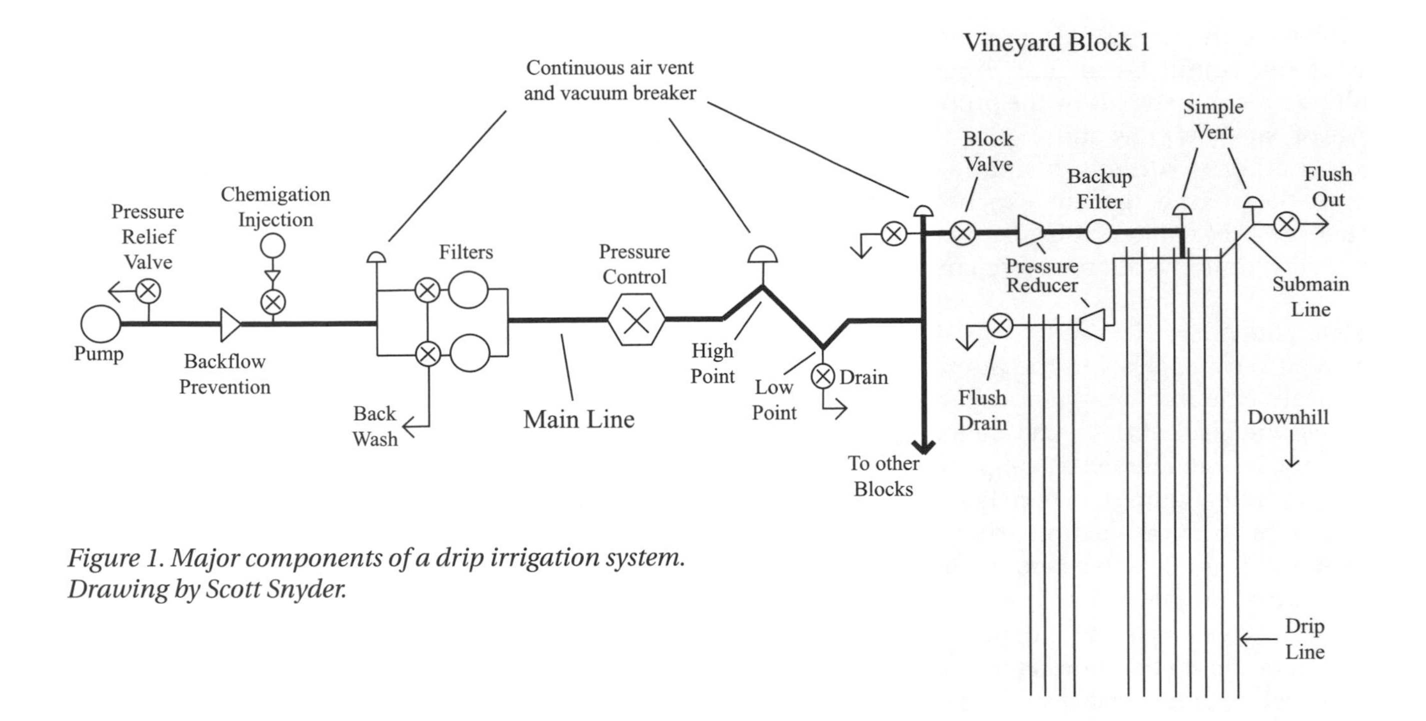

Figure 1. Major components of a drip irrigation system. Drawing by Scott Snyder.

leaves and debris keep growth down and thus minimize filtration load.

If water inflow is much less than irrigation output and losses, a pond must be of adequate size to accumulate enough water beforehand to complete an irrigation season. Remember to factor in evaporation loss when determining the required size of the pond; this can be estimated from "Pan Evaporation" data available from local weather stations. To increase volume, make the pond deeper in preference to wider. Some ponds leak enough to require sealing with bentonite or a plastic liner material.

If you are developing a new pond, it is a good idea to install a pipe in the bottom to supply the pump and allow draining. With proper fill and compaction, there is no leakage along the pipe. Configure the inlet so that water is not drawn off the very bottom. Put valves on the outside end for draining and for isolating the pump.

Ground Water. Water from wells and springs that is not influenced by surface water is called *ground water.* Since it is not visible, one cannot be sure how steady and reliable it will be. A storage tank or pond is necessary if steady water output is needed but the ground water source is intermittent. An intermittent source is indicated by the well pumping dry and requiring several minutes to refill. Electronic motor-load sensing is essential to signal a dry well and protect the pump. A totaling flow meter or an hour meter on the protected motor gives an indication of pumped volume.

If flow just has to be smoothed out or supplemented over a day or so, a tank may suffice. If average flow from the source is too low, the required tank size becomes impractical. In that case, only a pond can accumulate enough water for extended irrigation.

Ground water is intrinsically low in organic matter. It may be very pure and clean, or carry some amount of both dissolved and solid inorganic matter. Wells may pull in enough particulates to need a centrifugal separator or automated back-flushing filter.

Dissolved minerals that precipitate and clog lines and emitters can be a major problem. They must be treated to either stay in solution until through the system or come out of solution before going into distribution. The hardness minerals calcium and magnesium can be kept in solution with acid or proprietary chemical injection into the water flow. Low levels of iron or manganese can be treated by the same method. Moderate to high levels are difficult and, fortunately, rare. They must be oxidized, precipitated, and settled or filtered out before distribution. Media filtration is best in this situation. Aeration and settling is a low-tech approach for moderate concentrations. It requires a large tank or basin and second pump for filtration and distribution.

Very high iron/manganese concentrations, or circumstances with space constraints, are best managed by standard potable water treatment. This involves injection of a strong oxidizer, followed by catalysis and filtration with specialized manganese greensand media filters. Consult an industrial or municipal water treatment specialist if mineral problems are serious. When chemicals are used to treat water, the upstream source must be isolated from all treatment materials by the proper placement and use of backflow prevention.

Whether treatment is simple or complex, the essential point is that the expensive investment in thousands of emitters must be preserved with clean water. Filtration is addressed separately below.

Pump Suction

The path water takes to the pump inlet is the first critical element of the system, especially if it is located above the level of the water source. This path, the *suction*, has requirements for pipe size and limitations on elevation. Pumps create only a fraction of atmospheric pressure as suction. Therefore, water must arrive at the inlet with a certain amount of pressure, called the net *positive suction head*. If the pump is positioned too high (more than 20–23 feet above the water) or the suction line is too small or leaks, it will work poorly or not at all. Avoid creating a high spot in the suction line, which restricts flow as it fills with air released from the water in the partial vacuum of the line.

An underwater ("flooded suction") pump positioned at the base of the storage tank or pond dam simplifies water intake by the pump. If the suction is not flooded, there must be a way to prime the pump. Non-flooded suctions generally have a foot valve to maintain prime. A foot valve is simply a check valve at the inlet end of the suction, with a coarse screen. Flooded suctions require a manual valve so the system can be isolated for servicing. An inline check valve is necessary, and it can be located on either side of the pump.

Pumps

Most irrigation systems are not gravity-fed, so one or more pumps are needed. The nature of the water source and the required flow and pressure determine the size and configuration of the pump. Centrifugal pumps in various configurations are used most commonly for irrigation.

Power source. Current options for the pump power source include utility or solar electricity or a gasoline, propane, or diesel engine. Selection of the power source depends on availability and cost, the nature of the irrigation cycles, and personal preference. For example, engine power is not well suited for a site that is unattended for long periods or uses an automated irrigation schedule.

Bringing utility or solar electric power to a site can be prohibitively expensive. Be sure to investigate utility rate charges for agricultural use, and particularly the "standby" charges that accrue for all the months that power use is low. Standby charges can be substantial, exceeding the actual energy use costs.

Solar power is another option for remote areas, but it is limited largely by the cost per unit of power. Fuel cells show good potential for practical and affordable power within a few years. They convert fuel and oxygen directly and efficiently into electric power, without combustion.

Engine-powered pumps are often the best choice, especially for higher horsepower applications. The larger flow of sprinkler frost protection generally requires engine power. Select an engine that fulfills system needs when running at no more than 75% of rated power, to ensure longevity of the engine. Also, be sure the rating factors include operating altitude and peripherals such as the fan, water pump, and alternator. The drawbacks of engine power include noise and requirements for attendance, refueling, and maintenance.

Pump Size. The appropriate pump size is determined from the required flow and pressure necessary to deliver water to the system. *Total dynamic head* is the cumulative term for the pressure needed from the pump and is described in trade jargon as "feet of head." Height is used as a pressure unit because irrigation system design is always working with the height difference between the water source and the delivery point. This is called *static head*, the pressure exerted strictly by the height of water above any given point. Whether the water is in a half-inch pipe or a lake does not matter. A water depth of 2.3 feet exerts pressure of 1 psi (pounds per square inch), thus "1 foot of head" equals 0.43 psi. Convert to consistent pressure units as needed, but conceptually they are the same.

Ignoring tiny, complicated factors, total dynamic head is the sum of static head, friction loss, and net operating pressure. In addition to static head, water moving through the system loses power to friction, which is manifested as pressure loss. Finally, the system needs net operating pressure differential through every step. This is a way of saying that pressure difference and friction both affect flow. Zero pressure difference results in zero flow. Selection of the pump should include an additional reserve margin of pressure to allow for error and degradation.

Determining Static Head and Friction Loss. Delivering a specific flow and pressure to a specific place requires information about the static head and friction loss to be overcome. Static head is determined by measuring the elevation change between the water source (including below ground level in a well) and the delivery point in the vineyard. Accurate determination of elevation can be a challenge. Topographic maps, altimeters, stereoscopic photography, and direct static pressure measurement are some of the tools available for measuring elevation. Educated guesswork and simple geometry may suffice in nearly flat situations.

Here is one method to directly measure static head on a hillside. Get a 1,000-foot roll of quarter-inch "spaghetti tubing" and some fittings from a drip irrigation supplier. Attach a valve and female garden hose connector to one end. Fill the roll completely with water and close off the other end with a pressure gauge, which becomes the bottom end. Keep the valve closed until the tube is laid out from the high point to the low point. Open the valve at the high point and read static head off of the pressure gauge at the low point. Coils and the lie of the tubing between the two ends do not affect the reading as long as the tube is completely full all the way. Again, each 1 psi indicates 2.3 feet of elevation change.

Elevation determinations from a USGS topographic map are inexact. Common, low-price altimeters give mediocre results. A surveying altimeter or other surveying techniques can be accurate, but their cost is higher and may only be appropriate for large or complicated sites. However determined, elevation change in feet is the same as static head in feet.

Calculating friction loss uses a complicated equation involving the pipe size and flow rate. Using reference tables or a special slide rule simplifies the process of determining unit friction loss with the equation. Friction loss is usually expressed as loss per 100 feet of pipe. Pipe friction loss is simply the multiplied product of the length of the pipeline and the unit friction loss. Additional losses from elbows and fittings are often approximated as 5–10% of the calculated pipe friction. Understand that the pipe size for calculation is the inside diameter, known from the nominal size and type (schedule 40, class 200, etc.). Remember to convert different pressure units to one consistent unit.

Velocity is a more tangible way to visualize water flow, although less important than cumulative friction. A rule of thumb is to keep velocity in the pipe under 5 feet per second to keep the total friction low. In practice, velocities are mostly well below 5 feet per second in long level or uphill runs. Major components like filters and special control valves generally have friction loss versus flow data available, along with other design data.

The operating pressure at the block depends on losses through the valve station, topography, and type of emitters. For a typical block irrigated downhill, specify at least 15 psi at every emitter. Total dynamic head is then calculated by summing the static head, 105% of friction in the suction and main line, friction across the filter, friction in the block valve assembly, and operating pressure needed at the block. The calculation of total dynamic head for one 6-acre block in our example is provided in Table 1.

Pump Selection. Pumps are selected primarily on the basis of the horsepower (hp) needed to deliver the necessary flow in gallons per minute at the required pressure (total dynamic head). Continuing our example, the 6-acre block requires a flow of 66 gallons per minute. The required pump horsepower can be estimated with the figures for flow and total dynamic head plugged into the general equation below:

$$\frac{\textbf{Flow} = 66 \text{ (gallons/min) x } \textbf{Pressure} = 177 \text{ (feet of head)}}{3{,}960 \text{ (unit cancellation factor) x } \textbf{Pump efficiency} = 0.60} = 4.9 \text{ hp}$$

Pump efficiency is commonly about 60% for a medium or larger pump near its optimum performance. Smaller pumps and off-peak operation can be substantially less efficient. It is best to use actual performance data for prospective pumps, especially if pressure or flow will vary. See Figure 2 for an example of a pump performance curve that is appropriate for our example vineyard block. Notice that efficiency is highest around the middle of the pump performance range. Try to match high efficiency in the pump to the most common operating conditions of most blocks.

Select an appropriate pump with the approximate horsepower needed. Compare the flow and total dynamic head for each block with pump performance curves to select a pump that can best meet or exceed these demands. In other words, the point where the demanded head and flow meet on the graph, for each block, must be on or below the curve of the pump performance. How much below is the size of the reserve margin. Some pumps are better at maintaining pressure over a range of flow, so choose a pump with a flatter curve if your block size varies more than elevation. Other pumps are better at maintaining flow over a range of pressure; choose a pump with a steeper curve for larger elevation changes or fluctuating water level in a well.

Table 1. Calculation of Total Dynamic Head

Static Head		*Friction Loss*			*Operating Pressure*	*Total Dynamic Head*
Pumping Depth	*To top of Block*	*Filter*	*Pipe (105%)*	*Valve station*	*15 psi x 2.3 ft/psi*	
60 ft	40 ft	19 ft	15 ft	8 ft	35 ft	177 ft

Pump Arrangement. The pump can be run continuously and directly into the distribution system if the combination of flow and pressure needed for each block is fairly close to the output of the pump. Frequently however, there is at least one vineyard block that needs much less than full output. Continuous pumping to a small demand wastes power and could overheat the pump or overpressure the system. The latter problem could be handled with a pressure-reducing valve to maintain needed flow at a safe pressure. The wasted power might warrant a more complicated strategy, either cycling the pump off and on (with conventional pressure switch and tanks) or slowing it down (with a variable frequency drive on a larger pump). Larger pumps are awkward at small, intermittent tasks like supplying domestic water. Another well or a storage tank filled by the irrigation pump may meet such needs best. On the other hand, a small source filling a storage tank is a real time saver for vineyard spraying.

Consider using a second pump in series to get water to upper blocks if the elevation range of the vineyard is large. This could be mandatory if the total dynamic head is greater than the safe working pressure of the system components. The maximum operating pressure of many filters and other components typically ranges from 80 to 150 psi. Locating the filtration much higher than the pump would also reduce the pressure at the filter. Be careful about matching flow capabilities of the two pumps in such a situation, and consider the effect of filter backflushing. Failsafe controls to stop both pumps in case of interrupted flow must be included.

Likewise, two pumps in parallel might be better for a wide range of block size. Be careful about matching output pressures with parallel pumps, or one may "shut off" (overwhelm and stop) the flow of the other. Equally important is being sure the filter operates properly with the different flows. Parallel pumps may be run directly to the system or with a pressure control setup. The latter uses pressure tanks and a lead pressure switch on the first pump and a lag pressure switch to start the second pump at slightly lower pressure.

Electronic Controllers

It may be convenient or necessary to have automated controls for your irrigation system. Electrically powered systems can be set up with automatic operation fairly easily with programmable electronic timers. Choose one with enough features and stations to cover present and future needs. Also be sure the maximum run time is enough for your planned drip cycles. System equipment such as the water pump, booster pump, chemical pump, and backwash timer can be run in conjunction with the block control valves.

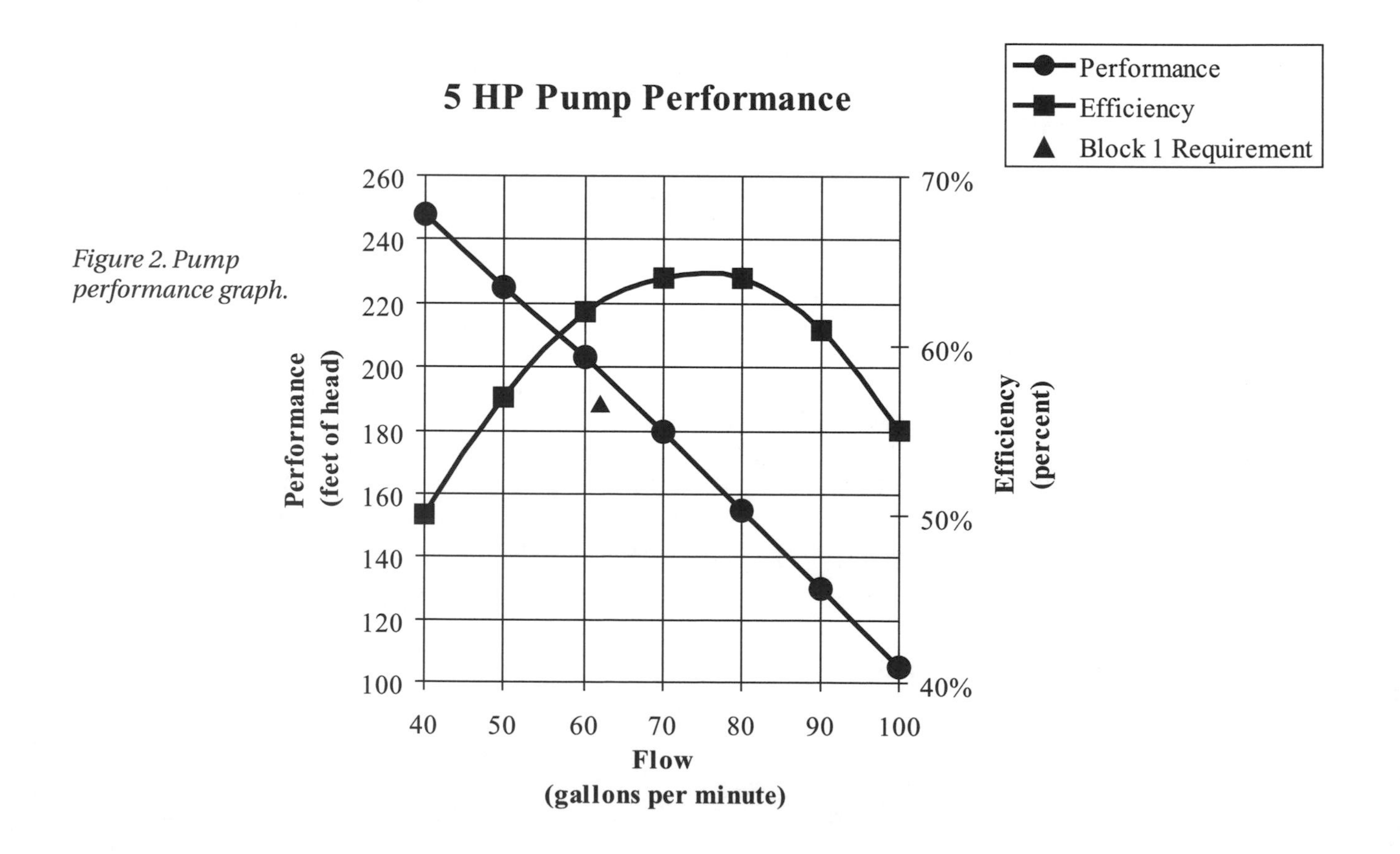

Figure 2. Pump performance graph.

Timers use a tiny amount of power, but it is not a good idea to take it from the pump's circuit. Use a different circuit, preferably one dedicated to the timer. The timer's electronics are sensitive to voltage spikes and transients like those created by motors. Aside from hardware damage by major surges, "dirty power" can scramble your programming entries in the timer memory. A lot is riding on proper functioning of the system, so provide the necessary protection. Modest surge protectors, either hard-wired or plug-in, are adequate.

Valves, Venting, and Other Devices

A pressure relief valve is prudent in any pumping situation except when discharging into an open body of water such as a pond. The valve must always be available to the pump and sized to pass enough water. If normal flow is blocked, the relief valve protects the pump and nearby pipe from damage by overheating and overpressure. The relief valve can be spring-loaded or a "sustaining and relief" diaphragm valve. The opening pressure is above the highest operating pressure and several pounds per square inch below the shut off head pressure of the pump. The regulated diaphragm valve is preferable when the relief pressure is only a little above operating pressure. Downhill runs may need pressure-reducing valves to keep pressure within safe levels. Pressure-sustaining valves are less commonly used, but sometimes they are vital in maintaining upstream pressure. Low points need manual valves or automatic drains to protect pipes from freezing. Domestic water must be isolated from agricultural water. This is done with an "RP," a reduced pressure principle double check valve assembly. All water sources must be protected from backflow of injected chemicals. Water sources strictly for irrigation may be protected from backflow by the use of a "chemigation check valve." Simple check valves are used wherever reverse flow has to be stopped.

Fertilizers and other soluble agricultural chemicals can be delivered through a drip irrigation system. This topic is covered well in Trimmer et al. (1992).

Every high point in the line, from the pump to the filter and from the filter to the ends of the main line, needs a continuous-acting air vent/vacuum breaker. These allow air to escape even under pressure, and air to enter into a vacuum. Venting air steadily during filling reduces water "hammer" (momentum), which occurs with a sudden release of trapped air. Releasing air also lets water flow through the full cross section of the pipe. Breaking vacuum lets the system drain well and prevents collapse of thin pipe walls.

One item that is often recommended, although it is optional, is a flow meter. This meter measures actual flow, which is useful for comparison to system design values and to indicate required maintenance.

Filtration

Every microirrigation system must have at least one filter, no matter how clean the water seems to be. Too much is invested in thousands of emitters to risk getting them clogged. There is no perfect and trouble-free filter, but in a given situation some are less suitable than others.

Different degrees of filtration can be matched to the application; 200 mesh (74 micron) or 80 micron filtration is fairly standard for drip systems. Practically all the filter types are available in a range of sizes or assemblies for flows up to thousands of gallons per minute. Pressure drop across the filter increases as debris builds up, so it must be monitored to maintain full flow. Monitor pressure by installing gauges before and after the filter or a differential pressure gauge that compares the two sides. Systems may need to include valves to enable draining and maintenance of the filter.

There are a variety of different filter types for different needs:

Simple Screen and Disc. Simple screen filters have a fine mesh stainless steel or fabric element, usually cylindrical. Various clever additions to the basic element extend the run time between disassembled cleanings. These include swirling inlet water or increasing velocity with an insert to move debris to a flush valve. Another option for coarser debris is a cranked internal brush and flush valve.

Simple disc filters use a tight stack of flat plastic rings with a molded pattern of ridges. The stacked patterns criss-cross to filter through more depth than a screen filter. Discs are better than screens for retaining algae. The practical difference between these two types of simple filter is less dramatic than sometimes proclaimed.

Screen and disc filters need shutdown and disassembly for periodic cleaning. They are most suitable for manually operated systems, and for water with relatively small amounts of inorganic contaminants, such as clean ground water. Prudent system designs incorporate simple filters at block valve stations as last-chance backup filtration.

Backflushing Filter. Backflushing filters reverse clean water back through the filter element to remove contaminants. They are usually set up to clean themselves automatically, which is convenient for heavier contaminant loads and automated irrigation.

There is a minimum flow and pressure needed for backflushing. If there is not enough flow capacity for backwashing along with irrigating, the irrigation must be briefly stopped or reduced. This can be done with a pressure-sustaining valve positioned downstream from the filter. If continuous flow is critical, such as for frost protection or into a secondary pump, use a larger number of smaller filter units for a reduced backflow requirement.

Disposing of backwash water may influence selection of the type of filter. For example, media filters put out a lot more backwash than other filters. Backwash water is typically put back into surface water, away from the suction intake, or into a drainage. Backwash recycling units are available to filter out solids for disposal and pump water back to the system.

Media Filter. Media filters are small tanks in an array of two or more. They filter through a thick layer of sand or similar "media," so they are big and heavy for a given capacity. Media filters are more sensitive to variable flows and have the most extensive setup and maintenance procedures to be followed. The media must be checked and replaced periodically because the sharp edges degrade and media slowly gets flushed out.

The filter bed captures debris in three dimensions, collecting it up to a few inches deep and thus having a relatively large holding capacity. Each tank is backflushed with filtered water flowing up through the sand bed and out a backflush valve. Filtered water comes from the other tank(s) in the array. Backflush flow fluidizes (fluffs up) the media a few inches to thoroughly release debris. Media filters are often used to remove the organic materials from ponds and surface water. They are less suited for ground water with dense, coarse particulates but are the best choice for removing chemically precipitated matter.

Disc Filter. Another popular type of automatic backflushing filter uses two or more disc filter stacks. Flushing is similar to media filters. Disc filters with automatic backflush give fair to good performance with both organic and inorganic debris. They are compact compared to media filters, especially for higher flow configurations, and backwash volume is also lower.

Rotating Point of Suction. The other major type of backflushing filter uses a cylindrical screen filter element with a rotating apparatus that works like a vacuum cleaner. Dirt collects on the inside of the cylinder, and a central tube with slotted nozzles is positioned close to the screen. This is the "vacuum cleaner." To backflush, the flush valve opens this tube to atmosphere. Water moving inward through the filter and into the nozzles pulls the dirt away from the screen and out of the filter. A hand crank, electric motor, or water is used to rotate the scanner so the entire inside surface is cleaned.

Centrifugal Separators

Any filter can become caked over with debris when water contains a large percentage of inorganic particles. Centrifugal separators are used prior to filtration to separate out particles more dense than water. Water enters tangentially at the upper end of the separator, through a small inlet pipe at rather high speed. Water swirls around and downwards in a vortex, then up along the central axis. The centrifugal action slings heavier particles to the wall; gravity slides them down the wall and into a lower chamber. From there, they are flushed out periodically through a manual or automatic ball valve, with the separator under pressure.

Small and less dense debris, including organic matter, pass right through the separator. Water goes through regular filtration after the separator. It is important to operate separators within their designed range of flow to get good separation. Smooth incoming flow is best, so the inlet should be straight for a length of several pipe diameters. For the same reason, constrictions such as a throttle or modulating valve need to be downstream of the separator.

Water Delivery and Distribution

Main line. Preliminary planning and early work should result in a final version of the water supply, pump, and filter by the time an irrigated vineyard is laid out. The job of the main line is to deliver clean water efficiently and economically from the pump and filter to the vineyard blocks, at needed flow and pressure. It is the piping that "stays charged," or full of water.

The general idea for routing the main line is to keep it reasonably short and direct to both existing and future vineyard blocks yet accessible during and after construction. Minimize crossing under rows and favor avenues over row middles. The main line should go to or near the uphill end of each irrigation block to minimize uphill flow in submain lines and drip tubing. The main should deliver at least 5–10 psi over the desired pressure to each block, to counteract losses in the valve assembly. The location and depth of all piping should be safe from driven posts, earth movement, tractors, and underground work. Use fine, loose material for backfill next to pipe, and keep rocks away. Imported fill may be needed around pipe in rocky sites. Air vents at high points and drains at low points improve performance and protect piping.

The valve assembly that makes the transition from main line to submain at each block location may be above ground or in a box below ground level. The valve station may include a flushing valve, air vent, pressure gauge, or drain valve in terminating the main line. Additional components that are part of the block are mentioned below. The assembly is generally in a row close to the upper end of a block and should be located either between vines or just beyond the end post. These locations make it safer from tractors and implements yet accessible from the avenue. Some blocks, notably those with irregular shapes or topography, may best be served with water delivery to somewhere in the middle of the block.

Note that the main line usually needs to terminate at an exact point in the vineyard layout. That means that trenching and pipe assembly are best done after vineyard layout is final and before driving trellis posts or planting. Installation is much more prone to error and change if done before vineyard layout stops evolving.

Limited curvature of trenches is acceptable for PVC pipe, as long as joints have cured for a few hours. The safe amount of bending decreases quickly with increasing pipe size. Common fittings are 45° and 90° elbows; other angles are accom-modated by closely pairing two fittings (45°, 90°, or tee) in a combination to line up with the intersection. Draw an "as-built" map after installation so the exact location of pipe is known and recorded. Repairs or modifications are inevitable, so you don't want to have to guess where the pipe lies in the future.

Submains. The submain begins with the manual or automatic valve to the block. Submains are the part of the irrigation system that controls and distributes water to each block and the drip tubes in each row. Other features of the submain may include backup filtration, vacuum breaking at high point, flushing at ends, and pressure reduction to that desired for the emitters. Pressure reduction may be incorporated in the supply valve or in a separate pressure-reducing valve. The latter may be located with the rest of the valve station or with the distribution piping and risers if the latter are not adjacent.

A typical operating pressure for emitters is 20 psi, although steep blocks might start at 10–15 psi at the top of the block so that the pressure is not too high lower down. Elevation differences over 70–80 feet within a block can require multiple pressure-reducing valves for subsections, individual rows, or even within each row. The distribution piping of submains sends water to, or near, the high point of each row of the block. The concept is to deliver water so flow goes downhill, counteracting pipe friction. The submain pipe is on a perpendicular or diagonal across vineyard rows, either between vine locations within the block or just outside the end post position.

Submain pipe size is determined by counting the number of emitters and calculating flow in each row. Row flows are added up from the ends back to the connection to the valve station, and pipe is sized accordingly. Normally, submain pipe size is decreased as flow drops to an appropriate rate in serving fewer remaining rows. Pipe size reductions should, however, be limited to just three or four sizes (i.e., 3 inch, 2 1/2 inch, 2 inch, 1 1/2 inch) so that flushing velocity is effective in the largest size.

The ends of the submain need a flushing valve but do not need the pressure relief valve often included in older installations. One or more vacuum breakers (simple air vents) at high points aids draining after shutdown and prevents pipe collapse. A small area of isolated or "point" rows is often just connected to a couple of adjacent rows to save a long run of trenching and pipe.

For each row, a riser comes from a tee in the pipe up to the drip tubing. The best material to use for risers is a special heavy-wall PVC hose called "IPS hose." It has the same outside diameter as the equivalent standard pipe size. The advantage of this material is its superior strength and flexibility, which protect it against breakage and mechanical damage above ground. The hose is solvent-cemented like other PVC, but a special deep socket on the tee and special flexible glue are necessary for reliable joints. The top end of the riser has a transition fitting to male garden hose thread for connecting drip tube fittings.

Tubing and Emission Devices. The vast majority of vineyard microirrigation systems use half-inch (nominal inside diameter) tubing and drip emitters. Drip emitters are appropriate in most conditions, but some soils have such poor percolation rates that not all the water from a dripper can infiltrate in one spot. Other sites have very shallow soil that prevents a reasonable volume of soil to be wet from one or two point sources. Such situations may be better served by microsprinklers that wet a larger surface area. Higher evaporative loss, increased weeds, and potential disease problems from wet trunks and higher humidity are factors to consider with microsprinklers.

Current practice is to place drip emitters close enough together to create a continuous wetted zone. This usually requires two smaller emitters per vine rather than one larger emitter. In any case, emitters should be located at least a few inches away from trunks to avoid wetting them.

Blocks with more than about 10 feet of elevation change within rows need pressure compensating emission devices to deliver uniform flows. Level terraces and flat blocks are suitable for non-compensating devices, which are less expensive. Drip tube is available with internal emitters preinstalled at standard or custom spacing. Compare total costs of internal versus external emitters. A worker should be able to install 120–150 emitters per hour under good conditions.

Compensating emitters have reasonably constant output within a pressure range of 10–60 psi. Output is nonuniform below 10 psi; above 60 psi is neither safe nor reliable. So, operating within or close to the 20–40 psi range is ideal. Flat blocks can be run at 20–30 psi, depending on row length; higher pressure is needed in longer rows to compensate for additional friction loss. Hillside blocks can start at 15 psi at the top, extending the downhill run within the safe pressure range. Install a pressure-reducing valve midslope at the point downstream flow is 0.1 gallon per minute or greater to keep pressure below from getting too high.

Tube and emission devices need to be reasonably distant from threats such as in-row cultivation, pruning, and harvesting. Tube with drip emitters is typically clipped to a wire around 16 inches above ground. Microsprinklers should be hung to give the desired coverage. Consider the option of connecting small tubing from the emitter to direct the water to the target.

Some people install drip tube stretched fairly tight to the wire. Loose polyethylene tube contracts by perhaps 1.25% or more in the colder temper-atures of winter. That may not sound like much, but it equals nearly 5 feet in a 375-foot row. Not allowing slack for that contraction creates a significant stress on fittings and end attachments. Instead, let the tube sag between each vine about a hand width (4 inches) at average vine spacing. Done uniformly, it looks fine. Sloping land requires tubes to be tightly clipped to the wire so they do not work downhill over time.

System Materials

Irrigation systems are mostly made of PVC, notably the basic fittings and pipe in various sizes and wall thickness. PVC is relatively easy to work with and performs well if proper techniques are used. PVC is usually assembled into a single sound structure with solvent-welded ("glued") joints. Glued PVC valves and fittings, rather than threaded or bolted, reduce labor, leaks, and other trouble. Care must be taken to keep primer and cement off of any moving parts, and be aware that glued joints are permanent. Long runs of buried pipe may be done with gasketed pipe instead of solvent-cemented. Unrestrained gasketed joints push apart under pressure, so trenches must have compacted backfill to prevent movement. Large, high-flow tee and elbow joints need concrete thrust blocking to hold them in place, regardless of type of joints. More complex components may utilize other plastics or metals, typically connected with pipe threads. Larger components tend to use bolted flange or grooved ("Victaulic") connections.

The PVC pipe sizing system is rather arcane, utilizing pipe "classes" and "schedules," indicating the pressure rating and inside diameter/wall thickness. Fittings and pipe outside diameter are consistent among the various materials. Two key design considerations are use of pipe with a margin of pressure capacity for the operating conditions, and use of heavier wall pipe where mechanical damage is possible. Thinner wall pipe is suitable for soft ground or good backfill but is more vulnerable above ground and in rocky soil.

It is foolish to put thousands of dollars of pipe in the ground with joints that leak or blow apart because of poor technique or skimping a few dollars on the proper applicator, primer, or cement. Printed and online information on proper methods of solvent welding is available from PVC cement and fitting manufacturers. An important caution: *do not* put pressurized air or gas in PVC pipe, even for testing. The energy stored in compressed gas can be released in shrapnel if the pipe is damaged or struck. Liquids are incompressible and create little explosive potential beyond water hammer, which can break things.

Summary: The Ideal Work Sequence

Begin with a site evaluation that includes an assessment of the need and possibility for irrigation. Determine the availability of water, flow rate, total volume, and quality. If an irrigated vineyard is feasible, begin vineyard planning and developing the water supply. Do preliminary design, including vine and row spacing, identify block and avenue rough locations, and select pump and filter. Order irrigation materials and, preferably, at least enough trellis to hang drip tube. Install the power, pump, and filter station, and do a final layout. Plan for and trench the underground portion of the irrigation system and install it. Backfill trenches and complete the installation of end posts and line posts for the trellis. Flush the underground system. Depending on conditions, timing, and preference, either install the support wire, drip tube, and emitters now, or plant the vines first. Prepare to inspect, repair, and maintain the system on a regular basis.

References

Schwankl, L., B. Hanson, and T. Prichard. 1998. *Micro-irrigation of Trees and Vines.* Publication 3378, Division of Agriculture and Natural Resources. University of California, Davis.

Trimmer, W. L., T. W. Ley, G. Clough, and D. Larsen. 1992. *Chemigation in the Pacific Northwest.* Publication PNW 360, Oregon State University Extension Service, Corvallis.

III

Vineyard Management

Managing a vineyard involves many diverse practices to produce the desired quantity and quality of fruit to meet the vineyard objectives. As stated earlier, there is no singular recipe for successful vineyard management. Every vineyard has unique site characteristics, production goals, and restraints that influence the selection and application of management practices.

The vineyard management section of this book covers all of the major aspects of managing a vineyard. The individual chapters introduce the practices that are available to the grower and discuss the considerations that should be made when deciding which practices to apply and how to apply them. Although the topics are covered separately, there is a high level of interaction among the vineyard practices, as well as interactions with the site and grapevines. Growers should understand these interactions and integrate their approach to managing the vines. Ultimately, managers must develop their own strategies for accomplishing the vineyard objectives.

This section begins with a discussion of methods of marketing grapes to wineries and some of the contractual considerations that should be made. Two chapters are devoted to sustainable and organic vineyard management, approaches to grape growing that have been partially codified into voluntary formal programs. Chapter 18 provides extensive coverage of soils and management of grapevine nutrition. Managing the water availability to grapevines has many viticultural implications and is practiced through the application of supplemental water by irrigation and the management of vineyard floor vegetation that competes with grapevines for soil moisture. Pruning is a fundamental viticultural practice that maintains the shape of the vine and renews the fruiting wood. Proper sunlight exposure of the grapevine canopy is a critical aspect of vine management, particularly for the production of high-quality fruit in a cool climate. Chapter 22 discusses the principles of canopy management and introduces the assorted practices that can be used to achieve the desired canopy characteristics. In the next several chapters, the challenges presented by diseases, insects, and mite pests are described, as well as the climatic threats of frost and winter freeze injury. Managing vineyard labor and ensuring compliance with government regulations offer challenges of a different sort to the vineyard manager, but challenges no less important to the overall success of the vineyard. The section concludes with chapters focused on the developing fruit crop and harvest. Crop estimation is an important tool for managing grape yields, and the practice of thinning is routinely used to regulate yields and influence fruit quality. It is important for the vineyard manager to understand how fruit develops and ripens, and to have the ability to monitor these processes as they occur. Harvest is the culmination of the season's effort to produce the grape crop. It requires advanced planning and considerable logistics to deliver fruit of good quality to the winery.

There is a general sequence to vineyard management practices that coincides with the natural progression of vine development through the seasons. Still, several cultural practices or vineyard activities may be needed at the same time, and the manager must quickly decide the best sequence of actions to take. A general plan should be prepared for the season, but the manager must always be flexible and ready to adjust to changing circumstances. Often, the effectiveness of cultural practices is highly dependent on their being conducted at a specific time or stage of grapevine development. Therefore, some of the most important skills of a vineyard manager are the abilities to set priorities, manage time efficiently, and organize the work effort.

15

Marketing and Contracts

Al MacDonald and Jesse D. Lyon

Growing good winegrapes accomplishes only half the objectives of a successful independent vineyard. The grapes must also be sold to a winery. Ideally, a marketing plan is developed in the early stages of vineyard planning. But marketing grapes is a continuous and dynamic process that requires knowledge of current market conditions and adaptability to changing situations. It also requires some business acumen and an understanding of contractual arrangements. This chapter introduces some of the important considerations for marketing winegrapes and grape sales contracts.

Marketing Data

In 1987 the Oregon Wine Advisory Board began funding annual surveys of the vineyards and wineries in the state. Survey data are collected and published by the Oregon Agricultural Statistics Service; reports from the most recent years are available online. The information drawn from these surveys should be of interest to new growers entering the grape market. The annual vineyard reports include the total number of acres of grapes planted and tonnage harvested. The numbers are further broken down by variety or county planted, average price for variety, and average production yields. High and low prices for selected varieties are also reported. A similar survey of wineries gives data on types of wine made and production levels for different varieties. Industry trends can be analyzed by reviewing past years' reports.

After reviewing these reports, the new grower should have a better initial understanding of yields to expect, average prices being paid, and which varieties are likely to be in demand. The reality of supply and demand also has an effect on the prices offered for grapes, and prices have been know to fluctuate by several hundred dollars per ton over a period of a few years. Today's high prices for one variety may attract more planting of that variety, creating an oversupply and lowering prices tomorrow. On the other hand, wine consumers may suddenly discover a previously overlooked variety, creating more demand and increasing the price for that variety.

Developing a Marketing Strategy

How does a new grower enter the grape marketplace? The answer depends on several factors, including the goals of the individual grower. There are now almost two hundred wineries in Oregon. One approach would be to call wineries, ask if they would like to purchase your grapes, coordinate bidding, and sell to the highest bidder. But wineries have different reasons for purchasing grapes, and each winery has unique needs and expectations. Some wineries provide a better fit with the vineyard characteristics and grower expectations than others. Growers, too, have different reasons for planting vineyards, and each vineyard has unique characteristics. Growers may find it helpful to determine their goals and assess their vineyard in order to target wineries that match their needs. It may be useful to consider the following questions:

- Do you intend to start a winery in a few years?
- Is your vineyard going to be your sole means of support?
- Is your vineyard part time (weekends)?
- Where is your vineyard located in relation to wineries?
- Do you wish to develop a long-term relationship with a winery?
- What style of wine will the grapes be used for?

For example, if you intend to start your own winery to someday use all of your grape production, but you wish to sell your grapes for a few years before building a winery, you may be interested in selling on the spot market each year and trying to capture the highest price. If, on the other hand, profits from your vineyard are to supplement a retirement income, you may wish to establish a long-term contract with a nearby winery, perhaps basing the price on a percentage over your cost of production to protect you from wide fluctuations in prices. New growers should be realistic in evaluating their vineyards, both from a budgeting and a marketing perspective. What are the expected yields and quality compared to other vineyards?

How do you ensure that you will have a good business relationship with the winery? The following are items to consider when dealing with a winery.

- If you are selling grapes to a winery for the first time, you may want to ask the winery for other grower references. Call these references and ask about payment histories and their experiences in dealing with the winery.
- If the winery does not provide a written contract, ask if the winery will accept one written by the grower.
- Discuss with the winery prior to delivery the number of tons expected, price, terms, and any other conditions the winery or you may desire.
- As a grower you will be expected to deliver clean sound grapes in a timely manner.

Contracts

One of the best ways of ensuring a good business relationship with a winery is by operating under a contract. There are two basic types of contract, oral and written. With an *oral contract,* you call or visit the winery, talk, and ultimately come to an agreement to sell your grapes at a specified price. This type of contract works best with parties who get along well, trust each another unconditionally, and both have good memories. Such an agreement is, however, very expensive, if not practically impossible, to enforce if either party fails to perform as the other intended, or if anything (even outside the control of either party) goes wrong. A *written contract* reduces to writing the terms and conditions of a sale or purchase so that both parties can clearly see what is expected of them. As the Oregon wine industry continues to grow, an increasingly sophisticated approach to grape contracting is warranted. Even between parties who get along and trust each other, a well-written contract is like a good fence between neighboring property owners; it defines their boundaries, protects the interests on each side, and preserves the relationship.

A written contract is strongly recommended for all grape sales to out-of-state wineries. Growers should give serious consideration to review of any such contract by their lawyer. Particular attention should be paid to clauses dealing with venue, which selects the state and county in which the provisions of the contract will be enforced. Growers will also be responsible for directly paying any tonnage taxes to the State of Oregon.

The "exchange of promises" in any grape contract should be designed to fulfill four basic purposes: (1) to communicate mutual understandings of fact; (2) to define all of the principal business terms between the parties; (3) to allocate responsibilities and obligations between the parties; and (4) to address contingent outcomes. A good contract defines the business relationship on all four of these levels, and it does so in a manner that meets both parties' basic objectives with clear and concise language. The basic elements of such a contract are reviewed below.

Mutual Understandings

Mutual understandings of fact are often addressed in a set of "recitals," background provisions that precede the substance of the written agreement. A few short statements about the parties and the nature and locations of their businesses and a brief summary of their objectives for the contract can help avoid fundamental flaws that might prevent a true meeting of the minds on the terms that follow. Recitals are also helpful to any third party (such as a business associate, estate administration, or judge) who later needs to put the agreement in context.

In a recital, the vineyard usually is referred to as the seller and the winery referred to as the buyer. This is a technical specification, but it is important to determine exactly what parties are involved with the agreement and to distinguish their relationship from a joint venture vineyard development, grape crop lease, or other legal relationship that may be found in alternative winery and vineyard transactions.

Principal Business Terms and Responsibilities

Although the principal business terms between a buyer and seller may seem to be straightforward or commonplace, careful drafting of these terms is absolutely necessary. Incomplete statements of both parties' expectations regarding the variety, source, pricing, quantity, and quality of grapes, or about the duration, exclusivity, or comprehensiveness of the contractual obligations, can lead to uncomfortable and sometimes costly haggling if a discrepancy is later revealed while the contract is being performed.

Good drafters think creatively to produce principal business terms that are sufficiently flexible to capture the value that both parties desire from the relationship defined by the contract. For example, a broad range of alternatives for pricing units (e.g., per ton, acre, or bottle) and quality incentives (e.g., per quantitative criteria, label designations, or awards) are already in place in Oregon. Careful drafters do not, however, sacrifice clarity to achieve flexibility. Questions such as "What items are included or excluded from the price?" and "How is the quantity or quality determined?" should be answered in the written agreement before a grower and winery ink their contract.

Product. The product should be specified explicitly. Are you selling a specific grape variety (Pinot noir, clone, and designated block, all of the grapes from the vineyard)? What is the quantity you are selling (e.g., number of tons, all the grapes from x number of acres, truck loads)? Are there provisions for excess tonnage, or for unexpected shortfalls? Does the seller have the right to withhold tonnage for open market sales?

Price. Price is usually a function of supply and demand and awareness of current market conditions is essential to both parties making a fair deal. Quality is also a major factor in determining a fair price. High-quality grapes as well as wines will always be in demand, and the price should reflect this. Different wineries have different standards of quality, and the grower should be aware of these standards when negotiating with a winery. Sugar, acid, and pH are the three most often used quality variables. Does the contract give a different price for different levels of these measurements? Who performs the measurements? If a dispute arises, is there a provision of third-party measurements? Growers should be aware that field samples may differ from tank samples. There may be a conflict if the contract states that the winery decides when to pick, yet prices are based on minimum measurements. Are the minimum standards based on your past field records reasonable?

Other quality standards used in the industry include such items as percentage of material other than grapes (M.O.G.; this usually includes leaves, canes, etc.), rot, mold, mildew, bird damage, and berry to stem ratio. Is there a schedule for price adjustments based on acceptable quality ranges? Will your load of grapes be rejected if minimum quality standards are not met? Is the stated price inclusive of grape taxes? Is there any bonus premium to be paid if a vineyard-designate wine subseqently wins certain awards or achieves a stated rating?

Over the past few years there has been some interest by wineries and growers in *bottle contracts,* an agreement in which the final price paid for grapes delivered is based on the retail (or wholesale) price of the bottle of wine made from the purchased grapes when it is released. The contract may state, for example, that the final grape price per ton will be the retail bottle price times a multiplier of 100. Thus, if the wine made from the purchased grapes retails at $20, then the grower would be owed $2,000 per ton ($20 x 100).

Often a base price is paid to the grower in the year the grapes are delivered, with the balance to be paid upon release of the wine. Using the above example, if a base price of $1,000 per ton is paid to the grower in the year the grapes are delivered, the balance of $1,000 per ton is paid when the wine is released.

If you are a grower negotiating this type of payment schedule with a winery, you should consider several factors. Is a base price stated in the contract, and is it high enough to cover the costs of production? The base price may be all you receive. Are your vineyard grapes being used exclusively in the production of this wine? Other vineyard grapes may add or detract from the value of the wine. How soon after delivery does the winery expect to release the wine? You may also want to consider what the effects of current interest rates have on your deferred payment. In the above example, suppose the average price being paid in the year you sell the grapes is $1,500.00 per ton and you have deferred $500.00 per ton in current income to receive a $500.00 per ton bonus in three years. With higher interest rates this may be only slightly more profitable and you would not have use of the cash. Knowing your own situation and the track record of the winery you are dealing with becomes more important in this type of payment schedule. If you are offered this type of contract, consulting competent legal and financial advisors is recommended.

Recently there has been a strong interest by wineries in lower yields per acre. To fairly compensate growers for reduced yields, wineries establish *per-acre contracts* with growers. As an example, suppose you have a 6-acre block of premium grapes that a winery wishes to be cropped at 2 tons to the acre, amounting to 12 tons of grapes. If the average price per ton were $1,500 and your average yield for that block was 2.5 tons per acre, you would expect to gross $22,550 for the entire block. At 2 tons per acre you would receive only $18,000 for the same block. To compensate the grower for the lower crop, the winery may offer to pay the grower $3,750 (2.5 tons x $1,500 per ton) per acre for the grapes, resulting in a gross of $22,550 for the entire block. The winery is thus able to control crop levels to their advantage given unexpected conditions at the vineyard. If the vintage is late and the winery wishes to thin to less than 2 tons per acre, then the grower can still expect to receive the agreed-upon amount per acre. If it is a warm year and the winery believes the vineyard can ripen a higher crop level, the grower still receives the same amount, and the cost per ton decreases for the winery.

Transportation. Included in the transportation clause are items such as who is responsible for delivery, timetables, delivery containers (what size and who provides), weight slips, and when the winery accepts title to the grapes (if you have 10 tons of grapes on a truck worth $1,500 a ton, you may want to check with your insurance agent to see if your load is insured). When does title and risk of loss transfer from the seller to the buyer?

Payment. Total cash price within 30 days of delivery would be ideal from the grower's point of view, but cash flow seldom allows wineries to offer this. If the contract is for installment payments, is there an interest charge to the winery, or a penalty clause for late payments? If a payment is late, does the full amount become due and payable? These things should be stated in the contract, and the grower must be willing to enforce this provision of the contract.

Labeling. In order to create demand and recognition for your grapes (and perhaps your own future wine label), you may wish to negotiate terms for the use of your own vineyard name(s) for vineyard-designate wines. Reliable viticultural area designation may also be important to the winery or grower for accurate labeling and marketing purposes.

Vineyard and Grape Management. Sales contracts may provide for the extent and nature of the winery's access to the vineyard, and for the degree of cooperation or control to be allocated to each party for production management issues (such as the use of pesticides and fertilizers, compliance with the LIVE program, picking methods, harvest dates, and field sanitation). Wineries and vineyards may wish to designate certain practices and data as "confidential information" not to be disclosed beyond the terms of their own transaction.

Any baseline expectations should be spelled out. Well-written representations and warranties about operational details, fiscal and managerial accountability, and financial wherewithal can be drafted to capture the rights and obligations upon which the basic business terms are conditioned.

Contingencies

Many contracts fail to address adequately unintended contingencies that may arise down the road. Some of these issues are related to the potential downsides of the agreement if things go awry, such as the risk of loss during shipment, the buyer's right to reject substandard products, or the seller's rights in the event of delayed payment. Common contingencies may revolve around business changes, such as the parties' relative rights to amend, assign, terminate, or renew the contract. Other conditions you may want to include in a contract would be references to warranties, successors, third-party arbitration, terms of the agreement (one year or long term), or anything else the two parties need to agree upon. Contract sections innocently headed "Miscellaneous"—and drafted in incomprehensible legalese that is all too common, but not at all necessary—can determine who wins and who loses costly disputes that may develop later.

Buyers and sellers who rely on worn-out forms and business assumptions could miss valuable opportunities and face business costs and liability they are unprepared to bear. Buyers and sellers who negotiate and use professionally crafted agreements can facilitate clear and creative business relationships that drive profitability and allow them to manage, insure, or hedge known risks with a greater degree of comfort.

Collecting Payment

Under amendments to the Oregon Agricultural Produce Lien (ORS 87.705-87.736) adopted in 2001, a non-waivable lien in favor of the selling vineyard is created upon delivery of grapes to a purchasing winery. Until and unless the vineyard is paid in full, the lien continues for 45 days from the date final payment is due. If the winery fails to pay within that time and the vineyard files a lien with the Secretary of State's records office (county filing is no longer appropriate), the lien is extended to 225 days from the date that final payment was originally due. When payment is made in full, the purchasing winery may request that the selling vineyard file a certificate of lien satisfaction; failure to do so within 10 days of request may subject the collecting vineyard to $100 damages due to the winery. An unsatisfied lien may be foreclosed by lawsuit filed at any time before the lien expires. A successful Agricultural Produce Lien plaintiff is entitled to recover reasonable costs and expenses for filing and recording the lien as well as attorney fees at trial and on any appeal. All of these time deadlines and filing requirements are strict, and you should fully understand them well in advance of filing a lien.

If the sales terms do not provide for payment beyond 30 days from the date grapes are delivered and accepted by the purchasing winery, a selling vineyard may have enforceable payment trust rights in the winery's inventory and sales proceeds under the federal Perishable Agricultural Commodities Act (PACA). The seller must provide a complete and accurate notice of intent to preserve PACA's statutory trust benefits to the non-paying buyer within 30 days of the date payment was due. PACA rights may be enforced in federal district court. Again, these deadlines and notice requirements are strict, and you should understand them well before filing a lien.

A grower also has the option of suing a winery under ordinary contract law for payment of grapes delivered. Suit in small claims court is appropriate if the amount is less than $2,500.

Example Contracts

Figures 1 and 2 are sample contracts that have information of the kind that should be included in any contract. Many wineries have their own contracts, which may differ from those presented here. There is no standard form contract appropriate for every deal between any grape grower and winery; the best agreements rely on a mix of industry norms and terms tailored to the unique needs of the buyer and seller at issue. These illustrations are examples only and are not intended as a substitute for sound legal advice.

The important differences between the per-acre contract and other contracts are the treatment of production and volume management. The per-acre contract states the winery intention for an expected crop level and has provision for alternate pricing if yield is lower than the agreed-upon level. If the lower than expected crop level is a result of conditions other than thinning, the grower would expect to receive less than the agreed-upon price per acre. For example, suppose the agreed-upon price per acre were $4,000 for a 2-ton per acre crop level. This would work out to $2,000 per ton to the grower, but because of a cool wet spring the actual crop without thinning results in a harvest crop level of 1.5 tons per acre. A provision might state that, if the crop level falls below 2 tons per acre, the grower will receive $2,000 per ton delivered. The net result to the grower would be a gross of $3,000 per acre instead of the $4,000 originally agreed upon. However, if it were a very late year and the winery did not think that the crop level of 2 tons per acre would yield the quality desired and instructed the grower to thin to 1.5 tons per acre, there would be no penalty to the grower. In this manner, the grower and winery share in the result of weather conditions that may affect the crop level.

Market Economics

Vineyard economics are covered elsewhere in this book, but you should be familiar with your own costs of production to determine a fair price for your grapes. Are your average yield predictions realistic? There can be dramatic differences in yield from year to year due to weather, disease pressure, bird damage, and so forth. Are your costs of production in line with other vineyards? Some vineyard operations may increase quality, but at a cost in excess of any extra compensation a winery may offer.

The economic picture in growing grapes in Oregon has narrow margins. Fortunately, most of us are growing grapes for reasons other than getting rich. To continue doing what we enjoy, however, most would agree that a return on investment sooner rather than later is reasonable, not to say necessary.

The best hedge against discouragement over the current balance sheet would seem to be serious attention to developing better cultural practices, lowering production costs, and establishing a sound financial plan with accurate record-keeping. Then agree to sound grape sales contracts with trusted buyers from whom payment will be collected.

Resources

Oregon Vineyard and Winery Reports, published annually by Oregon Agricultural Statistics Service, Portland. http://www.nass.usda.gov/or/vinewine.htm

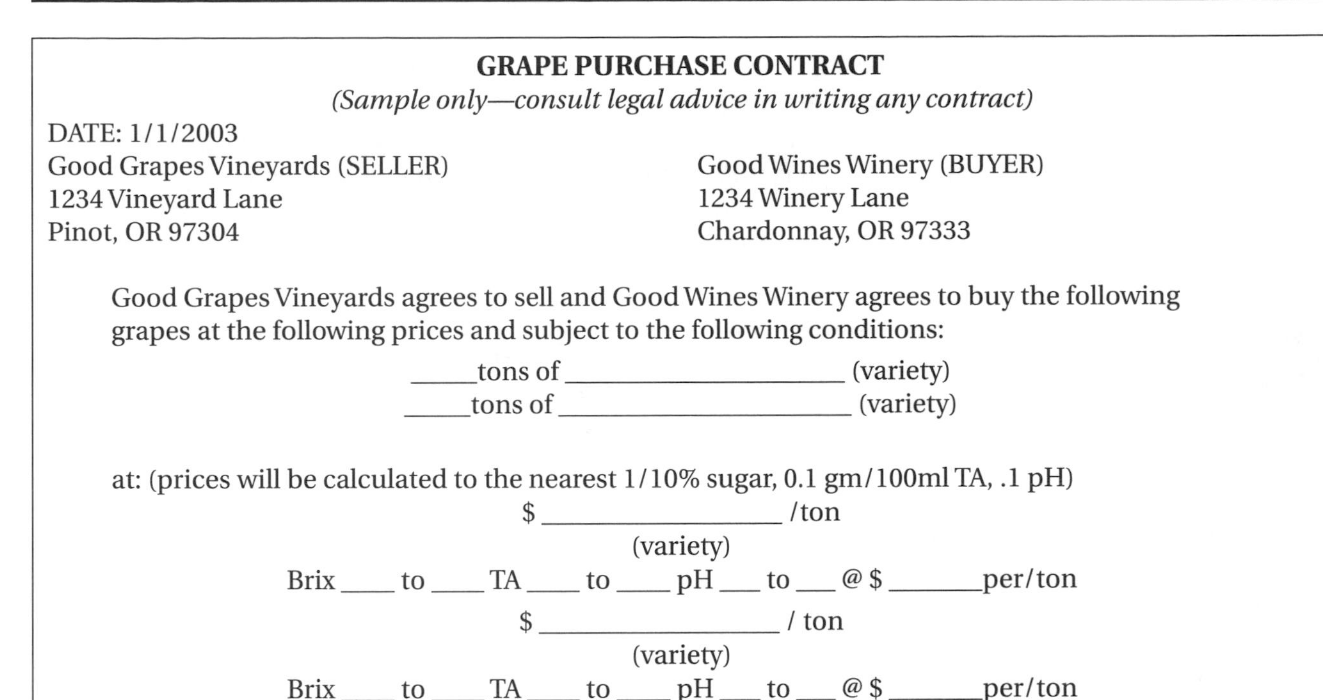

GRAPE PURCHASE CONTRACT

(Sample only—consult legal advice in writing any contract)

DATE: 1/1/2003

Good Grapes Vineyards (SELLER)
1234 Vineyard Lane
Pinot, OR 97304

Good Wines Winery (BUYER)
1234 Winery Lane
Chardonnay, OR 97333

Good Grapes Vineyards agrees to sell and Good Wines Winery agrees to buy the following grapes at the following prices and subject to the following conditions:

_____tons of ____________________ (variety)
_____tons of ______________________ (variety)

at: (prices will be calculated to the nearest 1/10% sugar, 0.1 gm/100ml TA, .1 pH)

$ _________________ /ton
(variety)
Brix ____ to ____ TA ____ to ____ pH ___ to ___ @ $ _______per/ton

$ _________________ / ton
(variety)
Brix ____ to ____ TA ____ to ____ pH ___ to ___ @ $ _______per/ton

Grapes that fail to meet any of the quality criteria for a given category will fail to qualify for pricing bonuses of that category. Grapes that fall below the minimum criterion may be refused by the Buyer or a negotiated price may be offered. This provision shall apply to all varieties of grapes covered by this contract.

The above prices are based on less than 2.5% material other than grapes (M.O.G.), rot, or mildew. If there is 2.5–5% M.O.G., rot, or mildew, the dockage will be 10%. If there is more than 5% M.O.G., rot, or mildew, the winery may reject the grapes or negotiate a new price.

Other Conditions of Sale:

1. Seller agrees to transport the fruit to Buyer's winery.
2. Grapes will be delivered in __ by ___ totes. If totes must be rented by the Seller, the Buyer will pay 1/2 the rental price. The off-loading of the containers at the Buyer's location shall be the responsibility of the Buyer upon acceptance, and upon off-loading of the containers the grapes become the property of the Buyer.
3. Buyer will measure sugar, total acid, and pH. Seller may observe the measurements and obtain another determination from a qualified laboratory in case of doubt as to the validity of winery measurements.
4. Seller will provide Buyer with gross and tare weights from a licensed weigh station.
5. During the season, Seller agrees to keep Buyer informed of maturity of the fruit. Seller will pick grapes in a timely fashion upon direction of the Buyer and begin harvest within 48 hours from the time specified by the Buyer.
6. Payment

______% of net price within 30 days of delivery
______% of net on or before __________
______% of net on or before ___________

A late fee of 1% per month will be charged on payments beyond the due date.

7. Other conditions: (specify)

_____________________________ Seller ____Date
_____________________________ Buyer ______Date

Figure 1. Grape Purchase Contract.

PER-ACRE GRAPE PURCHASE/SELL AGREEMENT

(Sample only—consult legal advice in writing any contract)

This agreement is made this _____ day of ______, 200__, by and between Good Grapes Vineyard. hereafter referred to as Seller, and Good Wines Winery, hereafter referred to as Buyer.

Variety ____________________

Quantity. All of the fruit from rows _____ to ______ (Block_____) of the Seller, located at _________________________.

Crop Level. It is the intention of Buyer to hold the crop levels to about _____ tons per acre. Therefore, Buyer has the right to state specific thinning instructions and Seller will execute those instructions to the best of its ability. To assist in these thinning determinations, Seller will estimate cluster weights by lag phase cluster samples. Note that different pricing applies if the crop load, without thinning instructions/actions, falls at _____tons per acre or below or the harvested Brix measurement of the entire lot falls below _____degrees Brix.

Price. Buyer shall pay to Seller $___________ per acre for each crop year, with the exception that if the crop harvested per acre falls at or below _____ tons per acre Buyer shall pay to Seller $________ per ton delivered. If the sugar level of the entire harvested lot falls below _________ degrees Brix the price shall be negotiated.

Payment terms. The total annual purchase price shall be paid as follows:

One-third due and payable 30 days after delivery.
One-third due and payable on or before ________.
One-third due and payable on or before ________.

Sampling. Buyer may take samples of grapes from the specified acreage at any time in order to measure the maturity of the fruit on the vines. Seller upon request will take and provide grape samples based on Seller normal sampling techniques. Buyer agrees to hold Seller harmless against any sampling error.

Picking. The decision as to when to pick will be made by Buyer. When notified Seller will commence picking as soon as possible, but at least within 48 hours.

Chemical. Seller will not use any chemicals on either the grapes or the vineyard floor that are not labeled and approved for use on grapes in Oregon. Cultural practices of the vineyard shall be consistent with good horticultural practice in order to produce the best fruit possible.

Applicable Law. This agreement shall be governed and interpreted in accordance with the laws of the State of Oregon and be enforceable in the county the vineyard is located.

Assignment. This agreement shall be binding upon the parties making it and their successors and assigns.

Life. This agreement is intended to continue in force for an extended period of years, with the possibility of renegotiations of terms and conditions or termination as follows:

On an annual basis, following harvest and before the end of that year, either party may initiate renegotiations of any or all terms and conditions of this Agreement. If the negotiation is successfully concluded, then a written modification to the Agreement will be prepared and signed by both parties.

If the negotiation is not successful there are three possible results: 1. Both parties will continue operating under this Agreement as then currently in effect, or 2. If both parties agree to terminate the Agreement, then it will cease to be in effect after one more crop year, or, 3. If one or both parties disagree on termination, the Agreement will continue in effect only for two more crop years.

The result of any termination action will be documented by letter and signed by both parties.

Integration. Seller and Buyer agree that this written agreement supersedes all prior oral understandings, contains the entire agreement between the parties, and forms the sole basis of governance of this agreement. All oral understandings not expressed in the Agreement are void. Amendment to this Agreement shall be made in writing signed by both parties to the Agreement.

Attorney Fees. In case suit or action is instituted with respect to this agreement or to enforce any of the provisions of it, the losing party agrees to pay such sum as the trial court may adjudge reasonable as attorney's fee, to be allowed the prevailing party in said suit or action, and if an appeal is taken from any judgment or decree of such trial court, the losing party further promises to pay such sum as the appellate court shall adjudge reasonable as prevailing party's attorney fees on such appeal.

Termination. Either party by written notice to the other party, given not later than the 1st day of April of the 1st year in advance of the harvest to be affected, may give notice of termination of the contract. The harvest following such termination notice shall be reduced by 25%, the next by 50%, the next by 75%, and the next by 100%

In witness, the parties have here set their hands and seal the day and year first above written.

______________________________ Seller ____Date
______________________________ Buyer ______Date

Figure 2. Per Acre Grape Purchase/Sell Agreement

16

Sustainable Viticulture in Oregon

Al MacDonald and M. Carmo Vasconcelos

Viticulture, reflecting a similar trend in other crop production systems in many parts of the world, has increasingly emphasized the importance of sustainability of production practices to the long-term viability of the wine industry, natural resources, and local communities. The USDA-funded Sustainable Agriculture Research and Education (SARE) program has defined sustainability in the following manner:

Sustainable agriculture refers to an agricultural production and distribution system that:
- Achieves the integration of natural biological cycles and controls,
- Protects and renews soil fertility and the natural resource base,
- Optimizes the management and use of on-farm resources,
- Reduces the use of nonrenewable resources and purchased production inputs,
- Provides an adequate and dependable farm income,
- Promotes opportunity in family farming and farm communities, and
- Minimizes adverse impacts on health, safety, wildlife, water quality and the environment.

The emphasis on sustainability of viticulture practices has lead to the development of formal programs designed to teach and encourage adoption of preferred practices to achieve specific goals. The preeminent body promoting development and certification of sustainable agriculture programs on a global basis is the International Organisation for Biological and Integrated Control of Noxious Animals and Plants (IOBC).

The IOBC defines integrated production (sustainable agriculture) as a system that produces high-quality food and other products by using natural resources and regulating mechanisms to replace polluting inputs and to secure sustainable farming. Emphasis is placed on a holistic system approach involving the entire farm as the basic unit; the central role of agro-ecosystems; balanced nutrient cycles; and the welfare of all species in animal husbandry. The preservation and improvement of soil fertility and of a biologically diversified environment are essential components. Biological, technical, and chemical methods are balanced, carefully taking into account the protection of the environment, profitability, and social requirements (IOBC/WPRS, 1993).

Oregon LIVE

In 1997, a group of Oregon winegrowers decided to investigate what could be done to increase sustainable practices in Oregon vineyards. The Swiss "Integrated Production" program was used as a model for developing guidelines customized for Oregon conditions. After a few years of testing the prototype program in approximately twenty-five vineyards in the Willamette Valley, an organization was created to pursue development of a formal, voluntary certification program that would meet international standards for integrated production. Oregon's LIVE (Low Input Viticulture and Enology) program was started in 1999, guided by the SARE definition of sustainability and the international standards set by the IOBC. In 2001, LIVE became the first sustainable viticulture program outside of Europe to receive international certification by the IOBC.

Goals and Objectives

One of the motivational factors for development of the LIVE program arose from the recognition that Oregon vineyards have fewer insect and disease problems than other grape production regions of the world. Thus, one of the goals of the program is to maintain and enhance this present advantage. Another goal is to encourage continual improvement of the sustainability of vineyard practices. The objectives of the LIVE program are as follows:
- To see the vineyard as a whole system.
- To create and maintain viticulture that is economically viable over time.
- To maintain the highest level of quality in fruit production. Integrated production should not require any compromise of quality standards.
- To implement cultural practices and to solve problems in such a way as to minimize the use of

off-farm inputs such as agricultural chemicals and fertilizers, with the goal of protecting the farmer, the environment, and society at large.
- To encourage farming practices that promote and maintain high biological diversity in the whole vineyard.
- To encourage responsible stewardship of soil health, fertility, and stability.

Program Requirements

Participation in the LIVE program is a contractual agreement with an annual membership fee. Attendance is required at an introductory information course and annual training courses designed to keep members informed of the latest research and practices applicable to sustainable viticulture. LIVE provides technical guidelines comprising the set of rules and optional choices that must be complied with to receive certification (LIVE Technical Guidelines, 2002). A scorecard system is used to evaluate compliance. Participants in the program are given specific guidelines on preferred production practices, and a handbook provides educational materials and additional resources to enhance learning by the producer.

The technical guidelines and vineyard scorecard are divided into six main sections: Floor Management, Disease and Pest Management, Soil Health and Vine Nutrition, Irrigation, New Plantings, and Administration. Within each section, specific practices are itemized. Certain practices are required and others are prohibited, but many have a variety of options with variable point scores. Positive points (10–20) are given to solutions and actions aimed at improving grape quality, diversifying the vineyard ecosystem, and reducing chemical inputs (pesticides, fertilizer, fuel, etc.). Negative scores are assigned to less preferred practices, suggesting the need for modification. An "Unacceptable" score is given for use of prohibited practices or for failing to complete a required practice. Compliance cannot be certified unless and until unacceptable scores are corrected. One Unacceptable score causes disqualification from certification status for the current growing season. Additionally, the grower must achieve at least 50% of the maximum number of positive points to receive certification. Evaluation is based on grower-submitted records of fertilization, pesticide use, and the scorecard of management practices. Growers are subject to unannounced inspection at least once a year for verification.

Vineyard Guidelines

Floor Management. The major focus is on managing the floor to provide favorable habitat for beneficial insects. The objective is to preserve the present situation of no significant insect pest problems, other than phylloxera, in Oregon vineyards. Botanical diversity in the cover crop and management practices that favor beneficial insects are encouraged within and around the borders of the vineyard. Beneficial insect populations are further encouraged by reductions in pesticide applications, and by using the least persistent formulations and targeting their use with a good IPM (integrated pest management) program. Soil erosion is prevented or reduced by using wide tractor tires, appropriate cover crops, and organic soil amendments wherever possible.

Disease and Pest Management. Disease pressure for powdery mildew and botrytis bunch rot is frequently high, but canopy management practices can contribute to effective disease control and reduce the need for fungicides. Growers are encouraged to use disease forecasting, pest monitoring, and threshold levels to determine if or when application of pesticides is necessary. Only pesticides from an approved list may be used. Reduced fungicide applications are rewarded with more points on the scorecard, and rotation of fungicides to prevent pathogen resistance is encouraged. Maintenance and calibration of all pesticide application equipment are required to ensure accurate applications.

Soil Health and Vine Nutrition. Practices are encouraged that improve or sustain soil structure and nutrient retention. Fertilizer applications are based on soil and tissue analysis and should correct deficiencies and maintain adequate levels in the soil. Nitrogen applications must be based on replacement of amounts removed by the grape crop and applied when most available for uptake by the vines. Addition of compost, organic fertilizers, or other organic matter is rewarded with positive points.

Irrigation. Irrigation is generally discouraged for mature vines more than three years old but is permitted in drought conditions or when vines are exhibiting water stress symptoms. Irrigation scheduling and quantity of water applied should be based on soil moisture-monitoring devices or grapevine physiological indicators.

New Plantings. New vineyard plantings should be laid out to minimize soil erosion potential. Grape varieties, rootstocks, and other planting systems that are likely to produce regular yields of high-quality grapes, and hence economic success, should be selected. Use of varieties, clones, and rootstocks that are resistant to diseases or pests is rewarded with points. Additional points are awarded for pre-planting adjustment of soil pH to within an optimum range.

Administration. An annual membership fee must be paid to cover the cost of administering the program. Complete records must be provided to LIVE on all applications of fertilizers, pesticides, and irrigation water. Vineyard records are the basis for compliance evaluations and for public acceptance of the program. They also assist the grower in self-evaluation of the vineyard and for future planning. Records are also required for sprayer calibration, irrigation and insect damage monitoring, and pest risk assessments. Participation in the introductory information course and annual training courses is mandatory.

Winemaking Requirements

The official "LIVE" label is granted to wines that fulfill the conditions for grape growing and winemaking. Wines with the LIVE label must be produced 100% from grapes originating from a certified LIVE vineyard. The wines must respect the regional appellation requirements. Certified LIVE wine cannot be blended with uncertified wine and retain certification. Chaptalization (sugar addition) cannot exceed 3 kg/hl (1.7%) volume alcohol unless an exemption is granted by the LIVE organization for that particular vintage. Total SO_2 content cannot exceed 120 mg/l, but exceptions are provided for barrel-fermented wines and wines with residual sugar, which must still conform to state legal restrictions. For all other criteria, the federal, state, or regional legal restrictions apply.

The wines must be evaluated by an independent tasting panel, must be clean, and must have varietal character. A wine analysis conducted by an independent specialist must accompany the wine sample for tasting. It must include percentage alcohol, titratable acidity (TA), and total and free SO_2. For all other criteria, the regional legal restrictions must be respected. The number of labels distributed by LIVE is based on the quality of wine produced.

Conclusion

The LIVE objectives and scorecard should be seen, not as an endpoint, but rather as the beginning of an ongoing process of improving the sustainability of vineyard practices. The Oregon vineyard is a dynamic entity, presenting fresh challenges requiring innovative solutions. New pests will require thoughtful responses that preserve the integrity of the program objectives. Better solutions to old and chronic problems need to be encouraged. The LIVE program is periodically reviewed and revised as necessary. Finally, it should be emphasized that participation in this program is entirely voluntary. Although vineyards are visited periodically to monitor compliance, the commitment of the participants is assumed and relied upon for the success of this program.

References and Additional Resources

LIVE Technical Guidelines. 2002. Low Input Viticulture and Enology, Inc., P. O. Box 102, Veneta, Oregon 97487.

IOBC/WPRS. 1993. Integrated production: principles and technical guidelines. *International Organisation for Biological and Integrated Control of Noxious Animals and Plants Bulletin* 16(1).

Sustainable Agriculture Research and Education program. http://www.sare.org.

17

The Organic Approach

Doug Tunnell and Edward W. Hellman

Time: Early 1990.
Place: Portland area wine shop.
Event: Conversation between an organic grower and his friend, a wine merchant.
"Organic, huh," the merchant says. "Yeah, when I was in the produce business, the organic produce was what we called all the stuff that fell off the back of the truck."

There was a time when scars and bruises were widely considered to be about the only things that distinguished "organic" produce from the rest of the fruits and vegetables in the market. But that has changed dramatically over the past two decades. The way in which foods of all kinds are produced has come under increased consumer scrutiny. From the standpoint of food and worker safety and environmental impact, agricultural chemical use is being examined as never before. One result has been increased governmental regulation, such as the Oregon Pesticide Use Reporting System, which requires applicators and farm operators to report all chemical inputs on a monthly basis to state officials. Another result has been a whole new interest in organic products.

Concurrent with the changes in consumer and regulatory attitudes has been the steady growth of the organic sector of American agriculture. According to *Organic Consumer Trends 2001*, "Retail sales of organic products have grown steadily for the past ten years, showing a compounded annual growth (CAG) of 22.74% over that period. Growth rates are similar over the past five year (22.61%) and three year (24.72%) periods. Assuming steady growth at a conservative rate of 20%, retail sales of organics in 2001 are projected at $9.3 billion. By 2005, sales are expected to reach nearly $20 billion" (Natural Marketing Institute, 2001). Organic products are no longer perceived as an isolated market sector on the fringes of American agriculture. They are rapidly moving into the mainstream, and with that shift has come some greatly needed regulation of the organic industry.

The long-awaited National Organic Standard became official in 2000, providing for the first time a single, nationwide definition for the term "organic." The National Standard details all the methods, practices, and substances that can be used in producing and handling organic crops and processed products. It establishes clear labeling criteria and specifically prohibits the use of genetically modified organisms, ionizing radiation, and sewage sludge for fertilization.

The National Standard also establishes a system designed to certify that producers and products are in compliance with this body of regulations. The system relies on a network of state or private agencies accredited by the USDA to inspect any farm, processor, wholesaler, or vendor that produces or markets a product as organic. For growers of organic produce, this entails annual on-site inspections by representatives of the certifying agency. These can be exhaustive, with inspectors checking such things as farm outbuildings, equipment, chemical inventory compared to supplier's invoices, and even sales receipts. The inspector may collect plant tissue to be analyzed for chemical residues by an approved laboratory. The objective is to provide a seamless paper trail that accounts for all farm products, practices, and inputs. The certifying agency is legally accountable to the USDA. Its job is to guarantee that any product labeled and sold as organic was produced in compliance with the National Standard.

Certification is at the core of the USDA's definition of organic products and processes. Even if a particular crop or product is produced using established "organic" methods, the U.S. government no longer allows it to be labeled and sold as organic without a certificate of approval from one of the state or private certifying agencies.

The oldest certifying agency in Oregon, Oregon Tilth Certified Organic (or Tilth), was among the first group of agencies accredited by the USDA as a certifier for the National Organic Program. Oregon Tilth's standards and procedures manual has been updated with the National Organic Standard at its core. This includes the National List, a general listing of all sprays and fertilizers approved for use in an organic vineyard. This list is the "bible" of approved inputs. Despite the claims of product manufacturers, it is the grower's responsibility to determine whether or not any given spray or material is on the list. Even the material chosen for trellis posts falls within the guidelines. If unlisted products are used, the grower and the product lose organic certification.

The foundation of the organic claim is the following broad definition of organic farming issued by the USDA:

> *Organic farming is a production system which avoids or largely excludes the use of synthetically compounded fertilizers, pesticides, growth regulators and livestock feed additives. To the maximum extent feasible, organic farming systems rely on crop rotations, crop residues, animal manures, legumes, green manures, off farm organic wastes and aspects of biological pest control to maintain soil productivity and tilth, to supply plant nutrients and to control insects, weeds and other pests.... The concept of soil as a living system is central to this definition. (Lampkin, 1990)*

This chapter presents the major considerations in planning, planting, and caring for an organic vineyard in Oregon. The obvious difference from conventional production practices is the strict avoidance of synthetic compounds for fertilizers and for disease, pest, and weed control. But the more profound distinction is in the grower's approach to the soil as a living system.

Soil Tilth

It is not by accident that the oldest and most widely known organic certification agency in the State chose to call itself Oregon Tilth. Tilth can be defined as the condition of soil under cultivation. The term generally refers to the physical characteristics of the soil as they relate to cultivation: ease of tillage, water percolation and drainage, and root penetration. But, these soil characteristics are influenced by organic matter and the actions of soil organisms. Therefore, the starting point of developing an organic program for any vineyard is the notion that soil is more than just a medium for anchoring vine roots in place while they draw water and nutrition toward the leaves and fruit. It is the understanding that the soil is alive, containing large numbers of microorganisms (bacteria, fungi, algae, molds protozoa), nematodes, worms, insects, and of course, living plant roots. There is much yet to be learned about the relationships among the diverse organisms interacting in the soil ecosystem, but organic growers work from the assumption that a healthy soil ecosystem promotes healthy plant growth (Sullivan, 2001). The amount of organic matter in the soil is considered a critical factor for soil health, since it influences tilth and provides the major food source for many soil organisms.

"This idea of the soil as a living system is part of a concept which maintains that there is an essential link between soil, plant, animal and man. Many people involved with organic agriculture believe that an understanding of this is the prerequisite for sustaining a successful farming system" (Lampkin, 1990). The organic approach thus entails much more than the simple choice of products from an approved list of farm inputs or decisions regarding cultivation or weed management in a vineyard. It is a worldview. It sees in the complexity of a natural system the fact that every intervention by a grower has far-reaching consequences, many of which may be beyond our everyday perception of what is going on in the vineyard.

This acknowledgement radically influences the ways in which an organic grower approaches problems. In general, it leads to the conclusion that in nature there may well be no simple, quick fix. When faced with a decline in soil fertility, for example, the organic approach would eschew the application of a single, processed, nitrogen-rich fertilizer and call instead for the more cumbersome and time-consuming application of well-made and seasoned compost.

The starting point of the organic system is that soil fertility entails much more than the balance of nitrogen, phosphorus, potassium, and a handful of micronutrients. In addressing plant productivity, the organic grower may also choose to encourage vine growth further by eliminating competition from weeds and grasses in the vineyard. But here again, the approach assumes that the application of herbicides as a strip spray in the vineyard has far-reaching consequences. In the organic view, the cost of choosing to apply a strip spray as measured against the goal of developing healthy, living soil is simply too great. The spray may solve one season's weed problem, but it sows the seeds of greater future problems by disrupting the microflora of the soil. The result is again the organic grower's decision to undertake the more difficult and time-consuming task of weed control by mechanical in-row cultivation or hand hoeing.

Farming a certified organic vineyard is not necessarily any more difficult than farming a conventional vineyard. But it is a very different project, requiring a distinctly different mindset. An organic vineyard never looks as crisp, clean, and neatly trimmed as its conventional counterpart, but consider all the costs of "neatness." The objective is to create an environment rich with biodiversity that is mostly self-sustaining, providing its own nutrients, disease resistance, and predator populations. No compromises in fruit quality are necessary. A properly managed organic vineyard is fully capable of producing the highest-quality fruit.

Cover crops. The most simple, economical, and thus widespread technique for building soil tilth is through the use of cover crops. All responsible farmers protect the topsoil from erosion by winter rains by planting crops whose roots hold valuable soil in place. Organic farmers often take that practice a few steps farther, using crops that fix nitrogen or provide many hundreds of pounds of organic matter for the topsoil. These "green manure" crops can be planted and tended economically with readily available materials and equipment. The techniques are as many and as varied as the crops that can make up the "manure." Variable crop blends are used to achieve different goals for different sites and soils. Some add significant amounts of nitrogen, others feed the microbial life of the soil. There are some excellent resources on the general characteristics of cover crops (Sattell, 1998; Sullivan and Diver, 2001; UC SAREP Cover Crop Database, 2002) and their specific use in vineyards (Ingels et al., 1998) that provide insights into potential crops and their management.

Compost. The application of compost is traditionally considered the organic grower's best means of "giving back" to a healthy soil. There are several organically approved fertilizers that address both specific and general soil nutrient deficiencies. But compost encompasses a time-honored holistic approach by which winegrowers have replaced the nutrients taken from the vineyard in the form of fruit at harvest. It is common to return the pomace (seeds and skin) by-product from winemaking back to the vineyard. But few vineyards today have their own source of livestock manure, which historically provided the raw material for homegrown compost.

The National Organic Standard defines compost as plant and animal material with a carbon–nitrogen ratio of between 25:1 and 40:1 produced in a static aerated pile system that maintains a temperature between 131°F and 170°F for three days, or material that was produced in a windrow composting system in which the pile maintains those temperatures for a minimum of fifteen days and has been turned at least five times. Any other material containing animal waste is considered raw manure and is subject to strict limits on use; for example, manure products cannot be applied in an organic vineyard within ninety days of harvest. Sewage sludge (biosolids) is specifically prohibited in the federal organic program.

Commercially produced compost from municipal waste (leaves, grass cuttings, Christmas trees, etc.) is becoming more widely available, and growers have reported some positive results in incorporating these materials into vineyard soils in Oregon. But be aware that transportation costs can be prohibitive. It is also critical to have commercial compost analyzed by an appropriate laboratory to define its nutritional and organic matter content and to detect any undesirable chemical compounds. Certification programs usually require that all off-farm products be analyzed and found to be within approved standards. Refer to the policy and procedures manual of the certifying agency before any composting project is undertaken.

Compost must be applied in large amounts to have a significant impact on hillside soils. Select a spreader with care to match the hopper size to the tractor horsepower, vineyard layout, and terrain. Often, refilling the spreader requires many trips, shuttling material from a central pile into the vine rows. Consider also that, for optimum results, growers probably need a long-term (five to ten years) composting plan. Several composting resources are available for further information on the subject: Cornell Composting, 2002; Rynk, 1992; Sustainable Agricultural Systems Laboratory, 2002.

Weed Control

An organic vineyard's aesthetic and goal regarding weeds differ from those of many conventional vineyards. It is not to conquer all the other lifeforms surrounding the vines. The objective is to control and manage them and, if possible, to use them to the vineyard's advantage. In the high-rainfall regions of western Oregon, the struggle to control weeds and native grasses is perhaps as great a challenge as in any growing region on earth. For an organic grower, weeds require a multifaceted approach that demands greater attention to selection of the right equipment and a larger allocation of time than in a conventionally farmed vineyard. Therefore, cultivation costs are generally higher for organic vineyards.

The vineyard alleyways can be managed with standard equipment: flail mower, rotovator, or disc. But the in-row area requires special handling. If left unmanaged, the in-row strip fills with broadleaf weeds, blackberries, and tall grasses that grow into the vine canopy and fruit zone. Abundant in-row weed growth inhibits airflow beneath the canopy, increasing the risk of frost and providing more favorable conditions for grape fungal diseases.

Over the years at Brick House Vineyard, the first author has employed many different methods and materials to manage in-row weeds. These are briefly described below, along with some of the limitations of each method. Be aware that not all are appropriate for all vineyards or, indeed, for the same vineyard in all years.

Black Plastic Mulch. The use of black plastic mulch in the vine row continues to be a widely practiced technique in many new vineyards, both organic and conventional. The plastic strip provides an effective weed barrier in the first year of use and helps retain soil moisture around the roots of young vines. Weed control on the outer margins of the plastic strip is, however, a particular problem. In addition, discing or tilling the alleyway usually throws soil up on top of the plastic, providing a place for grasses to grow and greatly complicating removal of the plastic after a year or two. The plastic strip also rules out the possibility of cross-tilling to control weeds in a newly planted vineyard before the first trellis wire is installed.

Sawdust Mulch. Sawdust mulch can be an effective approach to weed control in a newly planted organic vineyard. In 1995, the first author planted 11 acres of grafted vines using alder sawdust mulch. A 3–4-inch thick circle of sawdust was applied around each small plant. By August, the clay soil beneath the mulch was still moist and soft enough to penetrate with an index finger. The sawdust was good at inhibiting grass and weed growth around the plants. Cross-tillage with a narrow gauge tiller was possible in the first year because the mulch surrounded only the vines. Sawdust also eventually contributes to the organic matter in the soil. Another advantage of sawdust over black plastic is that there is no removal cost. But this method is generally most appropriate only in the planting year. Once the trellis is installed, precluding cross-tillage, it would be necessary to apply an expensive, continuous strip of sawdust to manage the entire in-row area.

In-row Mowers. Several types of in-row mowers are currently available. The fundamental design of these machines provides one or more retractable cutting heads, positioned outside the wheelbase of the tractor, which move in and out between the vines. Their effectiveness varies greatly depending on the design and configuration of a vineyard. These mowers were primarily developed for use in orchards, where the spaces between trees is much greater than the typical vine spacing in a vineyard. They are especially limited in close-spaced plantings. The mower head has little room to maneuver between vines planted at less than four-foot intervals. They can also be tricky to operate on sidehill slopes. It is highly recommended that you test drive one before you buy.

Weed Eaters. Yes, the garden-variety weed eater can be an effective way to clear out the strip beneath the wire. Sometimes it is the last resort when bermudagrass gets completely out of hand. Its superior maneuverability is obvious, but the labor cost for larger vineyards can be prohibitive. Labor costs can be reduced, however, by using a conventional tractor-drawn mower to cut the outer edges of the in-row area, leaving only a narrow strip to be mowed with the weed eater.

In-row Cultivators. These have been around for many years, and at least a half-dozen brands are commercially available. The basic concept is a mechanical hoe that slips between the vines and just under the surface of the soil, cutting weeds and grasses until it reaches a grape plant. The blade is quickly retracted when an aboveground sensor touches the vine trunk. The concept is nearly ideal, but all of these devices have their shortcomings. Grasses can be a particular problem; if cultivation is delayed too long, the grasses can be tall and thick enough to trip the sensor, rendering in-row cultivation an exercise in futility. Like the in-row mowers, it is best to test drive one before you make the decision to purchase.

Propane Burners. A fairly recent development has been the modification of propane burners as a vineyard tool for in-row weed control. Vineyard weed burners are mounted on a tractor's three-point hitch and aimed directly at the base of the vines. As the tractor passes up the row, the heat explodes the cell walls of the green growing plants. Woody tissues of the grapevine trunk are unaffected. Within a few minutes, burned plants show obvious damage, and two or three days later the strip appears as brown and lifeless as a conventional vineyard after treatment with herbicide. The burners are effective at killing a variety of troublesome broadleaf weeds, notably blackberries, field bindweed, and poison oak. They are also good for suppressing grasses, but the effect is limited to top burn. Grass roots remain undamaged, so growth is stunted for a time, as if the in-row strip had been mowed.

Several versions of weed burners have been adapted for the western Oregon climate, which is an important consideration. The local models are designed to work within the typical temperature and humidity range of our vineyards. Growers report the most success when the burners are set to ride 6–10 inches above the ground, high enough to aim down from above the tops of weeds. A common strategy is to run a "burn and return" schedule, making four or five passes through the vineyard from early spring until the grapevine canopy is formed. A burner with four jets (two on each side) supplied by a 100-gallon tank can treat about 14 acres of in-row strips when operated at 20 pounds per square inch at a speed of about 3 miles per hour.

Buds and young, green grape shoots are surprisingly difficult to damage with heat. Some growers who hoped to suppress sucker growth on their grapevines have been disappointed by their persistence, even after multiple passes. But precautions are in order once a canopy begins to fill out. Growers have reported damage to young grape leaves when the canopy traps excess heat from the burners as it rises from the in-row strip. Additionally, there is a significant fire hazard if burners are used later in the season when the soil has dried and there is an abundance of dry plant material.

Disease Control

Disease control in other organic crop production systems often begins with the use of resistant varieties. Unfortunately, the most valued winegrape varieties grown in Oregon are highly susceptible to several common diseases. Powdery mildew and botrytis bunch rot are the principal fungal diseases of concern here, and conditions for infection are frequently ideal in much of western Oregon. The uninitiated may think that organic practices preclude the use of all sprays to control diseases. But, in fact, no other piece of equipment is as critical to successful organic grape production as the vineyard sprayer, particularly in a moist, cool climate.

Most of today's new agricultural chemicals are "synthetically compounded" and thus outside the realm of products approved for certified organic growing. For disease control sprays, the organic grower must rely on "elemental substances" to protect the crop. Fortunately, elemental sulfur provides a highly effective control for powdery mildew. Recommendations for the use of sulfur are provided in the annual "Pest Management Guide for Wine Grapes in Oregon" (e.g., Connelly et al., 2002), published by the Oregon State University Extension Service. Powdery mildew control efforts must start early, and the first sulfur spray is recommended at about the 6-inch shoot stage. The Guide provides general recommendations for sulfur applications: short, 7-day intervals during the critical bloom stage, 10–14-day intervals at other times. Weather conditions and an abundance of powdery mildew spores at bloom time make it a critical period for controlling the disease.

Timing, along with good coverage, are the most important factors influencing the effectiveness of sulfur for powdery mildew control, but this cannot be accomplished by a preordained schedule. The grower must integrate many dynamic factors when making the decision to spray. Certainly weather significantly influences both the longevity and effectiveness of sulfur. Similarly, the grower must understand the life cycle of powdery mildew and the weather conditions that favor infection and disease development; see chapter 23 for this discussion. Finally, the vine growth stage, duration since the last spray, upcoming vineyard activities, and other variables must be factored into the decision of when to spray.

Much can be done to control botrytis bunch rot through canopy management techniques. Pulling leaves in the fruit zone exposes clusters to sunlight and air movement, thereby facilitating rapid drying after a rain. Other practices that contribute to an open, well-exposed canopy, such as shoot thinning, shoot positioning, and tipping, also makes the canopy microclimate less favorable for development of bunch rot. Several approved spray materials, including copper and certain oils, can provide slight control, but they are most effectively used in combination with canopy management practices.

Conclusion

The National Organic Standard promises to raise the level of quality control for organic products to a new high. While debate on precise products and practices may continue for years to come, the fact that a common, enforceable set of guidelines is now in place will contribute to the integrity of the organic claim.

Whether or not certification presents any clear price or profit advantage for organic growers remains an open question. It is most often the case that the organic approach is a matter of personal preference rather than a business/bottom line strategy. That being said, it appears that, although planting and managing an organic vineyard require greater labor input, overall long-term costs are generally comparable with conventional practices. An organic operation requires less fertilizer, chemicals, and the incumbent equipment to handle synthetic inputs safely. Furthermore, market research indicates a growing demand for organic products.

There is also little doubt that planting and managing an organic vineyard in Oregon are getting easier. The growth of organic farming has stimulated development of new technologies designed specifically for the practices employed by organic growers. It is worth noting that many of these are imports from Europe and other countries where organic products are in even greater demand. As imports, these technologies often sell at a premium and can be expensive to maintain. The release of several organically approved mildew eradicants has been a key development of the past few years. The development of many new, approved foliar amendments and spray technologies gives the organic grower several choices in designing a management program. New equipment and products have given organic growers more alternatives than ever before.

Ten years ago, efforts to grow commercial quantities of premium-quality wine grapes on a strict organic regimen in the Willamette Valley were widely considered to be fraught with risk and of questionable long-term viability. But the consistent successes of a handful of certified organic grape growers over the subsequent decade have largely put these doubts to rest. Even in very cold, wet seasons, appropriate applications of organically approved fungicides like elemental sulfur have proved to be sufficient to consistently suppress disease.

But this is not to say that all the best new products and equipment ensure success. The organic approach hinges on a series of decisions that may be totally site specific and beyond the realm of any simple recipes or prescriptions. Careful attention to the precise mesoclimate of a vineyard site and diligence in the timing of applications are of the utmost importance. Organic vineyard management in a damp spring or early summer in the Willamette Valley benefits greatly from the sustained presence of the owner or manager in the vineyard. Rainfall patterns can vary greatly from one side of a ridge or valley, especially in the high disease pressure month of June. Knowing precisely how much rain fell at what interval in a spray program intended to protect a rapidly growing canopy can be the deciding factor between success and failure. In the organic approach, timing is especially critical.

Although advances in equipment and materials have provided organic growers more alternatives than ever before, there is a distinct lack of independent research on these new methods and products. When it comes to how well organically approved products work in our specific climate, far too much is left to the trial and error of commercial growers, vineyard managers, and the claims of manufacturers. Continued advancement of organic viticulture in Oregon would be greatly facilitated by a sustained, systematic research program specifically designed to provide practical guidance to growers.

References and Additional Resources

Ames, G. 1999. *Organic Grape Production.* Appropriate Technology Transfer for Rural Areas, Fayetteville, Ark.

Connelly, A., J. DeFrancesco, G. C. Fisher, J. W. Pscheidt, M. C. Vasconcelos, and R. D. William. 2002. *2002 Pest Management Guide for Wine Grapes in Oregon.* Publication EM 8413, Oregon State University Extension Service, Corvallis.

Cornell Composting. 2002. Cornell Waste Management Institute, Cornell University. http://www.cfe.cornell.edu/compost/Composting_homepage.html

Ingels, C. A., R. L. Bugg, G. T. McGourty, and L. P. Christensen (eds.). 1998. *Cover Cropping in Vineyards: A Grower's Handbook.* University of California Publication 3338, Division of Agriculture and Natural Resources, Davis.

Lampkin, Nicolas, 1990, *Organic Farming.* Farming Press Books, Ipswich, United Kingdom.

National Organic Program. 2002. http://www.ams.usda.gov/nop

Natural Marketing Institute. 2001. *Organic Consumer Trends 2001.* Natural Marketing Institute, Harleysville, Penn.

Oregon Tilth. 2002. http://www.tilth.org

Rynk, R. (ed.). 1992. *On-Farm Composting Handbook.* Publication NRAES-54, Cornell University Cooperative Extension, Ithaca, N. Y.

Sattell, R. (ed.). 1998. *Using Cover Crops in Oregon.* Publication EM 8704, Oregon State University Extension Service, Corvallis.

Sullivan, P. 2001. *Sustainable Soil Management.* Appropriate Technology Transfer for Rural Areas, Fayetteville, Ark.

Sullivan, P., and S. Diver, 2001. *Overview of Cover Crops and Green Manures.* Appropriate Technology Transfer for Rural Areas, Fayetteville, Ark.

Sustainable Agricultural Systems Laboratory. 2002. http://www.barc.usda.gov/anri/sasl/sasl.html

UC SAREP Cover Crop Database. 2002. http://www.sarep.ucdavis.edu/ccrop/

18

Soil Management and Grapevine Nutrition

Alan Campbell and Daniel Fey

Across Oregon and around the world, grapevines are grown in a wide variety of soils. Each soil has its own distinctive combination of properties that influence vine growth and the resulting qualities of fruit and wine. Vineyard soil management begins with understanding the important properties of vineyard soils. These properties include soil texture, structure, depth, drainage, organic matter, organisms, water- and nutrient-holding capacities, pH, nutrients, and fertility.

An understanding of soil properties is important in choosing and developing a vineyard site, as discussed in earlier chapters. In established vineyards, soil properties are used and managed to optimize vine growth and fruit development for consistent production of high-quality fruit, appropriate to the philosophy, resources, and goals of the grower.

Grapevines are less nutritionally demanding on soils than most horticultural crops, but deficiency or excess of any nutrient can alter vine growth and the ability to develop and mature good fruit. Vineyard nutrition management should complement the ability of a vineyard soil to provide nutrients and water to the vine, so that the vine can develop appropriately and ripen an optimal crop. Vineyard fertility can be managed to stimulate or limit grapevine vigor. This should be done in balance with decisions regarding grapevine training and canopy and crop management, discussed in other chapters. Together these management activities produce a healthy, balanced vineyard, capable of producing fruit of consistent quality, representative of the vineyard site.

This chapter is divided into four sections. In the first section, important properties of Oregon vineyard soils are outlined, with examples from soil series in each region of the state. Vineyard soil management is discussed in the second section. Soil fertility and grapevine nutrition are reviewed in the third. The fourth section contains a brief perspective on the place of soil and nutrition in vineyard management.

Important Properties of Vineyard Soils

Management of soils begins with an understanding of what they are, how they are formed, and how they function. Soils are complex ecosystems composed of both living and non-living components. The living components include bacteria, fungi, other microbes, and a variety of animals and plants. The non-living components include visible and microscopic rock fragments (sand and silt), smaller minerals (clays), new and old organic matter, water, and air. Soils are formed in place from parent materials of igneous, sedimentary, or metamorphic rocks or of mixtures of these brought from elsewhere by water, wind, or gravity.

Soils develop into characteristic sequences of horizontal layers (horizons) through the weathering action of living organisms and physical processes over long spans of time (Figure 1). The one or more upper, or "A," horizons are distinguished by accumulations of partially decomposed (active) organic matter. The one or more middle, or "B," horizons are enriched by organic and inorganic materials from above. The lower, or "C," horizon is much less altered by biological or weathering processes, but roots may grow into it from above.

Due to differences in parent material, climate, organisms, topography, and time, more than 800 distinct soils series are officially recognized in Oregon. As the grape-growing industry has expanded in western, southern, and eastern Oregon, vineyards have been planted and maintained on more than 60 of these soils, and we have begun to understand the advantages and challenges of each for producing grapes and wines. These vineyard soils are distinguished by their place in the landscape and the depth and properties of their sequence of horizons, known as their profile. Many properties of soil horizons can affect the yearly cycle of grapevine growth and fruit maturation; these are reviewed below individually. The text by Brady and Weil (2002) is a helpful introduction to the prop-erties of soils

Soil Texture and Weathering

The texture of each soil horizon is defined by the relative amounts of the three sizes of inorganic particles (sand, silt, and clay) and is named using the soil texture triangle (Figure 2). Clay particles are the

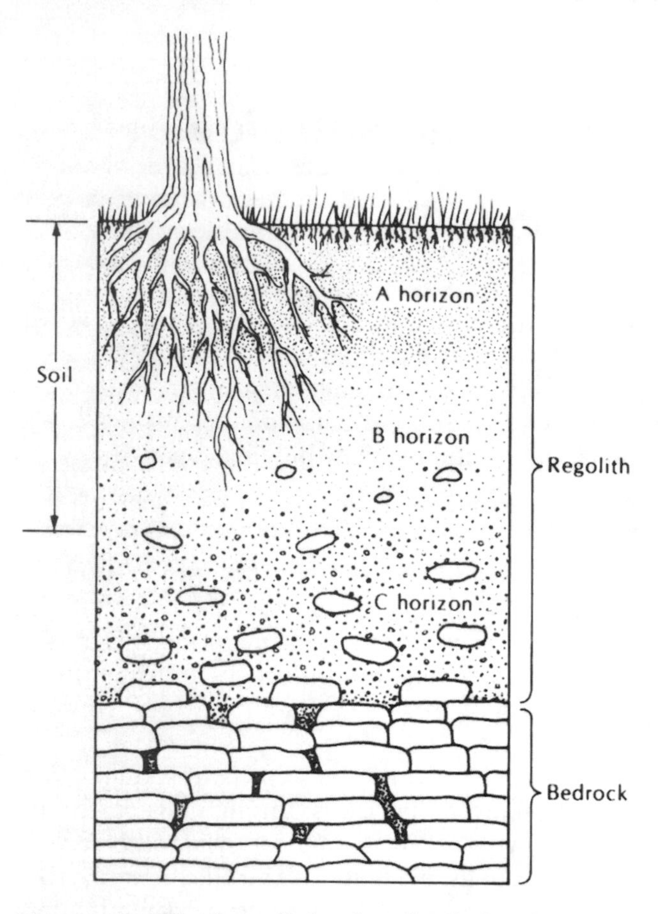

Figure 1. Profile of a well-developed soil, showing a typical sequence of horizons: the upper, A horizons distinguished by accumulations of partially decomposed organic matter; the middle, B horizons enriched by organic and inorganic materials from above; and the lower C horizon much less altered by biological or weathering processes. Below the regolith is the parent material and bedrock, from which the overlying soil has developed. Adapted from Brady (1990, fig. 1.3).

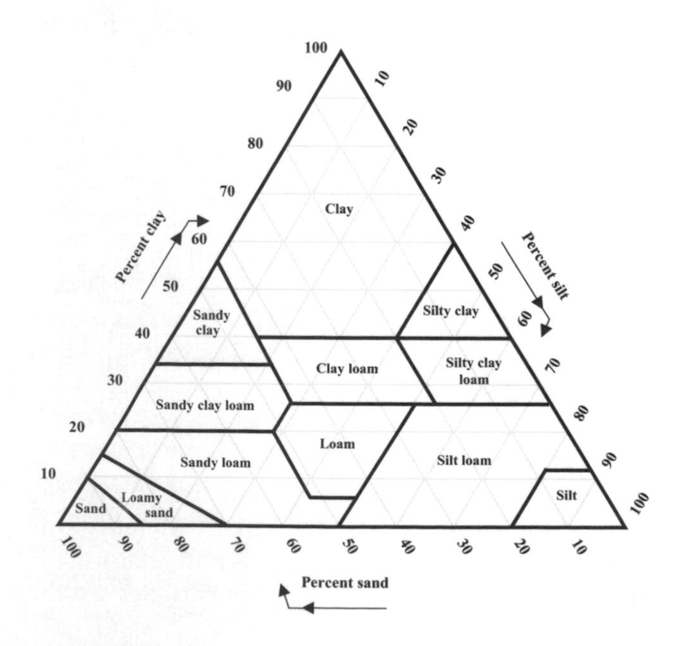

smallest in size, less than 0.002 mm in diameter, silt particles are between 0.002 and 0.02 mm in diameter, and sand particles are between 0.02 and 2 mm in diameter. Only sand particles are large enough to be seen individually. In the field, soil texture can be determined by feel, rubbing a moist soil sample between the thumb and fingers. Formation of a continuous ribbon indicates clay content. A smooth feel, with little stick or grit, indicates silt. A gritty feel indicates sand. A loam contains equal contributions from each size class. The amounts of sand, silt, and clay in each horizon, combined with organic matter, determine many of the viticultural properties of a soil discussed below.

Soil texture reflects both parent material and soil formation processes, such as weathering. Although parent materials vary, vineyards on stable hillsides in the Willamette, Umpqua, and Rogue Valleys are typically on older, well-weathered surfaces where clay formation has led to finer-textured topsoils and clay-enriched subsoils with reduced fertility. These more weathered soils, the *alfisols* and *ultisols*, are typically lower in nutrients and fertility than younger soils. Examples of well-weathered vineyard soils (with their surface horizon textures) include the Bellpine (silty clay loam), Jory (silty clay loam), and Willakenzie (silt loam) series in the Willamette, the Bateman (silt loam), Jory (silty clay loam), and Rosehaven (loam) series in the Umpqua, the Abegg (gravelly loam), Manita (loam), and Ruch (gravelly silt loam) series in the Rogue, and the Cherryhill (silt loam) and Oak Grove (loam) soils along the Columbia; see Table 1.

Many vineyards on hillsides and lower terraces and valley floors grow in younger, less-weathered soils, where clay, silt, and sand content and fertility reflect the composition of the parent materials. These soils with dark surface horizons, rich in organic matter, the *mollisols*, are typically higher in nutrients and fertility. Examples of these mollisol vineyard soils include Philomath (silty clay), Woodburn (silt loam), and Yamhill (silt loam) in the Willamette, Evans (loam) and Newberg (fine sandy loam) in the Umpqua, Carney (clay), Central Point (sandy loam), and Takilma (loam) series in the Rogue, Chenowith (loam) and Winchester (sand) along the Columbia, and Walla Walla in the Walla Walla AVA.

Soil texture and degree of weathering cannot be changed by cultural practices, but good soil man-

Figure 2. Soil texture triangle. Soil textural classes (outlined in bold lines) are defined by percentage of sand, silt, and clay (fine lines parallel to arrows). An average "loam" soil, for example, is composed of approximately 40% sand, 40% silt, and 20% clay particles. Adapted from Brady (1990, fig. 4.6).

Table 1. Properties of Common Vineyard Soils in Oregon's AVAs

	Parent Material	Soil Order	Typical Depth[a] inches	Drainage[b]	Topsoil Texture[c]	Subsoil Texture[c,d]	Topsoil pH[e]	Subsoil pH[f]	Water Capacity[g] inches	Relative Productivity[h] dry (irrigated)
Willamette Valley										
Jory	Basalt	Ultisol	60	wd	SiClLo	Cl	5.6	5.1	10.3	medium
Bellpine	Sedimentary	Ultisol	32	wd	SiClLo	Cl	5.6	5.2	5.3	low
Willakenzie	Sandstone	Alfisol	36	wd	SiLo	SiClLo	5.7	5.4	6.5	low
Woodburn	Lacustrine	Mollisol	65	mwd	SiLo	SiLo	5.9	5.8	13.0	high
Umpqua Valley										
Jory	Basalt	Ultisol	60	wd	SiClLo	Cl	5.6	5.1	10.3	medium
Bateman	Siltstone	Alfisol	63	wd	SiLo	SiClLo	5.6	5.2	11.5	medium
Evans	Alluvium	Mollisol	63	wd	Lo	Lo	6.2	6.4	11.3	med (high)
Rogue Valley										
Pollard	Mixed	Ultisol	61	wd	Lo	ClLo	5.6	5.4	6.9	low (high)
Manita	Mixed	Alfisol	58	wd	Lo	ClLo	6.2	5.7	9.8	low (high)
Central Point	Alluvium	Mollisol	67	wd	SaLo	SaLo	6.2	6.6	6.5	med (high)
Columbia Valley										
Oak Grove	Alluvium	Alfisol	78	wd	Lo	Cl	5.8	5.4	12.4	med (high)
Cherryhill	Sandstone	Alfisol	51	wd	SiLo	SaClLo	6.5	5.6	4.2	low (high)
Chenowith	Alluvium	Mollisol	88	wd	Lo	Lo	6.7	6.6	12.0	low (high)
Walla Walla	Loess	Mollisol	60	wd	SiLo	SiLo	6.8	8.8	11.2	low (high)[a]

[a] National Soils Information System, SSURGO 2.0 Database, Oregon Soil Series Data by county.

[b] wd = well drained, mwd = moderately well drained; from Official Soil Series Description, NRCS

[c] Sa = sandy, Si = silty, Cl = clay, Lo = Loam; from Official Soil Series Description, NRCS

[d] Finest texture of typical B horizons (C horizon for Evans series), from Official Soil Series Description, NRCS.

[e] Average value of typical A horizons, from Official Soil Series Description, NRCS.

[f] Extreme pH of typical B horizons (C horizon for Evans series), from Official Soil Series Description, NRCS.

[g] Calculated with values from Oregon Soil Series data by county, SSURGO 2.0 Database, National Soils Information System.

[h] Calculated from "component crop yield" table in Soil Survey for each county.

agement can increase soil organic matter, nutrients, and fertility and improve soil structure and tilth.

Soil Structure and Porosity

Soil structure and porosity are major factors in determining the distribution of grapevine roots within each horizon and down through the soil profile. Soil structure is created by the binding of sand, silt, and clay particles into aggregates by soil organic matter, with pores formed in the space between aggregates (Figure 3). Soil structure differs among horizons and soils due to differences in texture, soil organic matter, organisms, and soil formation processes. Structural aggregates used to characterize soil horizons vary in size and shape from 1/8-inch granules to plates, blocks, prisms, and columns inches across, and they may change for each horizon in a profile.

The size and extent of soil pores (soil porosity) determines the movement of water and air through the soil and so effects water-holding capacity, drainage, and potential for erosion. A well-structured soil horizon has a mixture of large and small aggregates, creating a network of large and small pores. Large pores permit oxygen diffusion, water movement, and root entry; small pores, and the aggregates themselves, store water and nutrients for root uptake. The distribution of pores sizes determines the ease of grape root growth through each horizon and the depth of grapevine rooting.

Maintenance of good soil structure and porosity is important in vineyard soils, especially in high clay and hillside soils. Jory (Willamette), Bateman (Umpqua), and Pollard (Rogue) series are examples of high clay content soils where high organic matter content helps form a strong granular structure that results in rapid water infiltration, good drainage, high water-holding capacity, deep grapevine root growth, and stable hillsides.

Soil Organic Matter

Soil organic matter benefits vineyard soils and vines in a variety of ways. Organic matter provides energy and nutrients to microbes and earthworms, and, through their activity, recycles nutrients to the soil,

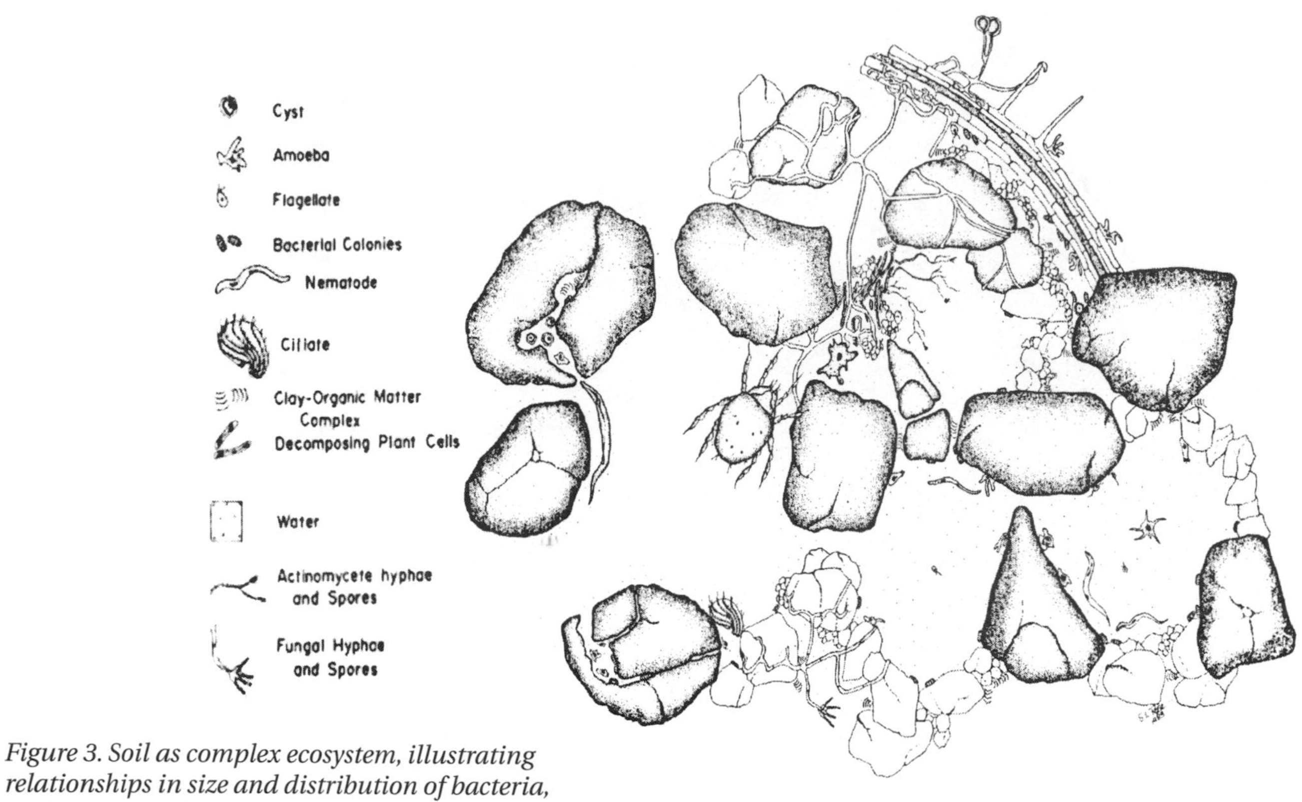

Figure 3. Soil as complex ecosystem, illustrating relationships in size and distribution of bacteria, actinomycetes, fungi, invertebrates, plant roots, organic matter, and clay, silt, and sand particles in a soil. Drawing by Sharon Rose.

forms and stabilizes soil aggregates, increases nutrient-holding and water-holding capacity, reduces surface crusting and erosion, and lowers bulk density. Maintaining soil organic matter is important to preserving the productivity and quality of vineyard soils.

Soil organic matter can be divided into three categories based on the relative susceptibility to microbial decay: active, slowly-cycling, and passive. Active organic matter is readily decayed, with a residence time of a few months to years, and consists mainly of high carbon–nitrogen ratio materials (15–30) that provide most of the nutrients to microbes and temporary benefits of coarse soil structure (macro-aggregates), water retention, and fertility. Slowly cycling organic matter has a residence time of a few to tens of years and consists mainly of finely divided plant tissues high in lignin; this is an important source of mineralizable nitrogen, phosphorus, and other nutrients for slow-growing fungi. Passive organic matter has a residence time of hundreds to thousands of years and includes humus and clay-humus complexes; it is responsible for most organic matter contributions to fine soil structure (microaggregates) and nutrient- and water-holding capacity. Soil organic matter decreases with depth and changes in composition from mostly active to slowly cycling to passive.

Soil organic matter differs among vineyards due to differences in texture, drainage, history of use, and soil organisms. Soils with finer textures tend to be higher in organic matter than coarser soils because of the higher organic matter input from higher productivity, slower organic matter decomposition related to poorer aeration, and greater organic matter protection from microbes through the formation of clay-humus complexes. Soils with poorer drainage tend to retain more organic matter than better drained soils. However, soils that are more frequently and deeply cultivated and from which more crop has been removed have less organic matter. Jory soils in the Umpqua and Willamette Valleys and Pollard soils in the Rogue are examples of fine-textured (silty clay loam), well-drained soils with high organic matter content preserved in clay-humus complexes to considerable depths. Jory soils may contain as much as 200,000 pounds of organic matter per acre.

Soil Permeability and Drainage

Adequate drainage is an important property of a vineyard soil. Grape roots die back if soil remains water-saturated following winter rains, and new root growth is inhibited by excessive soil water in the spring. Vines with root systems thus restricted to surface horizons are more stressed by summer drought and more stimulated by late summer rains. Good drainage is also important on hillsides, where poor water infiltration and drainage can result in severe erosion from heavy rains.

The permeability of each soil horizon in a profile determines how rapidly water drains. Soil permeability is determined by the size of soil pores, with large pores (>0.5 mm) allowing rapid movement and smaller pores (<0.1 mm) inhibiting movement.

Over time, poor drainage creates characteristic colors in a horizon that can be seen at any time of the year, such as mottles of yellow or gray due to local deposits and depletions of iron and manganese. Soil series are divided into drainage classes based on the depth to these redoximorphic features: well drained (>36 inches), moderately well drained (18–36), somewhat poorly drained (12–18), poorly drained (6–12), and very poorly drained (<6).

Most Oregon vineyards have been planted in well-drained and moderately well-drained hillside soils (Table 1). Soils with reduced drainage are more common in western Oregon, where they form from fine sedimentary rock and valley floor alluvium. Examples of moderately well-drained or somewhat poorly drained soils, planted to vineyards, include the Dupee, Hazelair, Helmick, and Woodburn series in the Willamette, and the Brockman, Carney, Coker, and Medford series in the Rogue. Drainage of these series may be improved by tiling and water diversions, and sometimes by ripping, the use of deep-rooted cover crops, or incorporation of organic matter by earthworms. Some soil series that have been avoided for vineyard development because of poor drainage include Bashaw, Hazelair, Panther, Waldo, and Wapato.

Coarse textured soils with high permeability and excessive drainage are also found in all viticultural regions of Oregon, but these are most common in the south and east. These soils formed from coarse parent materials deposited by water in floodplains, such as the Camas soils in the Rogue or Newberg soils in the Rogue, Umpqua, and Willamette Valleys, or by wind, such as the Winchester soils along the Columbia, or by weathering of coarse granitic rock, such as the Siskiyou soils in the Rogue. These soils have low water-holding capacity and require irrigation for satisfactory grape production.

Soil Depth, Rooting Volume, and Available Water

Soil water and nutrients are supplied to the grapevine through a branching network of roots distributed among the soil horizons. The depth and volume of the rooting network and the properties of each horizon determine both the water and nutrient reserves available to the vine. The pattern of water and nutrient supply provided by a soil through the growing season is a critical factor in controlling vine growth and fruit quality.

Most grape roots (and most nutrients) are found within the top 2 feet of soil, in the A and B horizons. But grape roots explore for water and nutrients to greater depths as a vineyard becomes established. Where conditions allow, grape roots can grow through the B horizons into the C horizon, parent material, or fractured bedrock beneath, to depths of 10 feet or more.

Soil series are assigned to depth classes based on the distance to the bottom of the B horizon. In a "shallow" soil, the B horizon ends within 20 inches of the soil surface; in a "moderately deep" soil, between 20 and 40 inches; in a "deep" soil, between 40 and 60 inches; and in a "very deep" soil, more than 60 inches from the surface. In the Willamette Valley, for example, Witzel, Nekia, Gelderman, and Jory, respectively, are shallow, moderately deep, deep, and very deep soils, all formed from weathered basalt.

Available water-holding capacity (AWHC) is a measure of the water held in the A and B horizons of the soil profile between field capacity and permanent wilting point. An estimate of water available for vine growth can be made by summing the AWHCs of each horizon explored by grape roots (Table 1). This underestimates available water in vineyards where roots extend into the C horizon or into fractured bedrock beneath.

Mature vines with deep roots in a large volume of soil have a larger annual reserve to supply needed water and nutrients and result in more consistent vine development and fruit quality. Effects of vintage are less. Vines with shallower root systems, whatever the cause, occupy a smaller volume of soil with lesser reserves and are more affected by annual variations in rainfall and water use. Effects of vintage are greater. Within this context, the texture and structure of each soil horizon can also have an important impact on rooting density and the rate and pattern of water movement into the vine.

Soil management practices can alter soil organic matter content, structure, and porosity, as well as soil fertility and water-holding capacity. These in turn influence the distribution of grape roots and the growth and development of vines and fruit.

CEC, Base Saturation, and Soil pH

Cation exchange capacity (CEC) is an important measure of a soil horizon's potential fertility. CEC is the ability of a soil to hold positively charged nutrients (cations), such as potassium, calcium, magnesium, and ammonium, as well as the non-nutritive cations sodium, aluminum, and hydrogen. Clays and organic matter (humus) are the main contributors to CEC, with less-weathered clays making greater contributions. CEC is typically measured in milliequivalents per 100 grams of soil (meq/100g) and varies fivefold, from 8 to 40 meq/100g, among Oregon's vineyard soils. The contribution of organic matter and highly weathered clays to CEC changes with pH, increasing as pH increases.

Base saturation is the percentage of a horizon's CEC that is actually occupied by base-forming cations (K^+, Ca^{++}, Mg^{++}, and Na^+); the remaining CEC is occupied by acidic cations (H^+ or Al^{+++}). Base saturation is a useful indicator of a soil's current fertility and its history.

The pH of a soil is a measure of the relative abundance of hydrogen ions in the soil water and reflects the balance of acidic cations and base-forming cations on the cation exchange sites. Soil pH is directly related to base saturation; as base saturation increases, so does soil pH. In Oregon vineyards, surface soil pHs range from 5.2 to 7.8 in the surface horizons, reflecting base saturations from 25% to 95%, and subsoil pHs range from 4.5 to 8.8, with base saturations from 15% to 100%. Lower pHs and base saturations are found in the wet Willamette Valley, and higher pHs and saturations in dry eastern Oregon (Table 1).

Weathering, Fertility, and Subsoil Nutrition

Cation exchange capacity, base saturation, and pH are useful tools for understanding a soil's fertility and the differences in fertility among Oregon's vineyard soils. Vineyard soils differ in their degree of weathering. In regions of high rainfall, weathering and leaching drain nutrients from a soil, and, though plants catch and return some nutrients to the surface horizons, over time the base saturation and pH of deeper horizons decrease (Table 1). In regions of low rainfall, weathering and drying move nutrients upward, and over time the base saturation and pH of surface horizons increase, and calcium carbonate may accumulate in subsurface horizons of high pH soils (pH 8.8 in Walla Walla).

The degree of weathering and leaching of a soil determines how nutrients, base saturation, and pH change with depth. In young, lightly weathered mollisols of western Oregon (such as Woodburn or Central Point), CEC, base saturation, and pH of horizons increase slightly with depth. In older, moderately weathered alfisols (such as Willakenzie or Ruch), CEC, base saturation, and pH of horizons decrease moderately with depth. In well-weathered ultisols (such as Jory or Bellpine), CEC, base saturation, and pH of horizons decrease significantly with depth. The base saturation of a typical mollisol is greater than 50% throughout its profile, whereas the base saturation of the B horizons of an alfisol are between 35% and 65%, and some of the B horizons of an ultisol are below 35%. Clearly, grapevines encounter very different chemical and nutritional environments as they grow deeper into these different soils.

Vineyard Soil Management

Vineyard soil management can be seen as a four-part process. Soil management begins with a map and an inventory of the important properties of each distinct soil area within a vineyard. Second, for each of these areas, management concerns are identified based on the soil properties, the grapes grown, the production goals, and the management philosophy of the grower. Third, management priorities are set and strategies and procedures are developed and carried out to meet specific goals. Fourth, the soil properties of concern are periodically reassessed to verify that the chosen management procedures are having the intended effect. Then the management cycle begins again.

The intrinsic properties of a vineyard soil such as soil texture and depth cannot be changed by cultural practices. But vineyard soil management should aim to maintain or improve the dynamic properties that are important to the productivity and quality of a vineyard soil. These include organic matter and organisms, porosity and drainage, appropriately balanced nutrient and water availability, structure and tilth, and other physical, chemical, and biological properties that encourage healthy grape roots and healthy vines. Vineyard soil management should also use strategies and procedures that are compatible with other vineyard operations and are appropriate to the philosophy, resources, and goals of the grower.

Mapping Vineyard Soils

Every vineyard should be mapped for the soil properties that are important to grape roots and vines. The map and description of soils found in each county's soil survey can provide a useful base for your vineyard map, though older maps are less useful for hillside soils. Soil survey information is available from the Natural Resources Conservation Service (the NRCS was formerly known as the Soil Conservation Service) of the USDA in printed form or via the Internet.

The base map information from the soil survey can be verified and refined by inspection of the actual soils. Professional services from certified soil classifiers are available for accurate classification and detailed mapping of soils and their critical properties.

Growers can also learn a great deal from systematic use of a 3 1/4-inch soil auger, a relatively inexpensive and convenient tool for examining soils to a 5-foot depth. With determination and practice, a soil auger can be used to inspect variations in the depth and properties of soil horizons across a vineyard. Intrinsic properties such as soil texture, structure, porosity, drainage, and permeability can be assessed with some learning and persistence. Many dynamic properties, such as recharge of soil water by winter rains and water use by cover crops and vines, can also be monitored with a soil auger.

From professional surveys or soil sampling, each grower should develop a series of maps of vineyard soils that shows variations in slope, soil texture, depth, drainage, water-holding capacity, and other important properties. These maps can be used to guide management decisions and to record management actions, such as areas sampled for soil and tissue tests, applications of fertilizer and amendments, or cover crops planted. These vineyard maps can be redrawn and modified as concerns, test results, and actions progress over time.

Managing Organic Matter

Maintaining an appropriate level of soil organic matter is important for maintaining the structure, fertility, and health of vineyard soils. This requires regular additions of organic matter to the soil, including on-site organic matter, such as prunings, leaves, and cover crops, and also imported organic matter, such as straw, wood chips, and composts.

Cover crops can add up to several thousand pounds (dry weight) of new organic matter per acre of vineyard soil each year, while cycling nutrients. Imported organic matter, such as straw, has been added in single applications to new or established vineyards at rates as high as 40 tons per acre. Compost has been added in the vine row at rates as high as 20 tons per acre. These additions raise the proportion of active organic matter, feed soil organisms, enhance the macro-aggregate structure of the soil, and increase the total nutrient supply. A small fraction of active organic matter eventually becomes slow-cycling and passive organic matter, improving the fine, microaggregate structure of the soil and increasing its nutrient-holding capacity (CEC).

Reduced tillage helps build and maintain all categories of organic matter, compared to conventional tillage. Nitrogen additions help increase stable soil organic matter by increasing productivity and by supplying the 10% nitrogen characteristic of humus. Additions of calcium (as lime) also improve the conversion of organic matter to stable forms. In the end, the maximum level of organic matter easily maintained at a given site is determined by soil texture, organic matter and nutrient inputs, environmental factors, and management practices.

Soil organic matter levels and related properties should be measured periodically, in each management area, to verify that management practices are having their intended effects.

Managing Cover Crops

Cover crops are grown in vineyards for many reasons, including addition of soil organic matter, preservation and improvement of soil structure, improvement of soil fertility, and control of grapevine vigor, as discussed in other chapters. Cover crops have different impacts on these soil properties and on grapevine growth, depending on the cover crops grown and their management.

Legumes are grown to add nitrogen to vineyard soils. Legumes are composed of tissues high in nitrogen (carbon–nitrogen ratios below 20) that are broken down rapidly in the soil, increasing available nitrogen and soil organisms, but with less persistence as active organic matter. Grasses, because of their lower nitrogen content (carbon–nitrogen ratios 20–100), are broken down more slowly than legumes and generally reduce available nitrogen while increasing active organic matter and macro-aggregate formation. Different mixtures of cover crop plants with complementary characteristics can be used to match the needs of a particular vineyard area.

Mowing, flailing, and tilling of cover crops release organic matter for decay, both by organic matter added at the soil surface and through dieback of roots within the soil. Timing of these operations can be used to control the impact of cover crops on nutrient availability and grapevine growth. For example, whenever cover crops are cut green and left as mulch with carbon–nitrogen ratios below 25, soil organisms digest them more quickly, releasing their nitrogen and other nutrients more rapidly into the soil. When cover crops are allowed to mature before cutting, with carbon–nitrogen ratios above 25 and higher lignin content, soil organisms decay them more slowly, releasing their nutrients more slowly and binding available soil nitrogen during the decay process.

Because cover crops take up nutrients and water in competition with grapevines, they can be used to reduce grapevine vigor on overly fertile soils. However, too much competition results in insufficient water and nitrogen for vine growth and eventual reduction in grapevine reserves, vigor, and health. Application of nitrogen (and other nutrients) to meet the nutrient requirements of cover crops can reduce their nutritional impact on grapevines.

Cover crop management can be used to control when water competition occurs and when nutrients become available in the soil. For example, frequent mowing reduces both competition and organic matter contributions from cover crops and spreads nutrient release over time. Infrequent mowing increases competition but adds more organic matter in larger nutrient pulses.

Managing Soil Organisms

Fungi and Bacteria. Grapevines, like all plants, depend on soil fungi, bacteria, and other organisms to decay organic matter, supply nutrients, build soil structure, and create beneficial soil conditions for growth (Figure 3). In turn, these soil microorganisms are supported by organic matter entering the soil from decaying roots, leaves, and shoots of grapevines and other plants.

Thousands of species of bacteria and hundreds of species of fungi are found in a square inch of typical Oregon vineyard soil. These organisms of the soil food web are not uniformly distributed but concentrated in the top few inches of soil and wherever organic matter is abundant: around roots, in litter, on humus, and on and within soil aggregates. The activity of soil organisms follows seasonal patterns, as well as daily patterns, with greatest activity in late spring when organic matter, temperature, and moisture are optimal.

Populations of soil bacteria, fungi, and other organisms can change rapidly in response to changing soil conditions. Bacterial populations change fastest, consuming simpler organic compounds released in the early stages of decay. Their numbers are increased by incorporation of fresh organic matter. Fungal populations change more slowly, consuming more complex and partially decomposed organic materials. Their numbers are most responsive to additions of fibrous plant residues high in lignin, such as compost, wood, and bark. Through their combined activities, bacteria, fungi, and other soil organisms convert organic matter to stable forms, supply nutrients, build soil structure, and create beneficial soil conditions. The *Soil Biology Primer* (Tugel and Lewandowski, 1999) is a popular introduction to soil biology.

Several species of fungi common to vineyard soils also readily form mycorrhizal associations with grapevine roots. Mycorrhizae aid in the uptake of scarce and immobile nutrients, but because of their natural occurrence little benefit has been found with further addition of these fungi to new vines or established vineyard sites (Schreiner, 2001).

Populations of bacteria, fungi, and mycorrhizae and their combined respiration are among the soil properties that can be measured to monitor the effects of management practices on the health of soil organisms in the vineyard.

Earthworms. Earthworms are the most visible soil invertebrates, and their activities produce many visible benefits to vineyard soils. Through their feeding and burrowing, earthworms and other invertebrates incorporate surface organic matter, lime, and fertilizers, improve porosity and soil aeration, increase water infiltration, improve soil aggregate stability, enhance nutrient availability, encourage grape root growth, and reduce the incidence of some root diseases.

Maintaining healthy earthworm populations is especially important to ensure good drainage in hillside soils threatened by erosion. Anecic earthworms, such as nightcrawlers, form deep, permanent vertical burrows from which they emerge to gather surface litter and move it down into the soil for later consumption. These vertical channels speed the downward movement of water following heavy rains.

Deep-burrowing earthworms are not native to Oregon but can be introduced if populations are low. Abundance can be noted by middens of inedible plant material (including grape leaf petioles) piled around burrow openings on cleared ground, or by flooding soil with dilute mustard solutions. Populations are increased by the food and protection provided by cover crops (especially if high in nitrogen and cut green), by organic matter mulches, and by reduction in depth and frequency of tillage. Populations are decreased by frequent and deep tillage and by reduction of food and cover. Earthworms can be moved from areas of high to low abundance by collection and release or by transport in soil. Once established locally, earthworms multiply and spread with appropriate soil and cover crop management (Werner, 1990).

Managing Permeability and Compaction

Compaction of soils by tractor tires is a common feature in Oregon vineyards. Tractor traffic compacts soils as it breaks down soil aggregates, moves soil particles closer together, and reduces soil pore space. Ruts and standing water between vine rows are the most visible signs. Movement of water and air into the soil is restricted in compacted zones and can lead to surface run-off and erosion. Grape root growth is reduced in compacted soil, and biological communities are altered by anaerobic conditions that result from waterlogging.

Recent studies have shown that compaction is not an immediate concern for most grape growers in western Oregon (Baham, 2000). Initial compaction is limited to the top 6 inches of soil, so the volume of soil compacted is small relative to the total volume of soil in lower-density vineyards. Compaction is probably more of a problem in dense vineyards, where tire paths are more numerous and closer to the vines. Compaction can be detected by comparing resistance to insertion of a short metal probe in high- and low-traffic areas.

Soils are most susceptible to compaction by traffic when they are wetted to field capacity or beyond. Tractor traffic should be avoided in wet winter and spring months, and trips and loads should be minimized in all seasons. Use of large low-pressure tires or crawler treads can greatly reduce compaction.

Cover crops can be used to reduce the impact of traffic on the soil surface and to rebuild structure and permeability in lightly compacted soils. Heavily compacted soils can be broken up with spading or appropriate tillage in dry weather.

Managing Soil pH, Calcium, and Magnesium

Grapevines are tolerant of a wide range of soil pH. Yet attention to pH is important when managing soil nutrition and fertility, because pH reflects the underlying nutrient status of a soil and directly effects the availability of some nutrients and the activity of many organisms with consequences for soil fertility (Figure 4).

The pH is a measure of the balance of acidic elements (hydrogen and aluminum) and basic elements (magnesium, calcium, and potassium) in a soil. It is directly related to base saturation: soil pH decreases as calcium and magnesium are replaced by hydrogen or aluminum on cation exchange sites, and these nutrients are lost by leaching or crop removal. The pH of Oregon vineyard soils ranges from 5.2 to 7.8 in the surface horizons and from 4.5 to 8.8 in the subsurface horizons (Table 1). The lowest pHs are found in vineyards on the older, well-weathered hillside soils of the Willamette Valley, many of which produce high-quality fruit. It is on these soils where application of calcium and magnesium should be most carefully considered.

Application of lime to raise soil pH, increase calcium and magnesium content, increase nutrient availability, and improve soil tilth is a standard practice in agriculture. Many agronomists recommend adjustment of the pH of vineyard soils to above 6.0 by the addition of lime. Others recommend addition of nutrients so that 60–65% of a soil's total CEC is occupied by calcium, 10–20% by magnesium, and 3–7% by potassium, for a total base saturation of 80–90%. These are long accepted fertility adjustments for the management of annual crops on lowland, mollisol soils (such as Woodburn, Evans, or Takilma), where crop removal and fertilizer additions regularly reduce surface soil pH and the native pH is above 6.0 and natural base saturation above 80%. These targets for pH or base saturation adjustment may not always be justified in vineyards on well-weathered, and base-depleted, hillside ultisols and alfisols (such as Jory, Bellpine, and Willakenzie), where pH and base saturation are naturally low.

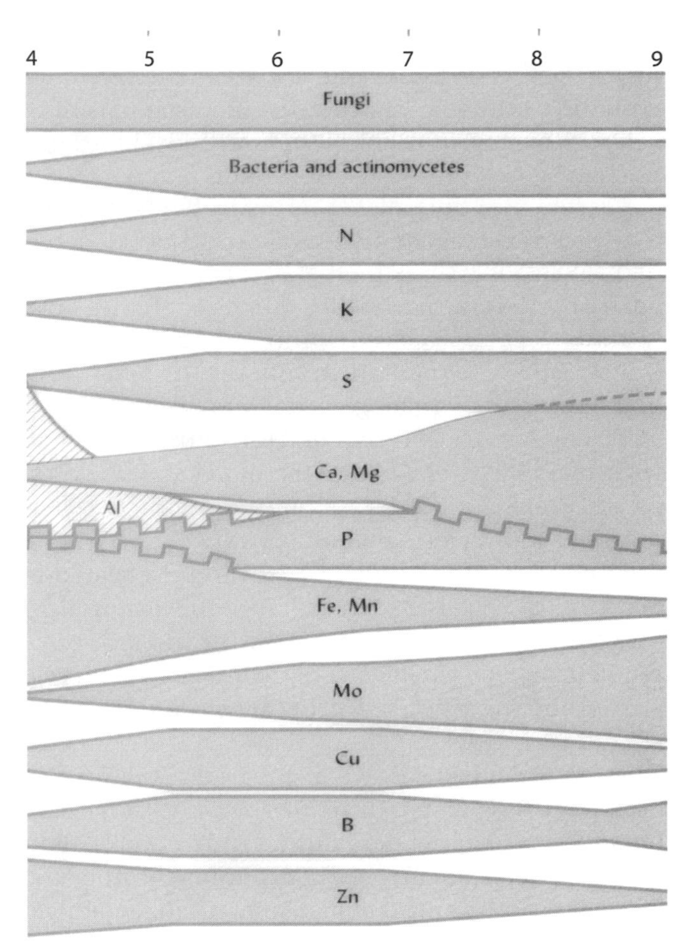

Figure 4. Relationships among soil pH, nutrient availability, and microorganism activity. The width of bands illustrates the relative availability of nutrients to plants and activity of microorganisms at different soil pHs. The notched lines between the phosphorus (P) band and the bands for aluminum (Al), iron (Fe), and calcium (Ca) represent reduced phosphorus availability at pHs below 5.5 and above 7.0, caused by binding with these metals. Adapted from Brady and Weil (2002, fig. 9.22).

Many hillside soils require a lot of lime to raise pH to above 6.0, or base saturation to over 80%, in the surface horizons, due to their high CEC and low base saturation. For example, to raise the calcium content of just the first foot of a typical Jory soil (with a pH of 5.6, a CEC of 30 meq, and a calcium saturation of 8 meq) to 60% calcium base saturation would require more than 10 tons of lime per acre. It is not clear whether the massive additions of lime, dolomite, and potash required to reach these targets would improve fruit quality and yield, and therefore be economically justified. They certainly would significantly alter the properties of the soils.

Liming is, however, often the most economical way to increase the fertility of many low pH soils. The availability of many nutrients that affect fertility is improved at more neutral pH levels (closer to 7), including nitrogen, phosphorus, potassium, sulfur, calcium, magnesium, boron, and molybdenum. Some nutrients, including phosphorus and potassium, are converted into unavailable forms by acidic soil and made more available by liming. A soil pH closer to neutral also increases the health of many beneficial soil organisms, including nitrogen-fixing and -recycling bacteria, earthworms, and other invertebrates. Liming to raise soil pH may be the best way to address these soil management concerns.

The amount of lime required for adjustment of the surface 6 inches of soil to a target pH can be determined from the SMP lime requirement test, often listed as "buffer pH" on soil tests (Hart, 1998). The amount of material applied needs to be adjusted to account for its calcium carbonate equivalent rating. The time required for reaction with the soil, once applied, depends on the degree of fineness or particle size of the material, indicated by sieve number or solubility rating.

Vineyard Nutrition Management

Grapevines are less demanding nutritionally than most horticultural crops. This is apparent from the range of soils on which they are grown and the general infrequency of deficiency symptoms in most Oregon vineyards.

A nutrient management program should complement the ability of each vineyard soil to provide nutrients and water to the vine, so that the vine can develop appropriately and mature an optimal crop. Nutrient elements should be applied only if the soil cannot supply them adequately to meet the management strategy of the grower.

Vineyard fertility can be managed to stimulate or limit grapevine vigor. Changes to vineyard fertility must be in balance with decisions regarding grapevine training, canopy management, and crop load, discussed in other chapters. Together these management activities can produce a balanced vineyard, capable of producing consistent crops of high-quality fruit representative of the vineyard site. Each grower needs to recognize when a limiting nutrient is beneficial or detrimental to production goals and vineyard health.

Essential Elements

Plant nutrients are the chemical elements that are essential to formation and function of normal plant tissues for both growth and reproduction. A grapevine requires 17 elements for normal growth. Three of these are captured from air and water through photosynthesis: carbon, oxygen, and hydrogen (with chemical symbols C, O, and H, respectively). The remaining 14 nutrients are absorbed through the root system from the soil. Six of these are called macronutrients because they are used in large quantities and have major structural or functional roles in the plant: nitrogen (N), phosphorus (P), potassium (K), calcium (Ca), magnesium (Mg), and sulfur (S). Eight are called micronutrients because they are required in much smaller quantities and have specific functions in individual enzymes or compounds: iron (Fe), manganese (Mn), boron (B), molybdenum (Mo), copper (Cu), zinc (Zn), nickel (Ni), and chlorine (Cl). Macronutrients each make up between 0.2–4% of a living tissue's dry weight; individual micronutrients are 0.5–150 parts per million of dry weight. A comprehensive introduction to plant nutrition has been published by Marschner (1995), and grapevine nutrition has been reviewed by Robinson (1992).

Limiting Nutrients

Grapevine development is limited if a nutrient is not available in sufficient quantities when and where it is required. Such deficiencies can result in characteristic changes in leaf or shoot color, shape, or growth, known as deficiency symptoms, although vine growth or fruit quality may be reduced before other symptoms become obvious (Gartel, 1993). In Oregon, deficiencies in nitrogen, potassium, boron, and zinc have been shown to reduce vine health, and deficiencies in phosphorus are suspected. Common causes, tests, and treatments for deficiencies of each of these nutrients are discussed below.

Nutrient deficiencies can arise from soil conditions such as a lack of the element in the soil, its presence in unavailable forms, or an inability for the nutrient to move rapidly to the absorbing root (low soil mobility). Deficiencies can also arise from plant conditions such as inability to take up the nutrient, to mobilize it, or to move it rapidly within the plant transport tissues (low plant mobility). Deficiencies may also arise from an interaction of soil and plant factors. Soil tests can be used to determine the availability of nutrients within the soil. Tissue tests can be used to determine the nutrient status of grapevines.

At low soil pH, aluminum (Al), an element that is not a plant nutrient, can also limit grapevine growth. Abundant in clay soils, aluminum becomes increasingly soluble and can become toxic as soil pH decreases below 5.3 (Figure 4).

Nutrient Mobility in Soils

Nutrients differ in their mobility in the soil, which affects their ability to be lost through leaching and the distance they can move to absorbing grapevine roots (Table 2). Boron (as borate) and nitrogen (as nitrate) are anions (negatively charged ions) and highly mobile

Table 2. Nutrients in Soil: Form, Mobility, and Analysis Guidelines

		Ionic Form	Ion Name	Mobility[a]	OSU Soil Test Guidelines[b] Low	Medium	High	Units
Macronutrients								
Nitrogen	N	NO_3^-	nitrate	high	<10	10-20	20-30	ppm
		NH_4^+	ammonium	medium	-	-	-	
Phosphorus	P	HPO_4^{-}	phosphate	low	<20	20-40	40-100	ppm
		$H_2PO_4^-$	phosphate	low				
Potassium	K	K^+	potassium	medium	<150	150-250	250-800	ppm
					<0.4	0.4-0.6	0.6-2.0	meq/100g
Sulfur	S	SO_4^{-}	sulfate	high	<2	2-10	>10	ppm
Calcium	Ca	Ca^{++}	calcium	medium	<1000	1000-2000	>2000	ppm
					<5	5-10	>10	meq/100g
Magnesium	Mg	Mg^{++}	magnesium	medium	<60	60-180	>180	ppm
					<0.5	0.5-1.5	>1.5	meq/100g
Micronutrients								
Boron	B	BO_3^{--}	borate	high	<0.5	0.5-2	>2	ppm
Zinc	Zn	Zn^{++}	zinc	low	<1.0	-	-	ppm
Iron	Fe	Fe^{++}	ferrous	low	-	nr	-	
		Fe^{+++}	ferric	low	-	nr	-	
Copper	Cu	Cu^{++}	cupric	low	<0.6	-	-	ppm
Manganese	Mn	Mn^{++}	manganous	low	<1.5	-	-	ppm
Molybdenum	Mo	MoO_4^{-}	molybdate	low	-	nd	-	
Chlorine	Cl	Cl^-	chloride	high	-	nr	-	

[a] From Marschner (1995). [b] From Marx et al. (1999).

in the soil and so can be removed easily through leaching or nutrient uptake. Potassium, calcium, and magnesium cations are moderately mobile in soil, and so they are lost through leaching only over long periods of time and move to roots most rapidly during periods of high water uptake. Phosphorus, iron, manganese, copper, and zinc are usually very immobile and so are not commonly depleted from soils, but they can be taken up only from areas immediately adjacent to the roots or their mycorrhizae.

Mobile nutrients applied to soils are readily available but may be quickly depleted. Immobile nutrients need to be applied in larger amounts or adjacent to roots to assure uptake, and some become unavailable over time. Mycorrhizae are important to grapevines for absorption of the immobile nutrients, especially phosphorus, and perhaps also copper and zinc (Schreiner, 2001).

Nutrient Uptake by Plants: pH and Nutrient Balance

Nutrients are imported into the plant from the soil by specific carriers that require oxygen, metabolic energy, and appropriate soil conditions for normal function. Low soil pH reduces the ability of grape roots to take up positively charged nutrients from the soil. As soil pH decreases below pH 6, some nutrients become less available for uptake, including nitrogen, phosphorus, potassium, sulfur, calcium, magnesium, boron, and molybdenum. Soil bacteria also decrease in activity with low soil pH and reduce their cycling of nutrients from organic matter reserves. For these and other reasons, liming of vineyard soils to raise soil pH is often recommended.

Potassium competes with two other macronutrients, calcium and magnesium, for uptake by grapes, and so the abundance of these two elements within the soil can affect availability of potassium. For this reason, deficiencies of potassium can result if low concentrations of soil potassium are combined with high concentrations or applications of calcium or magnesium. Deficiencies of magnesium can also occur for similar reasons after large applications of calcium or potassium to low-magnesium soils.

Nutrient Mobility in Plants

Nutrients differ in their mobility inside the grapevine, which affects where deficiencies appear whenever a grapevine's nutrient supplies become limiting. Most nutrients move upward easily in the xylem tissue, which transports water from roots to transpiring leaves. But not all elements are easily transported by the phloem tissue, which transports nutrients and organic molecules out of leaves or other stores when they are needed elsewhere in the vine.

Nitrogen, phosphorus, potassium, and magnesium are highly mobile in phloem and so can be moved out of older leaves to new tissues when they are in short supply in a grapevine. Thus deficiency symptoms for these nutrients are first visible, and usually more easily detected, as levels decline in older leaves. Sulfur, iron, manganese, copper, and zinc are somewhat immobile in phloem, and calcium and boron are least mobile. Immobile nutrients build up in older tissues supplied by xylem but may be in short supply in young tissues dependent on phloem transport. Thus deficiency symptoms for less mobile nutrients are usually first visible in new, rapidly growing, low-transpiring tissues such as young leaves, tendrils, flowers, and fruit.

As nutrient concentrations in leaves change during the season, they reflect both the vine's growth stage and nutrient status. Concentrations of mobile nutrients in leaves are usually highest in the spring, but they can change rapidly, decreasing in stages through the growing season as leaves expand and nutrients move. Concentrations of some less mobile nutrients increase during the growing season as they move upward in the transpiration stream and accumulate in leaves.

Table 3. Components of a Soil Test

Basic soil test includes:	*Units*
Organic Matter (humus)	percentage
ENR Nitrogen (estimated nitrogen release)	lbs/acre
Available Phosphorus	ppm or lbs/acre
P1 test, weak Bray (in western Oregon)	
P2 test, strong Bray (in eastern Oregon)	
Cation Exchange Capacity (CEC)	meq/100g, or $cmol_c$/kg
Exchangeable Cations	meq/100g, or $cmol_c$/kg
Potassium (K), Magnesium (Mg), Calcium (Ca), Sodium (Na), Hydrogen (H)	
Percent Base Saturation by Cation	percentage
Potassium, Magnesium, Calcium, Sodium	
Soil pH	
Buffer Index (SMP Method)	
Additional tests may include:	
Trace elements	ppm
Boron (B), Zinc (Zn), Manganese (Mn), Iron (Fe), Copper (Cu)	
Nitrate-Nitrogen	ppm or lbs/acre
Sulfur	ppm or lbs/acre
Excess lime (in eastern Oregon)	

Soil Analysis

Soil tests help characterize the soil in which grapevines grow and identify possible soil problems and nutrient deficiencies to allow for correction. Table 3 lists the components of a typical soil test. Baseline tests should be done on each soil management area in a vineyard, areas with distinct differences in soil type, depth, drainage, slope, and other important characteristics, including previous use. Soil properties and nutrient content do not change rapidly so, once baseline tests have been done, tests need not be repeated more frequently than every five years unless major application of fertilizers or lime are made or puzzling symptoms are seen. See Table 2 and Marx et al. (1999) for guidelines to soil test values from the Oregon State University Extension Service.

Soil tests in a vineyard should include samples representing both surface and subsoil horizons. Comparison of surface and subsoil tests indicates how greatly soil properties change with depth (distinguishing mollisols from alfisols and ultisols) and whether previous practices have significantly altered the native properties of the surface soil.

Surface soil samples should be collected at a depth of 0–12 inches. Ten or more samples randomly collected throughout a uniform management area can be mixed to provide an average sample for the area. Alternately, samples from distinct sites within the management area can be tested individually to indicate variability within an area, especially where previous uses (e.g., burn piles or cattle feeding) may have produced areas of high nutrient concentration. Be certain to use clean tools and containers for gathering, mixing, and shipping samples.

Subsoil samples should be collected, with a shovel or auger, at a depth of 2–3 feet to indicate soil conditions in the lower rooting zone. Again, subsoil samples may be mixed, but they should show less variability than surface samples. Tests on single samples taken from the most typical areas of a management unit can be used to give an estimate of the range of subsoil nutrition across an area.

Sampling sites for both surface and subsoil tests should be recorded on a vineyard map, so that the same sites can be resampled and retested if unusual results are found, and for greater consistency among tests in subsequent years. The same testing laboratory, or laboratories using similar testing procedures, should be used for each analysis to allow direct comparison of results between areas and over time. The Oregon State University Extension Service publishes a list of analytical laboratories serving Oregon, and each lab provides guidelines for handling samples to be tested (Hart, 2001).

Tissue Analysis

Tissue analysis allows monitoring of nutrient levels in growing tissues and provides the best indicator of the ability of a grapevine to obtain adequate nutrition from the soil. Patterns of nutrient concentration reflect the reserves of the grapevine and the balance of nutrients and water across the rooting profile. Absolute and relative amounts of nutrients help indicate if any one nutrient is limiting grapevine growth. Either leaf blade or leaf petiole tissue can be used to assess grapevine nutrition. Each has its advantages for particular nutrients or growing conditions, and both are used in different grape-growing regions of the world (Table 4).

It is important to sample tissues at a consistent developmental stage for results to be comparable to nutrition guidelines, or from vineyard to vineyard or year to year. Nutrition guidelines for grape leaves or petioles sampled at bloom or veraison are available from several regions. In Oregon, nutrition guidelines have been developed for petioles collected between July 27 and August 10, as vines approach or begin veraison.

Petiole tissues at bloom are probably better indicators of the status of nutrients with early-season deficiency symptoms, such as less mobile boron and zinc, but are less useful for nitrogen, which changes rapidly in concentration. Leaf tissues at veraison are probably better indicators of the status of the nutrients with late-season deficiency symptoms, such as the mobile macronutrients, potassium, phosphorus, and nitrogen, and relatively immobile calcium.

Long-term trends in nutrient status in a vineyard are more useful than single-season results for guiding soil management decisions, since yearly fluctuations in variables such as weather, soil moisture, and crop levels can cause short-term changes in grapevine nutrition. Nutrient analyses carried out regularly, together with a record of corresponding weather, soil, and crops, help establish the necessary context for informed nutrient management decisions.

Grape varieties, clones, rootstocks, and clone/rootstock combinations have been shown to differ in their patterns of nutrient uptake, transport, and use on the same soil and on different soils (Vasconcelos et al., 1997). For this reason, nutrient status should be analyzed and evaluated for each clone or clone/rootstock combination in each significantly different area of a vineyard. Comparison of tests from similar vines in similar vineyards in the same year can also help growers identify and assess nutritional concerns. The Oregon Vineyard Soil and Nutrition program began in 2002 to develop better information for soil and nutrition management, specific to each region, soil, and vineyard across Oregon, through analysis of shared soil and tissue test data.

At bloom, sample tissues are collected from leaves opposite the lower inflorescence of an average shoot, when 50% of the flowers have shed their caps, revealing the stamens. At veraison, tissues are collected from fully expanded leaves at the middle of a fruiting shoot, a consistent number of internodes above the cluster. Veraison samples are collected at the onset of berry color change, when berries in an average cluster have just begun to lose their green pigments and form the red or yellow pigments of maturity. It is important that compared samples be collected at a matching stage, using matching tissues and procedures.

Sample tissues should always be collected randomly from vines and shoots of average vigor that are spaced evenly through the management area, but not from vines at row ends or in outside rows. Samples should be collected away from sources of possible contamination such as dusty roads, driveways, or chemical mixing areas.

About 50 petiole or leaf blade samples should be collected from each management area, briefly rinsed with distilled water to remove dust, then blotted and air dried to remove surface moisture. Samples should be shipped in paper, rather than plastic, to avoid growth of mold. Be certain to use clean tools and containers for gathering, handling, and shipping samples. Analytical laboratories may provide additional handling guidelines.

Application of Nutrients

The method of nutrient application is chosen to match the nature of the materials and the purpose of the application. Broadcasting spreads the fertilizer material evenly over the surface of the vineyard soil. Between-row application applies nutrients for uptake by cover crops, whereas in-row application supplies nutrients for uptake by grapevines. Banding buries fertilizer in a narrow band adjacent to the vine row for rapid uptake in the region of highest grape root density. Foliar sprays apply nutrients to vine foliage for direct absorption into the leaves (Christensen, 1994). Fertigation supplies soluble nutrients in the vine row through drip irrigation.

Foliar sprays are appropriate for nutrient application where only small amounts are required (which is always the case with micronutrients) and where mobility of the nutrient in soil or plant is limiting. Foliar sprays may be used to correct a deficiency quickly in the growing season in which it is detected. Soil-applied nutrients are usually absorbed less quickly, particularly in nonirrigated vineyards.

Records should be kept of all fertilizer applications, untreated control rows should be noted, and responses to fertilizer application should be observed,

Table 4. Nutrients in Grapevines: Mobility and Tissue Analysis

				Bloom				7/21–8/10	Veraison			
				Petiole			Leaf	Petiole	Petiole	Leaf		
		Mobility[a]	Units	California[b]	Australia[c]	New Zealand[d]	France[e]	Oregon[f]	Australia[c]	Australia[c]	New Zealand[d]	France[e]
Macronutrient												
Nitrogen	N	high	ppm nitrate	350-2000	500-1200	570-1750	-	-				-
			% N	>0.5	-	0.8-1.0	3.0-4.0	0.66-1.50	-	2.2-4.0	1.5-2.8	1.5-2.5
Phosphorus	P	high	%	>0.15	0.2-0.46	0.21-0.50	0.6	0.11-0.35	-	0.15-0.30	0.16-0.25	0.15-0.20
Potassium	K	high	%	>1.5	>1.5	1.5-2.5	2.0	1.01-3.00	1.2-3.0	0.8-1.6	1.1-1.6	0.5-1.5
Sulfur	S	high	%	-	-	0.21-0.50	0.8	0.13-0.35	-	-	0.21-0.40	0.2-0.4
Calcium	Ca	low	%	-	1.2-2.5	1.4-2.5	0.5	1.26-3.00	-	1.8-3.2	2.0-4.0	1.5-4.0
Magnesium	Mg	high	%	>0.3	>0.3	0.31-0.80	0.4	0.46-1.25	-	0.3-0.6	0.2-0.5	0.2-0.4
Micronutrient												
Boron	B	medium	ppm	30-100	30-100	31-50	20-40	25-50	-	35-100	31-50	20-40
Zinc	Zn	medium	ppm	>26	>26	25-50	-	41-100	-	30-60	26-40	30-100
Iron	Fe	medium	ppm	-	-	31-100	40	31-100	-	-	40-100	100-200
Copper	Cu	medium	ppm	-	>6	5-20	-	6-20	-	10-300	18-34	5-20
Manganese	Mn	low	ppm	-	>25	25-200	20-40	61-650	-	25-200	41-100	20-60
Molybdenum	Mo	medium	ppm	-	-	-	-	-	-	-	-	5-10
Chlorine	Cl	high	ppm	<0.5	-	0.5-1.5	-	-	<1.5	<1.3	-	-
Nickel	Ni	medium	ppm	-	-	-	-	-	-	-	-	-

[a] From Marschner (1995, p. 97).

[b] "adequate" levels, from Christensen (2000, p. 4).

[c] "adequate" levels, from Robinson (1992, pp. 196-197).

[d] "adequate" levels, from Caspari (1996).

[e] "normal" levels, from Champagnol (1984, p. 154).

[f] "normal" levels, from Central Analytical Laboratory (2002), Oregon State University

comparing treatment levels or treated and untreated areas.

Managing Nitrogen

Nutrition. Nitrogen is the most commonly limiting nutrient in Oregon vineyards. It is responsible for 1–2% of the dry weight of a grapevine and about 0.2% of the weight of the crop (about 4 pounds per ton). Midseason petioles in Oregon are typically 0.66–1.50% nitrogen (Table 4). Nitrogen is an essential element in many critical plant compounds including enzymes, DNA, energy-transfer molecules, and chlorophyll.

The obvious symptoms of nitrogen deficiency are both a general decline in vigor (characterized by slower rates of growth and smaller leaves and shoots) and an overall yellowing of leaves and other green tissues (because of a deficiency in chlorophyll and other photosynthetic pigments).

Nitrogen is the most dynamic nutrient in the soil and vineyard, so careful attention must be paid to its management. Natural soil nitrogen reserves do not come from weathering of parent materials but from conversion of atmospheric nitrogen to organic forms by soil bacteria in a process known as nitrogen fixation. Nitrogen fixed by bacteria slowly accumulates in soil organic matter reserves over time and is a primary contributor to soil fertility. Each year, soluble nitrogen is released from soil reserves by microbial decay of new and old organic matter (a process known as mineralization) and becomes available for uptake by grapevines and cover crops. Nitrogen taken up by plants is incorporated into new organic matter and eventually recycled to the soil when that organic matter is again mineralized. Soluble nitrogen not taken up by roots can be lost from the soil, through leaching or volatilization back into the air, especially in high rainfall years.

Soil reserves of organic nitrogen can be very high in the active or slowly cycling fractions of many younger Willamette Valley soils, such as Woodburn, and reach levels of 20,000 pounds per acre in the passive organic matter fraction of some highly weathered soils, such as Jory. These soils display high fertility when first planted to grapevines because of rapid nitrogen release from reserves. Vineyard management practices that increase the breakdown of soil organic matter increase nitrogen supplies and grapevine vigor in the short term but eventually reduce nitrogen reserves and natural fertility. Soil levels of organic matter and nitrogen are typically much lower in the drier soils of southern and eastern Oregon.

Uptake, Transport, and Reserves. From bud-break to bloom, grapevines depend on nitrogen mobilized from internal reserves in the roots and trunk to support rapid shoot and leaf expansion. As midshoot leaves expand, current soil uptake becomes the primary source of nitrogen for continued growth. Growing fruit and later leaves accumulate most of their nitrogen from current uptake of soil nitrogen. If soil supplies become deficient, nitrogen is moved to fruit (and new leaves) from older leaves and other tissues supplied earlier in the growth cycle. Inadequate nitrogen movement to fruit can result in nitrogen levels in must that are insufficient for complete fermentation of resulting wines. But concerns over inadequate nitrogen for fruit development must be balanced with those over excessive nitrogen supplies that support too much vegetative growth.

At the end of the growing season, adequate nitrogen must be available for transport to internal reserves for support of the following season's early growth. Reduced internal nitrogen reserves result in reduced early vigor in the following year and a decline in potential yield and vine health. Conversely, increased nitrogen and energy reserves result in increased early vigor with a potential for excess vegetative growth, delayed fruit maturation, and reduced fruit quality and yield. For these reasons, careful attention must be paid to management of nitrogen in the vineyard.

Application. Most vineyards respond to nitrogen additions or release with increased growth. This is desirable in establishing a vineyard but requires observation and judgment in bearing vineyards. Young vines should maintain steady growth and thus usually require regular, small, nitrogen applications, since overwatering can leach nitrogen from around young roots.

In established vineyards, nitrogen is needed to support cover crops (if present) and increases in soil organic matter as well as to supply grapevines. Nitrogen reserves are cycled within the vineyard each year but are also lost through crop removal, leaching, and volatilization. Nitrogen reserves can be maintained or increased through nitrogen-fixing cover crops (legumes), mineral fertilizers, or organic amendments. Nitrogen application to cover crops for delayed release to the vines and increase in soil organic matter should be matched to the needs of the crop and soil. Nitrogen added in the vine row for direct uptake by vines may increase vine vigor in the current or following year. Common application rates for nitrogen in vineyards are 5–10 pounds of nitrogen per acre per year to maintain vigor, or 10–30 pounds to increase vineyard vigor slightly. These amounts of applied nitrogen are small compared to amounts fixed by legume cover crops, or contained in imported organic matter, and especially in comparison to total nitrogen reserves in the soil. Test applications of nitrogen at different levels and assessment of vine responses are advised.

Mineral fertilizers can contain nitrogen in any of three forms: ammonium (NH_4^+), nitrate (NO_3^-), and urea. Depending on the form, these differ in their cost, physical properties (including ease of handling, availability to plants, impact on soil pH, and potential loss through leaching or volatilization), and mode of application. Urea is the most concentrated and least expensive form of nitrogen, but it is also more difficult to handle and apply effectively. Ammonium sulfate is easier to handle and supplies sulfur, but is the most acidifying form of nitrogen. Mono- and di-ammonium phosphate both provide nitrogen as ammonium and phosphorus, but they are moderately acidifying. Calcium nitrate provides nitrogen as nitrate (the most available and leachable form), as well as calcium, and is the least acidifying, but it forms clumps at high humidity, making it more difficult to handle and store. Potassium nitrate is expensive as a soil nutrient but is sometimes used as a foliar spray, providing two nutrients.

Managing Phosphorus

Nutrition. The available phosphorus content of many Oregon vineyard soils is low compared to the seasonal needs of growing grapevines. Phosphorus is responsible for 0.1–0.4% of the dry weight of a grapevine and about 0.03% of the weight of the crop (about 0.6 pounds per ton). Midseason petioles in Oregon are typically 0.11–0.35% phosphorus (Table 4). Phosphorus is an essential element in many plant compounds, including the membranes of each cell, the DNA in each nucleus, and the energy-transfer molecules essential to metabolism.

Symptoms of phosphorus deficiency include reduced cluster number and crop yield, general decline in vigor, stunted leaves, and early yellowing and loss of basal leaves. If deficiency is more severe, red areas may appear at the edges of basal leaves and fill in between the veins, which remain green. Phosphorus deficiency can also reduce magnesium uptake and transport, resulting in magnesium deficiency symptoms.

Phosphate released from decaying organic matter is the primary source of available phosphorus in most soils. In low-phosphorus soils, grapevines depend on mycorrhizal fungi for phosphorus absorption and transport to exchange sites (arbuscules) in the root. Availability of phosphorus is affected by soil pH. Phosphorus is most available between pH 6 and 7; availability decreases at lower pH because of complexing with iron and aluminum, and at higher pH because of binding with calcium. Applied phosphorus can be transformed to unavailable forms.

Different extraction methods should be used to estimate available phosphate in low pH soils from western Oregon (Bray P1) and in high pH soils from eastern Oregon (Bray P2 or Olsen). Extractable phosphorus concentrations less than 20 ppm have been considered low in western Oregon soils, but levels as low as 2 ppm are found in seemingly healthy vineyards on well-weathered, hillside soils. Extractable phosphorus concentrations less than 10 ppm have been considered low in eastern Oregon soils (Table 3).

Application. Despite the low content and availability of phosphorus in many soils, no studies have been done on grapevine response to phosphorus application in Oregon. Several vineyards have experimented with regular additions of phosphorus to low-content soils. Applications of 50 pounds per acre of actual phosphorus in the vine row (below drip emitters) have improved growth and yield in Chardonnay vineyards in California, on clay soils with 1.9 ppm available phosphorus and petiole concentrations of 0.09%; such conditions are also seen in Oregon. Applications of 150 pounds per acre of actual phosphorus during site preparation have improved grapevine growth and yield on low-phosphorus soils in Australia. Some cover crops, such as legumes, with higher phosphorus requirements than grapes may benefit from phosphorus additions and increase phosphorus availability in the soil. Appropriate levels of phosphate to apply for optimal response need to be determined by vineyard trials.

Applied phosphorus can be transformed to unavailable forms and may then become available only very slowly. Phosphorus supply may be increased most economically by adjusting soil pH toward the 6–7 range, where reserves become more available.

Phosphorus fertilizers are derived from deposits of rock phosphate and processed into superphosphate, double superphosphate, and triple superphosphate, each with increasing phosphate content and solubility in water. Mono- and di-ammonium phosphate are roughly equivalent to the triple superphosphate in phosphate content and solubility and contain additional nitrogen that may increase phosphorus absorption. Rock phosphate itself contains only water-insoluble phosphates, which are not directly available to grapevines. Strongly acidic soils (pH <5.5) with high organic matter may, however, slowly convert rock phosphate to soluble forms for uptake, especially if it is applied finely ground or in colloidal forms, but rates of release are not clear. Large applications of phosphorus to soils low in available zinc can result in zinc deficiency.

Managing Potassium

Nutrition. Potassium is the nutrient required in the highest concentrations by grapevines, responsible for 1–2% of the dry weight of a grapevine and about 0.3% of the weight of the crop (about 6 pounds per ton). Midseason petioles in Oregon are typically 1.0–3.0% potassium (Table 4). Potassium, though an essential element, is not incorporated into organic compounds but remains in solution in the plant cell, where it serves essential osmotic, transport, and enzyme activation functions.

Foliar symptoms of potassium deficiency in grapes appear as color change (reddening in red varieties and yellowing in white varieties) beginning near the margin of older leaves and progressing inward, followed by browning at the margins or interveinal darkening, and leaf fall if severe. These symptoms can be intensified by low humidity and exposure of leaves to direct sunlight. Single-year deficiencies can result from the nutrient demands of heavy cropping or low soil moisture. Veraison tissue samples are the best indicator of grapevine potassium status.

Because potassium is involved in grapevine adjustment to water stress and in transport of sugars from leaves to fruit during maturation, effects of potassium deficiency on fruit maturation and quality are of considerable concern in dryland grape production. Potassium needs are highest late in the growing season when availability is lowest because of drying soil. Many vineyardists have experimented with late-season foliar applications of potassium to improve sugar accumulation in fruit, but with variable results.

In some areas, a high potassium level in harvested fruit is a concern because it can lead to loss of acidity and quality in wines. This problem appears, however, to reflect an accumulation of potassium in shaded leaves of overly vigorous vines and subsequent transport to fruit, and thus it may be corrected by proper canopy and vigor management.

Because potassium is not incorporated into organic molecules, active soil organic matter contains relatively little potassium for release through decay. Most potassium is supplied to plant roots through weathering of clays and release of potassium cations from exchange sites on the surfaces of clays and passive organic matter. Total soil potassium levels contained within clay minerals are typically quite high in Oregon soils, but only 1–2% of the total is on the surface of clay particles in available forms. Extractable potassium concentrations of less than 150 ppm or 0.4 meq/100 g are considered low in Oregon soils (Table 3).

Application. Temporary potassium deficiencies caused by heavy cropping or drought need not be corrected. But persistently low potassium levels in tissue and soil tests can be corrected by soil applications. Potassium deficiencies are sometimes quite local, so deficiency areas should be defined before treatment.

Potassium sulfate is the most common mineral potassium fertilizer for soils, but potassium chloride is also appropriate for use in western Oregon, where soil chloride levels are low. Because potassium is fixed rapidly in clayey soils, potassium is most economically applied in a concentrated band adjacent to the vine row, or in the irrigation stream, at a rate of 3–5 pounds per vine. When possible, optimal application rates for potassium, or any nutrient, should be determined by vineyard trials. Experimental controls are necessary for meaningful interpretation of results.

Post-veraison foliar sprays of potassium have been tried to produce an increase in late-season vine function and sugar accumulation. A variety of potassium compounds, including potassium carbonate, potassium thiosulfate, and various chelates, have been used, often with multiple sprays of 1 pound of potassium per acre. Responses have varied. Petiole or leaf samples should be taken to define the nutritional status of the vines before any fertilization program is initiated, to clarify potassium status (Christensen, 1994; Peacock, 1999).

Managing Boron

Nutrition. Boron is the second most commonly deficient nutrient in Oregon vineyards, in part because of its naturally low soil levels in areas of high rain fall. Midseason petioles in Oregon are typically 25–50 ppm boron (Table 4). Boron is critical in the formation of new cell walls and in the maintenance of cell membrane integrity.

Symptoms of boron deficiency typically appear early in the season in rapidly growing tissues. Tendrils, inflorescences, and internodes exhibit reduced elongation and may become locally swollen and necrotic. Pollen tube growth within the flower is apparently inhibited, so fruit set may be reduced with formation of small, seedless berries. Boron deficiency symptoms may be more pronounced on very acidic soils. Bloom-time tissue samples are the best indicator of grapevine boron status.

Transport of boron within the grapevine in cool spring weather can be problematic, even over short distances. Deficiencies can arise in new tissues on vines with adequate boron reserves. In vineyards where boron deficiency is a problem, a dormant spray in the fall and pre-bloom spray in the spring can usually supply adequate boron to buds and flowers. Sprayed boron also increases the boron status of the whole vine after cycling through the soil. Extractable boron concentrations less than 0.5 ppm are considered low in Oregon soils (Table 3). As with all nutrients, boron should be applied only when the need is indicated by tests or symptoms; excessive boron application can lead to boron toxicity.

Application. Boron is rightfully the most frequently applied nutrient in Oregon vineyards. When boron deficiency is diagnosed, spray applications are recommended, totaling 1 pound actual boron per acre per year. Boron is commonly applied as a dilute solution of 20.5% boron as borate. A post-harvest spray that wets buds, using up to 0.8 pounds of actual boron in 100 gallons water, is used to prevent stunting of early spring growth. A pre-bloom spray of 0.4 pounds actual boron per 100 gallons should ensure adequate boron for normal fruit set. Thorough coverage is critical, since boron mobility is low.

Managing Zinc

Nutrition. Zinc deficiencies are occasionally seen in Oregon vineyards, although, like boron, zinc is a micronutrient required in only small quantities by grapevines. Midseason petioles in Oregon are typically 40–100 ppm zinc (Table 4). Zinc is an essential component of several critical plant enzymes.

Symptoms of zinc deficiency include small, distorted and mottled leaves, reduced fruit set, and uneven fruit development and ripening ("hens and chicks" berries). Zinc, like boron, is relatively immobile in the grapevine. Large applications of phosphorus to soils low in available zinc can result in zinc deficiency.

Extractable zinc concentrations less than 1 ppm are considered low in Oregon soils (Table 3). Zinc deficiency is often quite local, and more common in sandy soils and in clay soils with high magnesium content.

Application. When deficiencies are suspected and verified by a tissue test, spray applications are recommended. A dormant spray of 5–15 pounds zinc per acre is applied in 100 gallons just prior to bud break. An additional early spring spray of 1 pound zinc in chelate form is applied in 100 gallons to young green foliage if deficiency symptoms have been severe. Neutral zinc (52% zinc) and zinc oxide (75% zinc) are both effective and economical as foliar sprays. Neither form is very soluble, and both require good agitation and flushing to keep sprayer lines clear. Again, good coverage is critical, since zinc mobility is low.

Managing Calcium and Magnesium

Nutrition. Calcium is an important element in both soil management and grapevine nutrition. It is responsible for 1–3% of the dry weight of a grapevine and about 0.05% of the weight of the crop (about 1 pound per ton). During the summer, calcium surpasses potassium as the most abundant nutrient in a grapevine, responsible for 2–4% of the dry weight of a green leaf. Midseason petioles in Oregon are typically 1.2–3.0% (Table 4). Calcium is a divalent cation with two positive charges critical to its role in plants. Calcium cross-links negative charges to strengthen walls and stabilize membranes around plant cells, and it also acts as a signaling molecule within.

Distinct symptoms of calcium deficiency are rare in Oregon but include brown borders on leaf margins progressing toward the petiole and bunch stem necrosis in developing fruit clusters. Low soil and tissue levels of calcium are associated with reduced shoot growth. On soils with subsoil pH below 5.0, calcium deficiency symptoms may be combined with symptoms of manganese and aluminum toxicity, such as stunted roots.

In soil, calcium's two positive charges may also cross-link negative charges on clays and organic matter to stabilize clay-humus complexes and soil microaggregates and to create and maintain soil structure and tilth. As calcium and magnesium occupy soil cation exchange sites, they displace hydrogen and aluminum and thus create and maintain the pH and chemical environment of the soil.

Magnesium is also important in both soil management and grapevine nutrition, but its functions are distinct from those of calcium. It is responsible for 0.1–0.3% of the dry weight of a grapevine and about 0.01% of the weight of the crop (about 0.02 pounds per ton). Midseason petioles in Oregon are typically 0.4–1.25% magnesium (Table 4). Magnesium's most important function is as the center of the green chlorophyll molecule that gives leaves their color and makes photosynthesis possible.

Distinct symptoms of magnesium deficiency are rare in Oregon but include yellowing (or reddening) between the veins of older leaves. As with calcium, low soil and tissue levels of magnesium are associated with reduced shoot growth. Magnesium deficiency symptoms due to manganese and aluminum toxicity may occur with subsoil pH below 5.0. Phosphorus deficiency can also reduce magnesium uptake and transport and result in magnesium deficiency symptoms.

Magnesium, like calcium, is a divalent cation, but it is much less effective in stabilizing clays and aggregates. For this reason, extractable calcium concentrations should be higher than magnesium concentrations in agricultural soils, with calcium–magnesium ratios between 6 and 2 in surface horizons for good tilth and structure. Calcium and magnesium together create and maintain the pH and chemical environment of the soil.

Application. Calcium and magnesium are commonly applied to soils to increase soil pH and to adjust the availability of each nutrient and the ratio between them. In agriculture, the term "lime" refers to materials derived from limestone and rich in calcium carbonate and containing up to 31% calcium. Limestone usually contains magnesium carbonate as well and is known as dolomitic limestone when magnesium concentration exceeds 11% (2:1 calcium–magnesium ratio). In either form, it is not the cation (Ca^{++} or Mg^{++}) but the negative-charged anion, the carbonate (CO_3^{-}), which reacts with H^+ to neutralize acidity and raise soil pH.

When no pH adjustment is desired, calcium can be added to soil as gypsum (calcium sulfate, 22% calcium), calcium nitrate (19% calcium), or triple phosphate (13% calcium), and magnesium can be added as Epsom salt (magnesium sulfate, 16% magnesium). See the earlier section "Managing Soil pH, Calcium, and Magnesium" for additional considerations in use of limestone materials for adjustment of soil pH and calcium and magnesium levels in vineyards.

Added calcium, magnesium, or lime should be incorporated into the soil when possible, because these materials move slowly from their site of application. If large amounts are applied, they should be well mixed to distribute their impact on soil properties. Healthy earthworm populations can increase the downward movement of surface-applied lime by maintaining drainage channels within the soil. Field trials should be used to determine appropriate levels of calcium, magnesium, and lime to apply for desired changes in soil properties, vine growth, and fruit quality.

Managing Grapes, Soils, and Nutrients

Every vineyard is different and must be managed for its unique combination of site, soil, vines, and management constraints. There is no single best way to manage the soils and nutrition of a vineyard, but good decisions require an understanding of the properties of soils that influence growth and development of vines. Although the intrinsic properties of a vineyard soil, such as texture and depth, cannot be changed by cultural practices, soil management can influence the dynamic properties critical to the fertility and quality of a vineyard soil.

Vineyard nutrition management strives to complement the ability of each vineyard soil to provide nutrients to the vine, so that the vine can develop appropriately and mature an optimal crop. Attention to vineyard fertility must be in balance with decisions regarding grapevine training, canopy management, and crop load. Vineyard fertility can be managed to stimulate or limit grapevine vigor. Nitrogen is the most commonly limiting and dynamic nutrient in Oregon vineyards, so careful attention must be paid to its management. Yearly attention must be paid to boron nutrition in many Oregon vineyards. Phosphorus, calcium, magnesium, and soil pH are likely more limiting to vineyard vigor than generally recognized. Potassium may limit vine function at the end of the growing season in some vineyards. Each grower needs to recognize when a limiting nutrient is beneficial or detrimental to production goals and vineyard health.

Winegrowers, winemakers, educators, and researchers across Oregon must continue to learn the impact of critical soil properties on vine growth and fruit quality. New ways to share information and identify and address common concerns will help growers manage each distinct vineyard soil more wisely for the production of high-quality grapes and wines.

Viticulture is new to Oregon's hillsides and valleys. With careful attention to soils and their management, grape growers and winemakers can leave a legacy of profitable and sustainable viticulture to future generations.

Acknowledgments

The authors wish to acknowledge the valuable contributions to this chapter by founding members of the Grape Underground: John Baham, Crop and Soil Science Department, Oregon State University; R. Paul Schreiner, USDA-ARS-Horticultural Crops Research Laboratory, Corvallis; and Andy Gallagher, Red Hill Soils, Corvallis.

References and Additional Resources

Baham, J. 2000. Soil compaction in western Oregon vineyards. Oregon Wine Advisory Board, *OSU Winegrape Research Progress Reports 1999–2000.* Oregon State University Agricultural Experiment Station, Corvallis.

Brady, N. C. 1990. *The Nature and Properties of Soils,* 10th ed. Macmillan, New York.

Brady, N. C., and R. R. Weil. 2002. *The Nature and Properties of Soils,* 13th ed. Prentice Hall, New York.

Caspari, H. 1996. Fertiliser Recommendations for Horticultural Crops. Horticulture and Food Research Institute of New Zealand. http://www.hortnet.co.nz/publications/guides/fertmanual/grapes.htm

Central Analytical Laboratory. Oregon State University. http://www.css.orst.edu/Services/Plntanal/CAL/calhome.htm

Champagnol, F. 1984. *Element de Physiologie de la Vigne et de Viticulture Generale.* Montpellier.

Christensen, P. 1994. *Foliar Fertilization of Grapevines.* NG6-94, University of California Cooperative Extension Service, Tulare County.

Christensen, P. 2000. *Use of Tissue Analysis in Viticulture.* NG10-00, University of California Cooperative Extension Service, Tulare County.

Gartel, W. 1993. Grapes. In W. F. Bennett (ed.), *Nutrient Deficiencies and Toxicities in Crop Plants.* APS Press, St. Paul, Minn.

Hart, J. 1998. *Fertilizer and Liming Materials.* FG 52, Oregon State University Extension Service, Corvallis.

Hart, J. 2001. *A List of Analytical Laboratories Serving Oregon.* EM 8677, Oregon State University Extension Service, Corvallis.

Hellman, E. 1997. Winegrape fertilization practices for Oregon. *Proceedings of the Oregon Horticultural Society* 88:215-220.

Marschner, H. 1995. *Mineral Nutrition of Higher Plants,* 2d ed. Academic Press, London.

McCarthy, M. G., P. R. Dry, P. F. Hayes, and D. M. Davidson. 1992. Soil management and frost control. In B. G. Coombe and P. R. Dry (eds.). *Viticulture,* Vol 2: *Practices.* 1992. Winetitles, Adelaide, Australia.

Marx, E. S., J. Hart, and R. G. Stevens. 1999. *Soil Test Interpretation Guide.* EC 1478, Oregon State University Extension Service, Corvallis.

Peacock, B. 1999. *Potassium in Soils and Grapevine Nutrition.* NG9-99, University of California Cooperative Extension Service, Tulare County.

Robinson, J. B. 1992. Grape nutrition. In B. G. Coombe and P. R. Dry (eds.). *Viticulture,* Vol 2: *Practices.* 1992. Winetitles, Adelaide, Australia.

Schreiner, R. P. 2001. Seasonal dynamics of roots, mycorrhizal fungi and the mineral nutrition of Pinot noir. Oregon Wine Advisory Board, *OSU Winegrape Research Progress Reports 2001–2002.* Oregon State University Agricultural Experiment Station, Corvallis.

Tugel, A. J., and A. M. Lewandowski (eds.).1999. *Soil Biology Primer.* NRCS Soil Quality Institute, Ames, Iowa.

Vasconcelos, M. C. C., S. Castagnoli, and J. Baham. 1997. Grape rootstocks and nutrient uptake efficiency. *Proceedings of the Oregon Horticultural Society* 88:221-228.

Werner, M.R. 1990. Earthworm ecology and sustaining agriculture. Components 1:4, University of California Sustainable Agriculture Research and Education Program. http://www.sarep.ucdavis.edu/worms/

19

Water Management

Allen Holstein and Edward W. Hellman

It is well known that water is a primary factor influencing vine behavior. Indeed, water management is considered to be a powerful viticultural tool for manipulating vine growth and, in some situations, improving fruit quality. Vineyard floor management practices, including the use of cover crops and cultivation, provide an opportunity to manage the soil water from rainfall. Supplemental irrigation is a necessity for grape production in the dry Columbia Valley and Walla Walla winegrowing regions, and in many areas of the Rogue Valley. In contrast, the high annual rainfall in the Willamette and Umpqua Valleys could lead to the conclusion that irrigation has little relevance in these regions. In fact, ample rainfall can contribute to excessively high vigor in grapevines and consequent problems. Certainly, winegrapes are grown successfully without supplemental irrigation in the Willamette and Umpqua regions. But a good argument can be made for the viticultural advantages of developing vineyard irrigation capability in Oregon's high-rainfall regions. The potential viticultural benefits of irrigation then must be weighed against the often-substantial expense required for system development.

Irrigation in High-Rainfall Areas

The issue of grapevine vigor in Oregon's high-rainfall regions requires closer examination. Vineyards in the Willamette Valley tend to exhibit a lot of early-season vigor, as opposed to late-season vigor in which vines continue to grow during the fruit-ripening period. The soils usually begin the growing season nearly saturated with water, and vines are often cropped at low levels. This combination can lead to high early vigor and succulent growth, with large leaves that consume more water over the season than would smaller, less succulent leaves. Consequently, high water use by grapevines, combined with that used by cover crops, usually dries the soil by the middle of August if summer rains are absent. A water deficit situation can subsequently develop that inhibits vine function (water stress) and may negatively affect crop yield or fruit quality. A similar scenario can develop in high-density vineyards or those planted on shallow soils with a relatively low total water-holding capacity. Thus, dryer than average summer conditions, in an otherwise high-rainfall region can create a situation in which some degree of water stress in grapevines is likely. It seems obvious, therefore, that the capacity to irrigate and avoid periods of damaging water stress can be beneficial to grapevines, even in the high-rainfall winegrowing regions of Oregon.

There is a fairly common belief, although usually vaguely defined, that some degree of grapevine "stress" contributes to high-quality fruit and wine. Generally this refers to water stress, but is sometimes interpreted quite broadly to include any factor, other than most diseases or pests, which limits vine size or fruit production. Unfortunately, such broad generalizations without consideration of vine physiology can lead to viticultural practices that unnecessarily limit crop production and may not improve fruit quality. Water stress, applied to an appropriate extent at the right time, can be used to manipulate grapevine growth and in some cases improve fruit and wine quality (Matthews and Anderson, 1988). Recent research has demonstrated the value of applying a controlled degree of water stress to grapevines by "regulated deficit irrigation" and other strategies (Dry et al., 2001). Utilization of these strategies in the Willamette Valley is complicated by the early-season abundance of soil moisture, although the use of competitive cover crops can reduce soil water content. The management strategy of the first author attempts to achieve mild water stress to stop vine growth earlier, enabling the vine to shift more photosynthates to support fruit ripening. Vines are not stressed enough to undermine productivity or fruit quality.

The ability to induce water deficits in grapevines is a powerful tool for restricting vine growth and must be used carefully to avoid excessive stress. Under circumstances of vigorous growth, a reduction in water availability can be used advantageously to restrict vine function, resulting in slowed growth.

Taken too far, excessive water stress can impair photosynthesis and other plant functions that support growth and development of the vine and fruit.

Irrigation to prevent excessive water deficits is important for maintaining yields and fruit quality under dry conditions, that is, dry seasons or dry sites with shallow soil and lower water-holding capacity. For example, 1999 was a dry season in the Willamette Valley, and nonirrigated vines in some vineyards turned yellow and dropped leaves in response to severe water stress before fruit ripening was complete. Irrigated vineyards generally produced better grape quality compared to vineyards with obvious excessive water stress. The personal experience of the first author affirms the value of irrigation for more consistent production of desired yield and quality of grapes. It is clear that, in some locations and some seasons, irrigation contributes to improved grape and wine quality.

Irrigation Management

Irrigation management is concerned with two questions: when to apply water, and how much water to apply. From the previous discussion it is clear that the answers to these questions depend on the irrigation strategy employed and the particular characteristics of the vineyard, including soil, climate, vine spacing, canopy size and exposure, scion, rootstock, and floor management.

Water management requires a thorough knowledge of the vineyard site characteristics and design features. Soil types often vary within a vineyard; water-holding capacity and effective rooting depth determine how much of the annual rainfall may be available to grapevines. Size and exposure of the grapevine canopy influence water consumption, and rootstocks may vary in their water uptake characteristics. The growth cycle and water consumption of cover crops must also be considered in the overall water availability equation.

Soil moisture content can be monitored with a wide variety of sensors. The use and merits of various soil moisture monitors are reviewed by Coggan (2002) and Selker and Baer (2002). The general methodology is to place sensors at multiple, representative locations in vineyard blocks, usually at two depths within the grapevine root zone. Sensors are monitored on a regular schedule, the frequency depending on the rate of soil drying, to determine when soils have reached a target level of dryness. Irrigation water is then applied, usually to deliver a predetermined quantity of water or, less precisely, until the deeper-positioned soil moisture sensors indicate an increase in water content.

Timing of irrigations can also be based on estimates of vine water status as measured by leaf water potential using a pressure chamber (bomb). Leaf water potential is a measure of the negative pressure (suction) of the sap flow in the xylem, caused by the transpirational pull as water evaporates and moves out of stomata. The pressure chamber method involves the slow application of positive pressure to a detached leaf until sap is forced out of the petiole, indicating that the negative pressure of the sap has been equalized. The pressure is displayed on a gauge and leaf water potential is expressed in units of bars or megapascals (mPa) (1.0 mPa = 10 bars). It is generally considered that grapevines are not under water stress at midday leaf water potential values of less than -10 bars, and mild stress occurs between -10 and -12 bars. Meaningful data require precise but simple methodology, which is described by Williams (2001).

The advantage of monitoring leaf water potential with the pressure chamber is that it provides the most direct indication of vine water status. Leaf water potential becomes more negative as water is held more tightly to soil particles as soil becomes drier. Higher transpiration rates also increase (more negative) the leaf water potential. Thus, the vine's response to soil and atmospheric conditions are integrated into the measurement of leaf water potential. Irrigation scheduling is based on a predetermined level of leaf water potential. Williams (2001) begins grape irrigation in California when the midday leaf water potential is at or more negative than -10 bars. He indicates that many commercial vineyards use a target of -10 bars for white wine varieties and -12 bars for red wine varieties.

Quantity of water applied can be based on replacing all (no deficit) or a portion (deficit irrigation) of the water used by the grapevine over a given period of time. An estimate of grapevine water consumption can be calculated based on the potential evapotranspiration (PET) in the equation

$$\text{Grapevine evapotranspiration (water use)} = ET_o \times K_c$$

The reference evapotranspiration (ET_o) is the estimated soil water loss from evaporation plus the water used by a grass reference crop. The daily ET_o value (provided by some weather stations) of the reference crop is converted into grapevine evapotranspiration by multiplying by the grapevine crop coefficient (K_c). Crop coefficients for grapevines vary through the growing season to reflect the development of the canopy and vine water demand as the season progresses. Because vine development rate varies from season to season, crop coefficients are usually associated with growing degree-days rather than calendar days. Grapevine crop coefficients are available for Washington (Evans et al., 1993), California (Williams, 2001), and other regions but have not yet been developed for Oregon.

Irrigation scheduling must also take into consideration the delivery capacity of the system. Often a vineyard is too large to irrigate all vines simultaneously, so a schedule must be developed that enables all vines to be irrigated with the desired amount of water at the proper time. Scheduling often requires the anticipation of water needs to prevent scheduling conflicts between different sections of the vineyard. Irrigation systems are generally designed so that vineyard blocks can be watered individually. This gives the manager the scheduling flexibility necessary to manage blocks of varying soil types or requiring different irrigation strategies.

There are presently no published irrigation management guidelines for Oregon vineyards, although Wample (1999) recently provided some general recommendations. Guidelines for other growing regions are available (McCarthy et al., 1992; Peacock et al., 1998; Wample, 1998) and could be adapted for use with appropriate consideration of Oregon climatic conditions, production systems, and a thorough knowledge of a vineyard's relevant characteristics.

Some common irrigation practices currently used in the Willamette Valley are as follows. Newly planted vines are usually irrigated throughout the first growing season. Applications of 1–2 gallons per vine, twice a week, are a common practice. Water applications are reduced late in the season to allow the vines to harden off (acclimate). Vines in their second and third years are generally watered based on their present size and desired vigor. Vines that did not establish well in their first year are often pruned back to two buds and treated as if they were just planted. Vines that grew well in their first year are watered a little less in their second year by delaying the start of water applications. Delaying irrigation avoids overwatering in the early season. Vineyards with well-established third-leaf vines are usually watered like mature vines; irrigation is not applied until growth has mostly stopped. Irrigation of young vines has enabled the first author to achieve rapid vine establishment and commercial yields in the second season without compromising yields in subsequent years.

For mature vines, most growers do not irrigate until shoot growth has stopped or nearly stopped. Watering during the active shoot growth stage encourages continued growth and subsequently higher water demand and excessive canopy problems. A common practice is to apply 2 gallons per vine, twice a week. This amounts to only 0.18 inches of water per week for 1,200 vines/acre. It is probable that more water, if available, could be applied without restarting shoot growth when dictated by higher evapotranspirational demand. It is desirable to have water capacity that would enable the grower to apply 1–2 inches over a four- to six-week period.

Water Management via Floor Management

Irrigation is not the only way to affect soil moisture status; floor management is also an important tool. This topic is covered in detail in Chapter 20, but a few summary comments are provided here.

Annual cover crops are primarily used for controlling erosion and have little impact on season-long soil moisture. Perennial cover crops compete with grapevines for moisture and thus can be used to reduce excessive vine vigor. There are numerous cover crop options with varying features, but in terms of relative competitiveness there is generally a greater difference between having a cover versus no cover than between types of cover crops. The degree of competition from cover crops can be manipulated by the extent of floor covered, mowing height, and other floor management practices.

Competition from perennial covers can be used to decrease early-season vine vigor, but if summer rains are negligible this is at the expense of late-season moisture availability. Often a perennial cover crop in every alleyway, or in some cases alternate alleyways, is too competitive for nonirrigated vines. A combination of floor management and irrigation practices provides the greatest flexibility in managing soil moisture.

Economics

A key question is whether the cost of an irrigation system can be recovered during the life of the vineyard. The answer requires analysis of the costs and benefits, some of which, such as improved fruit quality, are difficult to evaluate quantitatively. The costs of the water distribution system alone for a standard drip irrigation system can be $1.50–$1.75 per vine, depending on the vineyard design and site characteristics. Development of the water source (well, pond, etc.) and the delivery system (pump, filters, main lines, etc.) can require substantial costs and are very site-specific. It goes beyond the scope of this discussion to describe these costs. In some cases, landowners consider these costs to be associated with the value of the land. In other cases, costs of developing the water source and delivery system are directly allocated to the vineyard. How this accounting question is answered has a great influence on the irrigation costs to be recovered by the vineyard.

The ability to recover irrigation costs is influenced by whether the vineyard is independent or an estate vineyard associated with a winery. In the former case, it is the experience of the first author that distribution system costs are more than recovered by increased quantity of grapes alone. Whether the vineyard can recover the additional costs of water source development and the delivery system depends on the particular circumstances of the vineyard. Improved fruit quality may or may not be rewarded with a higher price. Overall, the cost-benefit analysis is more favorable for an estate vineyard, which has the ability to recover more costs through the sale of wine. Increased grape production leads to increased wine production, and higher grape quality could lead to higher wine value.

Conclusions

This chapter has reviewed how water management can be used to manipulate grapevine growth for improved quantity and quality of grapes. Water management can be achieved throughout a wide range of soil moisture; excessive soil water can be depleted by competitive cover crops, and irrigation can supplement inadequate soil moisture. Beyond these simple examples, a wide array of irrigation, vineyard, and floor management options exist that enable the vineyard manager to manage soil moisture and manipulate vine growth.

It is easiest to see the advantages of water management under the more extreme conditions. Depletion of excess water can be used to reduce grapevine shoot growth, creating a more favorable canopy microclimate for high-quality fruit production. Supplemental irrigation in a situation of severe soil water deficit maintains vine functions for full development of the crop and complete ripening of fruit. The unique aspect of water management in Oregon's high-rainfall regions is that the same vineyard can experience both extremes in one season. There are clear viticultural advantages in these regions to utilize techniques such as cover crops or close vine spacing to reduce excess soil moisture early in the season, in combination with supplemental irrigation later in the season. Water management in Oregon's dryer growing regions is oriented toward conserving the more limited soil moisture provided by rainfall and supplementing it with irrigation. The principal benefits are again derived from maintaining vine functions during the crop development and ripening periods.

References

Coggan, M. 2002. Water measurement in soil and vines. *Vineyard and Winery Management.* May/June, 43-53.

Dry, P. R., B. R. Loveys, M. G. McCarthy, and M. Stoll. 2001. Strategic irrigation management in Australian vineyards. *Journal International des Sciences de la Vigne et du Vin* 35:129-139.

Evans, R. G., S. E. Spayd, R. L. Wample, M. W. Kroeger, and M. O. Mahan. 1993. Water use of *Vitis vinifera* grapes in Washington. *Agricultural Water Management* 23:109-124.

Matthews, M. A., and M. M. Anderson. 1988. Fruit ripening in *Vitis vinifera* L.: responses to seasonal water deficits. *American Journal of Enology and Viticulture* 39:313-320.

McCarthy, M. G., L. D. Jones, and G. Due. 1992. Irrigation: principles and practices. In B. G. Coombe and P. R. Dry (eds.), *Viticulture,* Vol. 2: *Practices.* Winetitles, Adelaide, Australia.

Peacock, B., L. Williams, and P. Christensen. 1998. *Water Management and Irrigation Scheduling.* Publication IG9-98, University of California Cooperative Extension, Tulare County.

Selker, J., and E. Baer. 2002. An engineer's approach to irrigation management in Oregon Pinot noir. Oregon Wine Advisory Board, *OSU Winegrape Research Progress Reports 2001–2002.* Oregon State University Agricultural Experiment Station, Corvallis.

Wample, R. L. 1998. Water relations and irrigation management of wine grapes. In *Growing Grapes in Eastern Washington.* Good Fruit Grower, Yakima, Washington.

Wample, R. L. 1999. Irrigation management for high quality wine grape production: considerations for western Oregon. *1999 Proceedings of the Oregon Horticultural Society* 90:139-148.

Williams, L. E. 2001. Irrigation of winegrapes in California. *Practical Winery and Vineyard,* November/December, 42-55.

20

Floor Management

Ray D. William and Dai Crisp

Managing the vineyard floor encompasses the traditional challenges of weed control and erosion prevention but also presents the opportunity to influence grapevine growth by managing soil moisture. Recently, viticultural practices have increasingly considered the varied interactions of grapevines with other plant and animal life in the vineyard. Plant interactions can, viticulturally, be both positive and negative. Weeds can be problematic to grapevines, competing for water, nutrients, and sunlight. But cover crops, though also competing with grapevines, are important tools to help manage grapevine vigor, improve soil, reduce erosion, and facilitate equipment access to the vineyard. In some cases, a species considered a weed in one situation could, in a different circumstance, be a component of a cover crop that is purposely employed for beneficial effects in the vineyard. This chapter is meant to stimulate ideas and discuss floor management options for your considered integration into vineyard management.

The Vineyard Ecosystem

It is useful to view vineyards as more than just intensively managed grapevines; vineyards should be viewed as ecosystems that consist of numerous plant and animal species. The diverse members of the vineyard ecosystem sometimes compete with each other for resources, and they may directly or indirectly influence the growth, development, and reproduction of other species. Collectively and individually, the members of the ecosystem affect the local environment, including the soil and microclimate. Our primary concern, of course, is the grapevines. But to manage grapevines effectively we must be aware of the interactive relationships among the species that make up the ecosystem.

An important ecological process to understand is *succession*, whereby one plant species is succeeded by others until a *climax*, or stable, plant community is achieved over time. The species composition and distribution of plant communities are associated with resource availability: water, nutrients, light, and space. The common climax vegetation for many western Oregon hillside vineyards is native bentgrass sod intermingled with false dandelion, blackberry, Canada thistle, poison oak, and an array of winter annuals.

Planted perennial sods demonstrate a strong tendency to maintain the original sod and therefore may represent an introduced climax vegetation. However, most other weeds or planted cover crops, including wildflower mixes, rapidly decline through natural succession unless ample resources and cultural practices are employed to prevent that process.

Vegetation also contributes to foodwebs, above and below ground, involving numerous types of mammals, insects, mites, fungi, bacteria, nematodes, earthworms, and more. Fleshy roots such as false dandelion often attract gophers to inhabit the community along with deer, shrews, and field mice. Numerous insect and other arthropod species inhabit the ecosystem; in relation to grapevines, these species may be innocuous, injurious, or beneficial (e.g., predacious to injurious insects, mites, or nematodes).

Floor Management Objectives

Vineyard floors comprise two ecosystem zones, alleyways and in-row strips. Both are managed to achieve multiple objectives, but objectives and management practices often are different in the two zones. Alleyways represent the larger percentage of floor area and must support equipment and worker traffic. In-row strips represent a smaller area but have a greater potential influence on grapevines because of their immediately subjacent position.

Vegetation, or lack of such, on a vineyard floor influences grapevines, soil conditions, and characteristics of the entire ecosystem. Thus vineyard managers often have specific objectives when they select floor management practices. The major floor management objectives are introduced below.

Minimize Erosion. Most hillside soils in western Oregon are erosive, and vineyards are commonly located on slopes with rows oriented vertically to improve tractor safety and operator comfort. Erosion potential is therefore high and vegetative covers are essential, at least during winter, to minimize soil loss. Vegetation protects the soil surface from raindrop impact that dislodges soil aggregates, enabling them to move with water run-off. Protection also minimizes sealing of the soil surface from raindrop impact, resulting in improved water infiltration rather than run-off and erosion. Therefore, it is in the best interest of landowners, vineyard managers, and the environment to protect soils for long-term productivity and management.

Influence Grapevine Growth. Soil moisture availability has a large influence on growth of both the grapevine and vegetative cover. Competition for water can be detrimental to grapevines when water is scarce or beneficial when in excess. The type of vegetation, extent of area covered, and how it is managed determine the degree of competitiveness of floor vegetation. Soil moisture management is a major factor in achieving balanced grapevines—vines with enough, but not too many, shoots and leaves to properly ripen a crop of the desired quantity and quality. Choices among soil covers, mowing or tillage equipment, herbicides, and timing of practices provide ample opportunities for developing management schemes that fit your objectives for the site.

Improve Trafficability and Avoid Compaction. Many vineyard practices are time-sensitive, so the ability to work in the vineyard under wet weather conditions requires a stable surface. This is something a vegetative cover in the alleyways can provide. Vineyard equipment traffic in alleyways contributes to possible soil compaction, especially when soils are moist. Grass sods with their dense root systems minimize compaction by tractors or other vehicles.

Improve Soil Tilth. Soil chemical and physical properties are well known for many soils, but soil biology and the dynamic relationship among numerous soil-borne organisms are just now being studied. Soil tilth represents the blend of these properties and dynamic behavior of the soil system over years. In addition to protecting the soil surface, plant roots penetrate and form channels. The burrowing activities of worms and other animals also creates channels that improves soil aeration and may increase water penetration. Dead leaves and roots contribute residues or food sources for a multitude of soil-borne organisms that interact and influence soil tilth and biological productivity.

Enhance Biodiversity. Some managers strive to increase the biological diversity within their vineyards as a mechanism to reduce potential pest problems. The integrated production practices of Oregon's LIVE (Low Input Viticulture and Enology) program encourage biodiversity and the presence of flowering plants in the vineyard. The assumption is that the nectar and pollen produced by flowering plants nurtures beneficial organisms such as predator insects, parasitoids, or nematodes, which may prevent the occurrence of pest problems. The impact of biodiversity within vineyards on grape production is, however, mostly unknown at this time.

Improve Working Conditions. Vegetative covers contribute immensely to improved working conditions. Reducing dust with soil covers improves breathing by workers. Eliminating or preventing what ecologists call "safe sites," places for weed seeds to germinate and establish, reduces infestations of false dandelion and other weedy vegetation that contributes to gopher mounds and rough surfaces for tractors. Also, minimizing safe sites diminishes (but does not eliminate) establishment of noxious weeds such as Himalayan blackberries, poison oak, and thistles, resulting in improved disposition and productivity of workers and equipment operators.

Modify Vineyard Air Temperatures. The vineyard microclimate is influenced by the nature of the floor. Bare soil absorbs more sunlight energy and reflects the heat back into the crop canopy, creating slightly elevated temperatures. A vegetative cover, in contrast, creates a cooler vineyard microclimate and can increase the risk of frost injury. Close mowing or flailing, both in row middles and under the vines, can slightly increase temperatures and lower frost potential in early spring. Tall vegetation can also influence air drainage down a slope, trapping or slowing colder air as it flows downhill. Fruit ripening can be enhanced in fall from increased heat reflected from bare ground.

Floor Management Strategy

In-row. Strategies for managing in-row strips generally emphasize eliminating most competition with grapevines and minimizing erosion potential. This means that a relatively "clean" strip is maintained under the vine row with cultivation equipment or by the use of herbicides. Management widths can be adjusted to fit the need: wider to reduce moisture loss, or narrower to increase competition. Vegetation control efforts are often suspended in the fall to allow germination and growth of native vegetation, providing erosion protection over the winter.

Alleyway. Alleyway strategies are more varied, and the relative priority of the various objectives for the vineyard floor must be considered when formulating a strategy for each block. Erosion protection is usually among the highest priorities, but, since the period for eroding rains in Oregon mostly coincides with the dormant season of grapevines, it is possible to achieve other objectives while still addressing erosion concerns. Often the most important objective in floor management is to influence grapevine growth by manipulating soil moisture availability. A vegetative cover competes with grapevines for soil moisture, so the timing and degree of competition are critical characteristics of a floor cover that can be manipulated by the vineyard manager.

The competitive characteristics of the alleyway cover are carefully considered in modern viticultural practice, enabling managers to control the degree of competition with grapevines. For example, a site with relatively deep fertile soil tends to promote excessive growth of grapevines. Managing a competitive perennial cover in the alleyway helps reduce vine growth. On the other extreme, a site with a shallow soil with low water-holding capacity and no irrigation supports less growth of grapevines. A perennial alleyway cover in this circumstance would compete excessively with grapevines, resulting in unacceptably low vigor and undesirable stress to the vines. Little or no competition is desired in this circumstance, so perennial covers are generally avoided in favor of clean cultivation most of the season, with an annual cover planted in fall to provide erosion control. Similarly, grapevine establishment is enhanced when no cover is present during the first few years of the vineyard. In between these extremes, varying degrees of competition can be established by the selection of appropriate cover plant species and by mowing and tillage practices.

Alleyway management strategies often must be devised to achieve multiple objectives. For example, alternate rows can be cultivated to reduce competition while a vegetative cover is maintained in the other rows for equipment traffic.

Be aware that management strategies may need to change over the life of the vineyard. Mature vineyards that have been continuously managed with a permanent grass cover can, over time, experience a general decline in vine growth. Such vines can become unbalanced, with too small a canopy to support the crops typical of previous seasons. Under these circumstances, reducing competition from the cover and improving soil fertility could return the vines to their previous productivity.

Achieving Balance Improved Pinot Noir Fruit Quality

Dai Crisp

A block of Pinot noir exhibited excessive vigor on a site with too much ground water. During the first years of vineyard establishment, cultivation was used to keep alleyways and in-row areas nearly weed-free. As the vines grew larger and produced a deeper root system, soil moisture from the high water table became readily available, resulting in very high vine vigor. The first crops from these excessively vigorous vines produced fruit with undesirable vegetal and herbal flavors.

To bring the vines into balance, we allowed vegetation to grow in the alleyways to deplete excess ground water. The volunteer vegetation was primarily perennial fine fescue and bentgrass, with a mix of more than fifteen species including various flowering broadleafs. Mowing times were alternated; we allowed every other alleyway to grow high to maximize its water consumption. Alleyways were mowed with a rotary or flail mower, depending on the desired height. Cultivation was still used to keep the in-row area clean.

Competition from floor vegetation helped bring the vines into balance, which greatly improved fruit quality. The vineyard went from producing above-average to "reserve" designation Pinot noir wines. Soil tilth was also improved, and the less-dense canopy reduced the incidence of botrytis bunch rot. Although increased management time was required, fruit quality improved dramatically, increasing the value of the crop.

Management Practices

Type of Cover. Many vineyards simply manage the existing vegetation for the alleyway cover. Often, native bentgrass and false dandelion dominate western Oregon vineyards. Native species are adapted and tend to reseed and persist naturally, requiring less management time and resources than non-native covers.

Native covers can be shifted toward desired vegetation by managing seed production and resource availability. Autumn is an opportune time to fill niches with desirable species when moisture and nutrients become available. If seed of desirable species is present, light cultivation creates conditions for germination and establishment. Undesirable perennial weeds such as blackberries, thistles, or poison oak must be removed, either by application of a herbicide or by digging out by hand in early spring. Once established, the desired structure and function of the vegetation community can be maintained by avoiding cultivation and by minimizing disturbance from tractor tires.

Several improved varieties of perennial grass can be planted to replace the native bentgrass and false dandelion succession. Dwarf turf-type red creeping fescues such as 'Ensylva' offer disease resistance and low maintenance, but they require control of the creeping stolens to keep them out of the in-row area. Bunch-type dwarf fescues such as 'Covar' or

'Micklenberg' sheep fescue and 'Sierra' hard fescue offer low-cost options, although stems are tough and require sharp mowing or flailing blades. If established uniformly, fescues dominate the alleyway and exclude many weedy plants from invading. These sods have low maintenance requirements, tolerate traffic, drought, and low fertility, but are slow to establish. Vine root growth is limited within the sod root mat, thereby forcing grape roots to grow and deplete water from greater depths. A detailed discussion of perennial and annual cover crop species and varieties is found in Ingels et al. (1998) or on the University of California Cover Crop Resource Page website.

During establishment of a sod cover, weeds often infest the planting. Weeds can be minimized by quickly developing a full cover. One effective method is to plant a combination of fescue and a nurse crop such as 20% (by seed weight) of a dwarf perennial ryegrass, which provides quick cover. Adding the equivalent of 20 pounds of nitrogen per acre at planting contributes to fast establishment of perennial ryegrass. Low fertility in subsequent years favors the fescue over ryegrass, and few weeds can get established if the cover is full.

White clover sometimes becomes established in the turf after several years. Because clovers fix nitrogen from air, soil fertility is enhanced and may contribute to uneven vine growth across the vineyard.

Annual plant species are used as winter vegetative covers where alleyways are clean cultivated in the growing season to eliminate competition for soil moisture. Local winter annual weeds often are allowed to reestablish with fall rains, providing erosion protection. In some situations a more reliable alternative is to plant an annual cover crop in the fall.

Cereals, legumes, or mustard species are commonly used winter annual cover crops in western Oregon. The cereals include cereal rye, several varieties of oats, spring barley, and triticale. They are easy to drill and germinate quickly, and prostrate varieties such as 'Micah' barley or 'Celia' triticale suppress many annual weeds. Subterranean clover and annual bluegrass are winter annuals that set seed in June. Common vetch grows slowly in winter, with lush growth in late spring. Mixing vetch with triticale and low-stature barleys can provide nitrogen, but experimentation is required to achieve the desired blend. Some cereals can grow more that 3 feet tall, requiring adequate equipment to flatten and incorporate the residues. Alternatively, growth must be arrested with timely control in spring.

Mowing. Frequency and height of mowing influence the degree of competition from an alleyway cover. Close mowing decreases water consumption by the cover and increases water availability to grapevines. Mowing high and infrequently encourages moisture loss from the site. Mowing practices are often varied through the season. For example, a site with excessive soil moisture in the spring could be mowed high in early spring and mowed low later in the year as water availability equilibrates with vine requirements.

Mowing practices can also be adjusted to achieve other floor objectives. The first mowing is usually done prior to budbreak to reduce the cooling effect during potential frost events. Subsequent mowing can be timed to allow seed development of desirable plants while minimizing development of seed heads of late-maturing weeds. Alternate-row mowing of vegetation is recommended by the Oregon LIVE program to promote biodiversity and the presence of flowering plants throughout most of the season.

Tillage. The practice of tilling the soil to reduce or eliminate floor vegetation is applied to varying degrees to fit the need. Tillage options include the type of implement, the extent and depth of tillage, and adjustment of timing and frequency.

Conventional tillage equipment, such as a disc or spring-tooth harrow, is used to eliminate alleyway vegetation and the competition it provides. Other tillage tools are used to reduce cover competition to a moderate level. One such tool, the undercutter cultivator, has a very thin shank and horizontal blades that look like airplane wings. It undercuts the root system, slicing roots 3–4 inches below the surface with minimal disruption of the soil. Cover vegetation is not completely killed by this practice; it remains in place, reducing dust and erosion potential. Another moderate cultivation option is a power spader, which provides a digging action that leaves clumps of live vegetation with severed roots and slowed growth potential.

Within the vine row, a clean strip is often maintained with the use of specialized cultivation equipment such as a grape hoe or French plow. These implements have various types of trigger mechanisms that cause the cultivator to swing back out of the way when a sensor contacts a grapevine trunk or trellis post. In-row cultivators with a rotating head can be adjusted for ground speed and rotational velocity to either pulverize the soil or leave it rough and cloddy to slow erosion in winter. Wet soil areas often require rotary spading or mowing in the row later in spring to manage moisture and enhance work conditions. A flamer or weed burner can also be used in spring to slow the growth of in-row cover vegetation temporarily.

Cultivation of alternate alleyways is another method of enabling an intermediate degree of competition from vegetation. Tilling alternate rows conserves soil moisture for vines while retaining a

Vegetation Management Solves Problem of Three Vineyards in One

Dai Crisp

Every vineyard site offers its own set of challenges to achieving the vine balance that will produce the desired quantity and quality of grapes. One nonirrigated, hillside Sauvignon blanc vineyard exhibited almost the full range of vigor. The upper part of the hill had shallow soil; vines in this area were undervigorous and did not fill the trellis. The bottom of the hill had deep, very fertile soils; vines in this lower section were extremely overvigorous. The fruit from this very large, out-of-balance canopy tasted herbal, grassy, and vegetal with distinctive bell pepper flavors that persisted at maturity.

The vineyard row orientation was north–south from the top of the hill to the base. The floor was a volunteer mixture of grasses dominated by bentgrass. Cultivation was used to control in-row vegetation in all areas of the vineyard.

My solution to these variable site conditions was to manage the vineyard as if it were three separate blocks. The drier, upper section required early and intensive cultivation to minimize competition for soil moisture. Every alleyway was tilled with a chisel plow followed by a disc harrow. The reduced competition promoted increased vigor in these previously underperforming vines.

The midslope region had moderate water availability, so it was managed to somewhat reduce competition. Alleyways were treated in an alternating pattern: cultivation, mowed grass cover, cultivation, and so on.

The lower portion of the vineyard had a high soil water content, so intense competition from the alleyway cover was used to reduce vine vigor and bring the vines into balance. The grass cover was allowed to grow knee-high and produce seed heads before it was mowed high. The excessive vine vigor was dramatically reduced by the alleyway competition. The resulting canopy was much more in balance; combined with careful hedging and good leaf removal in the fruit zone, it produced fruit with melon and honey flavors and floral aromas that carried into the finished wine.

The trickiest part of this management strategy is to stay with the plan and avoid the temptation to perform an operation, such as tillage, uniformly down the entire length of the row. It requires discipline and diligence to raise the mower or cultivator at the appropriate time. At first glance the vineyard may seem messy or unfinished because of the varying cover treatments, but in our case the grapevines achieved the uniformity of balance we were looking for.

covered row to facilitate vineyard traffic. Coarse tillage is generally preferred because it reduces dust problems by leaving chunks of vegetation intact, which reinitiate growth with fall rains.

Vines growing without irrigation on dry sites or shallow soils often benefit from a higher degree of tillage that minimizes competition, such as cultivating every alleyway. This practice is generally most appropriate for newly planted vines, high-density vineyards, and vineyards on low-capacity sites.

The effectiveness of most tillage operations depends on timing. Tillage must always be timed with respect to soil moisture and soil type to prevent glazing or creating hard pans and compaction layers as a result of tillage implements sliding past wet soil. Tillage practices are usually timed to precede the maximum vegetative growth stage of grapevines in spring, allowing time for decomposition and release of nutrients. Disturbance can also be used to initiate a new cycle of plant succession in the vineyard floor. Vegetation adds organic residues to the soil when the cover is tilled or turned under. Any disturbance regime such as cultivation, flaming, herbicides, or other practices that kill or suppress vegetation contributes nutrients through sloughing of roots or decomposition of the entire plant.

Herbicides. Herbicides offer another option for managing weeds in alleyways or in-row strips. Although herbicides can be used alone, this practice is most effective when integrated with other practices described previously. An example might be use of an herbicide to control difficult weeds or to suppress growth of floor vegetation in early spring, followed by combinations of tillage or mowing to achieve desired results.

Herbicides can be applied before weeds emerge, often to bare soil in winter, or to actively growing vegetation. Postemergence selectivity is achieved with tolerance by grapevines to the chemical, or by application to avoid contact with the vine. One fairly common practice is to direct applications of a non-selective, translocated herbicide, such as glyphosate, to vegetation within the row strip under the vines. Two applications are usually adequate to control most of the vegetation during the wet spring and early summer. Herbicides can also be used to remove undesirable species selectively from in-row strips, alleyway covers, or elsewhere in the vineyard.

Opinions on herbicide use and preferences about personal safety and long-term impacts on the environment vary greatly among growers, scientists, and consumers. Criteria for choosing herbicides may include worker exposure and hazard, residual properties of the chemical in the vineyard, selective application and timing techniques for crop safety, sensitivity of the undesirable weeds to the chemical, or compliance with certification programs.

Maintaining Vegetation Makeup. Shifts in the species makeup of floor vegetation and weed populations are a natural consequence of vegetation management practices. When a management practice is repeated, plants that resist a specific practice or sequence of practices will dominate or succeed the colonizing species. Repeated cultivation provides favorable conditions for field bindweed and other deep-rooted perennials, although niches for desirable winter annuals are enhanced. Repeated mowing or flailing selects for prostrate or low-growing weeds or ground covers. Repeated use of the same herbicide, or of those with the same mode of action, selects for plants that tolerate the chemical or selects resistant biotypes that eventually dominate the site. Weed and vegetation management requires combinations of practices and constant modification to maintain the desired vegetation cover.

References and Additional Resources

Ingels, C. A., R. L. Bugg, G. T. McGourty, and L. P. Christensen (eds.). 1998. Cover Cropping in Vineyards: A Grower's Handbook. University of California Publication 3338, Division of Agriculture and Natural Resources, Davis.

William, R. D., D. Ball, T. L. Miller, R. Parker, J. P. Yenish, T. W. Miller, D. W. Morishita, and P. J. S. Hutchinson. 2002. Pacific Northwest Weed Management Handbook. Oregon State University Extension Service, Corvallis. http://weeds.ippc.orst.edu/pnw/weeds

LIVE Technical Guidelines. 2002. Low Input Viticulture & Enology, Inc., P. O. Box 102, Veneta, Ore. 97487.

Cover Crop Resource Page. 2002. University of California Sustainable Agriculture Research and Education Program, Davis. http://www.sarep.ucdavis.edu/ccrop/

Sattell, R. (ed.). 1998. Using Cover Crops in Oregon. Publication EM 8704, Oregon State University Extension Service, Corvallis.

The Extension Toxicology Network (EXTOXNET). 2002. Oregon State University, Corvallis. http://ace.orst.edu/info/extoxnet/

21

Pruning

Edward W. Hellman and Dick O'Brien

Dormant pruning is a critical component of the grape production system, providing the mechanism to maintain the training system, to select the fruiting wood, and to manipulate the potential quantity of fruit produced. After training of a young vine is completed and all of the "permanent" vine structures are developed, annual dormant pruning removes the previous year's fruiting canes or spurs (now two years old) and excess one-year-old canes.

The fruiting habit of grapevines dictates a pruning practice that encourages the annual development of new fruiting wood. Fruit is produced only on shoots growing from one-year-old canes. Therefore, healthy new canes must be produced every year to maintain annual production of fruit.

The training system is designed to encourage the production of new fruiting canes at specific positions on the vine—the arms. Pruning is used to remove unsuitable or extraneous canes selectively, retaining a small number of good canes. Canes are carefully selected to serve two functions: produce fruitful shoots in the coming season, and produce healthy shoots from which a good fruiting cane can be selected in the next dormant season. At each arm, these functions can be divided between two canes: a fruiting cane or fruiting spur (depending on the training/pruning system), and a renewal spur. Alternatively, a single fruiting cane or spur can be used at each arm, and one of the basal fruitful shoots is subsequently retained as a fruiting cane for the next season.

Timing

Dormant pruning of grapevines can be done at any time between leaf drop in the fall to budbreak in the spring. However, the logistics of completing the job in a specific time period and the availability of labor often influence the timing of pruning. Vine health considerations also enter into the timing decisions. Pruning in the fall may increase vine susceptibility to freeze injury compared to later pruning. Therefore, in regions where there is a significant risk of cold injury, it can be advantageous to postpone pruning until after winter's coldest temperatures. Postponing pruning also enables an assessment of cold injury and adjustment of pruning levels to compensate for injury losses. Later pruning commonly causes the vines to "bleed" sap from the pruning cuts, but this is not harmful to the vine.

Pruning Level

In addition to maintaining the vine's training system, pruning reduces crop production by removing fruitful buds. Varying the extent of dormant pruning is one method to influence cropping level. The term *bud count* (also node count or node number) is used to denote the number of dormant buds retained at pruning. Generally, bud count includes only buds with clearly defined internodes in both directions (Wolf and Poling, 1995); thus, basal buds are not included in the count. Basal buds, sometimes referred to as noncount buds, are not included in bud counts because frequently they do not produce shoots, and if they produce shoots they are often unfruitful.

Grape growers often prune vines with the intent to achieve a balance between fruit production and adequate, but not excessive, shoot growth. Increasing the bud count increases the number of shoots, which, if excessive, can lead to a crowded canopy and increased shading. Cropping levels are also increased when bud count increases, and the vine may not be capable of fully ripening high crop levels despite the increased shoot number. At very high bud counts the vine compensates for the large number of shoots with shorter shoot growth and fewer clusters per shoot (Coombe and Dry, 1992).

Excessive pruning—retaining too few buds—leads to an undercropping situation. Removal of fruitful buds reduces crop, but it also eliminates primary shoots. When there are too few shoots in relation to the vine's growth capacity, the vine compensates for the deficit by increasing the vigor of the remaining shoots, producing more extensive lateral growth, and stimulating shoot growth from secondary, tertiary, or latent buds. The consequence is often an excessively shaded canopy that provides a poor fruit-ripening environment.

Because pruning directly influences the number of shoots and the potential crop level, it is often the most significant annual management practice affecting vine balance. Consequently, the concept of vine balance is the basis of most pruning strategies. In some winegrowing regions, balanced pruning formulas are used to guide growers' decisions on the number of buds to retain. The bud count is based on an estimate of the weight of extraneous canes removed by pruning—the *pruning weight.* For example, the formula [20 + 10] indicates that 20 buds should be retained for the first pound of pruning weight and another 10 buds for each additional pound. Thus a vine with a 3-pound pruning weight would retain [20 + 10 + 10 = 40] buds. Wolf and Poling (1995) recommend a [20 + 20] pruning formula for Chardonnay, Riesling, Cabernet Sauvignon, and Cabernet Franc in the mid-Atlantic states. Balanced pruning formulas may not be reliable, however, when summer trimming of shoots is done as part of canopy management. Development of balanced pruning formulas requires research on the specific varieties and growing conditions of a region.

Another pruning strategy that utilizes the vine balance concept is to adjust pruning levels based on a visual accounting and assessment of cane growth from buds retained in the previous year. A balanced vine has strong, but not overvigorous, cane growth from all retained buds. If some canes are weak, correspondingly fewer buds should be retained for the next season. This may not fix the problem, however, and other vineyard factors should be investigated as possible contributors to low vigor.

If some canes were excessively vigorous, that is an indication that the vine's canopy may have been too large, perhaps because too few buds were retained the previous year. Again, other vineyard circumstances must also be considered, such as having vine spacing too close for the soil type, improper water management, or excessive application of fertilizer. The situation might be remedied by retaining a corresponding number of additional buds to accommodate the excess vigor, if there is adequate space on the trellis. If a crowded canopy is likely to be a problem, another solution will be necessary. One approach is to retain one or more "vigor diversion" or "kicker" canes in addition to the fruiting canes. Shoots are allowed to grow from these canes (diverting the excess vigor) until about midseason; then the entire cane is cut off at its base. Other vineyard management practices should also be considered to reduce excessive vigor, including the use of competitive cover crops.

Sometimes pruning level is approached simply from the perspective of how many buds are required to fill the allotted trellis space, or to attain a desired shoot density in terms of buds per unit area. This approach may be satisfactory if all the vines in the vineyard are uniformly in balance, but it ignores vine-to-vine variability that can often be significant. Therefore, this approach is unlikely to give satisfactory results if there is significant variation of vigor in the vineyard, or if the vineyard is generally out of balance.

The goal of achieving a balance between cropping level and shoot growth is often a challenge. Despite the powerful influence of pruning on crop levels, it may not be enough by itself to achieve the desired vine balance. Pinot noir grown in western Oregon often requires additional crop thinning during the season to ensure complete ripening of the fruit. This is usually not a consequence of inadequate canopy, but rather of the limitations imposed by the cool climate on the amount of fruit Pinot noir can ripen to high quality standards. Therefore, it is a common practice to cluster-thin Pinot noir at some point in the season to limit the crop to a size that can be properly ripened.

Pruning Practices

Pruning Cuts. All pruning operations should be conducted with well-maintained, sharp pruning tools. Pruning cuts on canes or spurs should be made at least 1 inch beyond the last retained bud. For cane pruning, it is common to make the cut directly through the next node beyond the last retained bud. Cutting through the extra node prevents it from producing a shoot, but the enlarged nodal region helps keep the tying material from slipping off the end of the cane. Ideally, cuts should be made at approximately a 45° angle, preferably with the lower end of the cut angled away from the bud.

Cane-pruning. The first step in pruning is to identify the fruiting canes for the next year. Desirable fruiting canes develop under conditions of good sunlight exposure, which is a function of the training system, last season's pruning level, and other canopy management practices. Good sunlight exposure promotes bud fertility and wood maturity. Fruiting canes and renewal spurs should be selected from positions close to the trunk head to prevent the arms from becoming too long, which causes a non-productive gap in the canopy above the head (Figure 1). The characteristics of desirable fruiting canes are (1) firm wood with brown periderm nearly to the tip; a sufficient number of healthy, fruitful buds; and without mechanical damage or visible disease infections; (2) round in cross section with relatively short internodes (3–4 inches) and moderate diameter (0.25–0.5 inch); and (3) well positioned on the arm (i.e., arising close to the trunk).

After good fruiting canes are selected (either one or two depending on the training system and vine spacing), another good, well-positioned cane is

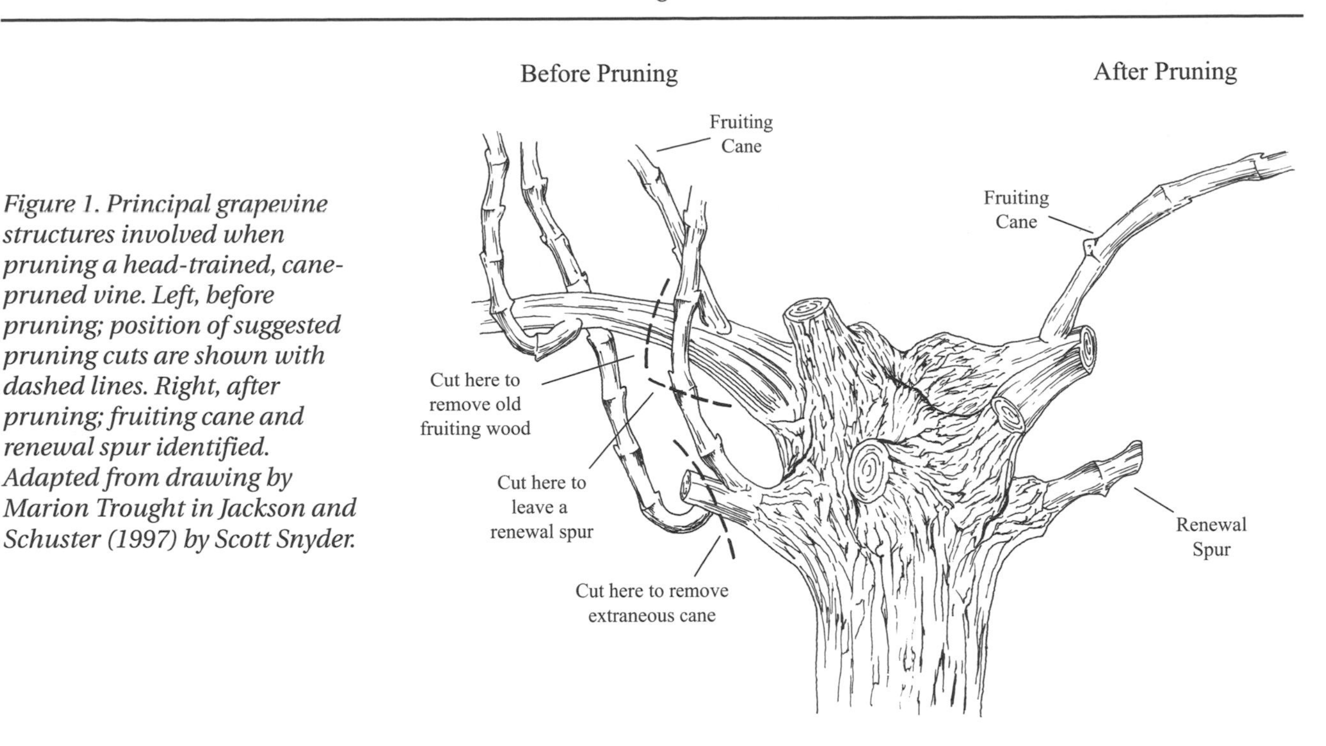

Figure 1. Principal grapevine structures involved when pruning a head-trained, cane-pruned vine. Left, before pruning; position of suggested pruning cuts are shown with dashed lines. Right, after pruning; fruiting cane and renewal spur identified. Adapted from drawing by Marion Trought in Jackson and Schuster (1997) by Scott Snyder.

selected as a renewal spur and pruned back to one or two buds. Periodically, it may be useful to retain a watersprout (during shoot thinning) that is closer to the trunk than the current renewal spur. At the next dormant pruning, the watersprout cane becomes the renewal spur. This practice keeps arm length from becoming excessively long. An alternate method does not retain a separate renewal spur. Instead, it is assumed that in the next dormant season a good basal cane from last season's fruiting cane can be selected as the new fruiting cane.

The remainder of last year's fruiting wood and all other extraneous canes, including suckers and watersprouts, are removed. Suckers are traced back to their source and cut back completely to remove all their basal buds. Fruiting canes are trimmed to a length that retains the desired number of dormant buds. The pruning cut is made through the next node (bud) beyond the retained buds, so that the enlarged portion of the node prevents the tie from slipping off. Next, all tendrils and laterals are removed, the cane is bent up onto the fruiting wire, wound once around it, and tied at the end.

Spur-pruning. Cordon-trained vines are typically spur-pruned (the practice is sometimes referred to as cordon-pruning). Just as with cane-pruning, the arm positions of cordons are established by the training process and all fruiting and renewal spurs arise from this area. The arms should be evenly spaced along the cordon and oriented in the proper direction (up or down depending on the training system). Select suitable canes for the new fruiting spur and renewal spur using the same criteria described for cane pruning. Remove the old fruiting wood from the previous season. The selected fruiting cane is shortened to create a fruiting spur with two to four buds, depending on the fruitfulness of basal buds and the desired cropping level. The renewal spur cane is cut back to one bud. Similar to cane-pruning, selection of canes for spurs should take into consideration the position of the cane on the arm. Select canes to maintain as compact an arm as possible and to maintain the desired spacing between arms.

Replacement of Arms. Over time, some older vine structures must be replaced, either because they have outgrown their area or because they have become injured, diseased, or are declining in vigor. For example, arms tend to increase in size as a consequence of cane renewal. When a cane that grew from the former fruiting cane is retained as a renewal spur, a stub of older wood from the fruiting cane remains. Extension of arms can be slowed by careful selection of well-positioned fruiting canes or spurs, and renewal spurs.

Arm replacement for cane-pruned vines begins during the normal practice of suckering and shoot thinning in spring, with the retention of a well-positioned watersprout arising close to the top of the trunk. In the subsequent dormant season, the watersprout cane is cut back to two buds and retained along with the fruiting cane or fruiting spur. The next year, the entire old arm is cut off just beyond the watersprout cane (now two years old), from which a good fruiting cane can be selected. Alternatively, if the watersprout is strong, it can been used immediately as a fruiting cane and the old arm removed. A similar process can be used to replace an unproductive or damaged cordon. A watersprout arising from near the trunk is selected and trained in the same manner that the original cordon was developed.

Pruning Logistics. It can be advantageous to divide the various steps of pruning into distinct operations performed at different times. For example, the old fruiting wood can be cut off first, usually retaining its basal cane in case it may be needed for fruiting or renewal. Suckers could also be removed as part of this operation. The brush (pruning waste) is then pulled from the vineyard or otherwise disposed of before the next operation. The remaining canes are now easier to see, making the process of selecting the fruiting and renewal canes easier. Extraneous canes are cut off and the fruiting canes trimmed and tied. A final brush disposal completes the pruning process.

It is helpful to throw the brush on the ground of every other row. Using alternate rows makes it possible to travel through the vineyard on the clear rows prior to brush disposal. Brush can be removed from the vineyard (called brush pulling) or chopped up on-site, which returns organic matter to the soil. The volume of brush in mature vineyards makes brush pulling more difficult, so chopping is preferred.

References

Coombe, B. G., and P. R. Dry. 1992. *Viticulture,* Vol. 2: *Practices.* Winetitles, Adelaide, Australia.

Jackson, D., and D. Schuster. 1997. *The Production of Grapes and Wine in Cool Climates,* Lincoln University Press,. Canterbury, New Zealand.

Wolf, T. K., and E. B. Poling, 1995. *The Mid-Atlantic Winegrape Grower's Guide.* North Carolina Cooperative Extension Service, Raleigh.

22

Canopy Management

W. Mark Kliewer and Ted Casteel

A grapevine canopy consists of the aboveground parts of the vine formed by the shoot system. This includes shoots (leaves, petioles, lateral shoots, and tendrils) and the fruit, trunk, and cordon or canes. Management of the canopy generally refers to the number and arrangement of shoots and fruit clusters. The objectives are to produce a continuous canopy for efficient fruit production and to create the appropriate environmental conditions for the intended grape yield and wine quality. Producing the desired canopy characteristics and environmental conditions actually begins with vineyard planning—matching the vine and row spacing, rootstocks, and the training system to the site capacity. After vineyard establishment, an array of vineyard management practices can be used to influence canopy characteristics.

It is critical to recognize that appropriate canopy conditions vary depending on growing region, grape variety characteristics and requirements, and the economic goals of the vineyard or vineyard block. In Oregon, intensive canopy management to provide good sunlight exposure of leaves and fruit is generally considered optimal for production of ultra premium Pinot noir wine. Intensive canopy management requires high labor inputs that can be economically justified only by an appropriately high market price for the grapes. Wine grapes of lower value do not bring adequate returns to recover the costs of intensive canopy management.

Vineyard canopies are said to be *continuous* when the foliage from adjacent vines meets and *discontinuous* when individual vines have canopies that are separate, that is, when there are gaps between vines. Discontinuous canopies are less efficient economically.

Dense or crowded canopies have excess leaf area resulting in many shaded leaves and clusters, whereas open or low-density canopies are not shaded. Good viticulture practices are aimed at achieving an appropriate balance between vegetative and reproductive development, since an excess of shoot vigor (canopy shading) generally has undesirable consequences on crop yield and fruit composition. Dense canopies are known to influence yield and quality negatively, affecting several processes including budburst, inflorescence initiation, fruit set, berry growth, and the balance between sugars, acids, and pigments, which collectively describe fruit quality.

The main key to good vineyard canopy management is proper utilization of solar radiation (sunlight). The benefits derived from exposing various parts of grapevine canopies to sunlight may be summarized by comparing the characteristics of canopy components in the interior of a dense shaded canopy with those in the interior of an open, well-illuminated canopy.

Buds from shaded canopy interiors have
- fewer cluster primordia/bud
- smaller cluster primordia, which develop into smaller clusters
- fewer shoots that develop from primary and secondary buds
- less shoot and grape production by basal buds
- poorer survival rate in cold winters

Shoots from shaded canopy interiors have
- less periderm formation (maturation of wood)
- less xylem and phloem development
- a lower level of carbohydrate storage
- less lateral shoot growth and maturation

Leaves from shaded canopy interiors have
- lower rate of photosynthesis and net assimilation
- shorter lives and a tendency to become chlorotic and abscise early

Fruits from shaded interior shoots generally have
- lower sugar and lower rates of sugar accumulation
- lower anthocyanin pigments and phenolic compounds
- higher titratable acidity, particularly malic acid
- higher pH and potassium
- more fungal disease problems

Canopy management includes a range of techniques that alter the position and number of shoots, leaves, and fruit, which in turn alters the canopy microclimate. Microclimate is the climate within and immediately surrounding a plant canopy. Climate values within a canopy can differ drastically from the above-canopy climate because of the presence of the canopy. This is especially true of sunlight, which can be less than 1% of ambient sunlight in dense canopy vines (Kliewer, 1982). Air movement within dense canopies may be less than 10% of above-canopy values. Humidity, temperature, and evaporation potential are also influenced by canopy density. The microclimate within dense canopies may predispose vines to a high incidence of fungal pathogens, especially botrytis bunch rot, and disease control is made more difficult by the inadequate penetration of chemical sprays. The interior region of a dense canopy does not dry out as quickly as an open canopy, and the higher humidity favors botrytis development.

The tools used to manipulate vine canopy characteristics of an established vineyard can be grouped into two categories: general vineyard practices that influence vine growth and thus canopy characteristics, and so-called canopy management practices that involve direct physical manipulation of the canopy. Irrigation, fertilization, and floor management are general practices that can have a great influence on vine growth. Canopy characteristics are influenced by several direct manipulations, listed below and discussed in more detail later in this chapter:

- dormant pruning and training: affects shoot density and location
- vigor diversion: retention of extra, disposable canes utilizes excess growth capacity
- shoot thinning and suckering: reduces shoot density
- shoot positioning: determines location and orientation of shoots
- hedging or shoot trimming: shortens shoot length
- leaf removal in the fruit zone: improves fruit exposure to light and air movement

Expected responses to several of the canopy management practices in Oregon vineyards trained to the vertically shoot positioned system are summarized in Table 1. A comparison of low- and high-density canopies on grapevines physiological processes are summarized in Table 2. It is readily apparent from Table 2 that open, well-exposed, low-density canopies favor production of sugars, anthocyanins, phenolic compounds, and other photosynthetic products important for high wine quality.

Table 1. Expected Responses of a "Normal" Pruned, Vertical Shoot Positioned Vineyard to Various Cultural Practices

Change of Cultural Practices	*Light to Cluster*	*Photo-synthesis per Unit Crop Weight*	*Tartrate (G/Berry)*	*Malate (G/Berry)*	*Potassium (G/Berry)*	*Wine Potassium*	*Wine Color*	*Crop Yield*	*Total Soluble Solids or Brix*
Scott Henry	Increase	Increase	No effect	Decrease	Decrease	Decrease	Increase	Increase	No effect or decrease
GDC Trellis	Large increase	Increase	Increase	Decrease	Decrease	Large decrease	Increase	Increase	No effect or decrease
Severe Pruning	Little effect or slight decrease	Decrease	No effect	Increase	No effect	Decrease	Increase or decrease	Decrease	Increase
Hedging (Topping)	Slight increase	Slight increase but total sugar in fruits less	No effect	Decrease	No effect	Increase	Decrease	No effect	Decrease or decrease
Moderate Water Stress	Increase	Little to no effect	Slight increase	Decrease	Decrease	Decrease	Increase	Slight increase	Increase
Removal of Secondary Shoots	Increase	Increase	Increase	Decrease	Decrease	Decrease	Increase	Decrease	Increase
Shoot Positioning	Decrease	Decrease or no effect	No effect	Increase	Decrease	Decrease or no effect	Decrease or increase	No effect or decrease	Slight increase or no effect
Leaf Removal (Fruiting Zone)	Increase	Increase	No effect	Decrease	Decrease	Decrease	Increase	No effect	Increase or no effect

Recognizing Excessive Canopies

There are several symptoms or ways to recognize vineyards that have high canopy-density problems:

- The interior part of canopy has a large number of yellow leaves; this condition usually does not become apparent until after veraison.
- Most of the fruit is not visible looking from outside of canopy.
- The fruit has high titratable acidity, malate, potassium, and pH.
- The fruit has low sugar, phenols, and anthocyanin pigments in skins.
- The fruit has low monoterpenes and other flavor compounds.
- The fruit and wine have high "herbaceous," or grassy, characteristics.
- The fruit has a high level of botrytis bunch rot.
- There are few canopy gaps.

Several methods are available for assessing vineyard canopies, as follows:

Yield/Pruning Weight Ratio. Perhaps the most informative and easiest way to determine if a vineyard has a good balance between crop yield and vine growth is to measure the crop yield at harvest and the weight of dormant wood removed at pruning time. Commonly, ten to twenty typical vines for each vineyard block are measured for this assessment. The ratio of yield to pruning weight gives a good indication of the balance between fruit and vegetative growth. The literature indicates that a ratio between 4 and 10 provide a good vine balance, but experience in Oregon for Pinot noir and Pinot gris indicates that the acceptable range for producing high wine quality is a narrow range of 3 to 5. Vineyards with values less than 3 are characterized by high shoot vigor, low crop yields, and excessive vine growth. Vineyards with high values (>6) have high crop yields in proportion to the amount of vine growth.

Average Cane Weight. Counting the number of dormant canes per vine before pruning, combined with measuring pruning weight, enables a calculation of the average cane weight (pruning weight divided by cane number). Optimal values vary with variety and climate. Representative values for low, moderate, and high vigor vines are as follows:

	Vine Vigor		
	Low	*Moderate*	*High*
Mean dormant cane weight (grams)	<10	20–40	>60
Yield/Pruning weight ratio	>10	3–6	<3

Balanced vines typically have indices associated with moderate vigor. Generally, vines pruned to ten to fifteen buds per pound of dormant pruning weight result in vines with a good balance between foliage growth and crop weight. Pinot noir generally requires leaving fewer buds per pound of pruning weight than most other cultivars for producing top-quality wines.

Point Quadrat Analysis. The method of point quadrant analysis (Smart and Robinson, 1991) simulates a beam of sunlight passing through a vine canopy. This should be done between veraison and harvest, when the canopy has filled out. The method entails using a thin metal rod about 3 feet long. The rod should be inserted at intervals in the fruit zone, horizontally or at an angle, and all contacts are counted as it passes through the canopy. Each contact with a leaf or cluster is recorded, as are areas where the rod makes no contact. A large number of inserts are needed to obtain valid results, perhaps 50–100 at about 3–4-inch intervals. The point quadrat tells us the proportion of leaves and fruits that are exterior or interior in the canopy. Once all the data are recorded, they can be used to calculate percentage of gaps (number of gaps divided by number of insertions times 100); leaf layer number (number of leaf contacts

Table 2. Comparison of Low- and High-Density Canopies on Grapevine Physiology

Physiological Process	*Low-Density Canopy*	*Interior of High-Density Canopy*
Transpiration	Higher rate of transpiration	Lower rate of transpiration
Photosynthesis	Higher rate of photosynthesis	Lower rate of photosynthesis
Respiration	Often high rate of respiration during the day due to sunlight warming	Daytime respiration rate is largely controlled by temperature. Respiration rate generally lower
Translocation	Exterior leaves are exporters of photosynthesis products	Little or no export of photosynthesis products from interior leaves
Phytochrome	Fruits and leaves are exposed to higher ratio of red to far-red light, which promotes anthocyanin synthesis	Fruits and leaves are exposed to lower ratio of red to far-red light and have lower anthocyanin synthesis
Water relations	Exposed leaves and fruits experience more water stress	Interior leaves and fruit experience less water stress

Table 3. Winegrape Canopy Ideotype, Shoot and Fruit Indices for Optimal Microclimate

	Optimal Value[a]	*Importance for Wine Quality*	*Justification*
Shoot length	14–18 nodes (2.5–3.5 ft)	High	Short shoots cannot ripen fruit properly; long shoots contribute to shade and high pH.
Lateral development	Restricted, less than 5–8 lateral nodes per shoot	High	Excessive lateral growth indicates high vigor and causes shade
Ratio, leaf area:fruit weight	~12 cm^2/g (6 ft^2/lb) range 6–15 cm2/g (3–8 ft^2/lb)	High	Low values are inadequate to ripen fruit, high values cause high pH
Ratio, yield:canopy surface area	1-1.5 kg/m^3 (0.2–0.3 lb/ft^2 for cool climates, 3 kg/m^2 (0/6 lb/ft^2) is a likely maximum figure for hot, sunny climates	High	Exposed canopy surface area required to ripen grapes
Ratio, yield: pruning weight	4–10	High	Indicates vine balance; low values mean excessive vigor, high values mean overcropping
Growing tip presence	Nil	High	Encourages fruit to ripen
Cane weight	20–40 g (0.7–1.4 oz)	High	Indicates desirable vigor level and vine balance.
Internode length	60–80 mm (2.4–3.1 in)	Moderate	Values vary with variety; indicates desirable vigor level.
Pruning weight	0.3–0.6 kg/m canopy (0.2–0.4 lb/ft)	High	Lower values indicate canopy too sparse, high values indicate excess values possible for nondivided canopies
Shoot spacing	12–8 shoots/m canopy length (3–6 shoots/ft canopy length)	High	Higher values indicate shoot crowding and excess shade; lower values indicate sparse canopies and potential low crop yields

[a] Indices are usually measured at or near harvest or after leaf fall.

divided by total insertions); and percentage of exterior clusters (number of exterior clusters divided by total cluster contacts). The optimum proportion of gaps should be in the range of 20–40% using the point quadrat technique (Smart and Robinson, 1991). Generally about five shoots per foot of canopy length produces an optimum canopy microclimate. Tables 3 and 4 summarize wine grape canopy ideotypes for microclimate characteristics and the indices considered optimal for most vineyards.

Visual Vineyard Scoring. Vineyard status just prior to harvest can also be assessed by visually scoring for the amount of lateral shoot growth, leaf size and color, shoot length and node number, and the number of growing shoot tips after veraison. All of these characteristics are related to the physiological status of grapevines.

Ideally, lateral shoot length in the fruiting region should be about five to eight nodes. Long lateral shoots are indicative of excessive vigor and contribute to fruit shading, which can have a negative impact on fruit composition.

Leaf size and color largely reflect the water and nutritional status of grapevines. Large, dark green leaves indicate excessive availability of nitrogen and water, leading to overly vigorous vines with canopy shading problems. Very pale leaves usually indicate nutrient deficiency, drought stress, or a water-logged condition. Ideally, exterior leaves on the middle portion of shoots during the post-veraison period should be a dull, medium green color, and leaf size should be average for the variety or slightly smaller.

Generally, shoots with fifteen to eighteen primary leaves are the ideal length. Shoots longer than this indicate excessive shoot vigor, whereas shoots with less than twelve nodes reflect inadequate growth and leaf area to mature fruit clusters fully. During the post-veraison period there should be few or no growing

Table 4. Winegrape Canopy Ideotype, Canopy Characteristics

	Optimal Value	*Importance for Wine Quality*	*Justification*
Row Orientation	North–South	Low	Promotes sunlight interception, especially for sunny regions
Ratio, canopy height: alley width	~1:1	High	High values lead to shading of canopy bases; low values mean inefficient sunlight interception
Foliage wall inclination	Vertical or nearly so	High	Underside of inclined canopies is shaded
Fruit zone location	Near canopy top	Moderate	Promotes high yield although fruit phenol levels can be excessive
Canopy surface area (SA)	~21,000 m^2/ha (92,000 ft^2/ac)	High	Low values associated with low yield potential; higher values not possible without excess shade
Ratio, leaf area:canopy surface area (LA/SA)	<1.5	High	Indicates canopy density; low values desired.
Shoot spacing	~15–18 shoots/m canopy (4.6–5 shoots/ft)	High	Gives about ideal canopy for moderate vigor vines in vertical canopies; values for non-positioned canopies can be higher

shoot tips. Shoots that continue to grow during this period slow sugar accumulation in fruits and delay the time of harvest.

Corrective Measures for Excessive Canopies

General Vineyard Management Practices

Vine canopy problems can be exacerbated by vineyard management practices such as excessive irrigation or nitrogen fertilization. In such cases, recognition of these as contributing factors and modification of practices may alleviate much of the overstimulation of growth. Withholding needed water or nutrition is, however, not a good method to reduce excessive vigor that is caused by inherent site characteristics or other factors. Water deficits at the wrong time or to an excessive extent can cause severe stress to grapevines and negatively affect fruit yield or quality. Similarly, grapevine nitrogen deficiency has been correlated to low must nitrogen levels and implicated in problem fermentations (Watson et al., 2000).

A better approach might be to utilize cover crops and floor management techniques (see Chapter 20) to provide timely, controlled competition for soil moisture. Judicious management of water, fertilizer, and floor vegetation can be effective and powerful tools for influencing canopy characteristics and may reduce the need for some of the direct canopy manipulation practices described below (see Table 5).

Canopy Management Practices

Pruning. The first opportunity to influence a canopy is at dormant pruning time; pruning level determines the potential shoot density of the canopy. This is really the best time to assess vine vigor. The skeleton of last year's growth can tell you things that are very difficult to see during the growing season. If there are many short canes, tighter pruning (leaving fewer buds) is a remedy. If the growth is excessive (big canes, wide internodes), additional buds are called for. If the wire is full and the canopy is congested, conversion to a divided canopy training system (see below) may be a good option if the spacing can accommodate it. Again, pruning can be your first and best time to achieve balanced vines and a balanced vineyard.

Vigor Diversion. Another canopy management tool created at pruning time is vigor diversion (or growth diversion). It is intended for vines in which the growth has been excessive and is often employed in Oregon vineyards. The idea is to leave extra or "disposable" canes at pruning time, sometimes called "kicker" canes. These are allowed to grow until midseason, diverting growth from the selected fruiting canes, and are then cut off. Ideally, this disposable cane should be held out of the way on the trellis so that it does not interfere with spray penetration or shade the permanent canopy. Bethel Heights Vineyard in the Willamette Valley has had great success with growth diversion in one of our problem blocks.

Shoot Thinning and Suckering. Most vines push (grow) shoots from bud locations other than those left intentionally at dormant pruning. These "non-count" shoots should be removed early in the season, after most of these unplanned buds have pushed but early enough so that they can be easily broken off. Because the fruitfulness of the emerging shoots influences thinning decisions, the practice is usually delayed until

Table 5. Corrective Measures for Existing Vineyard Canopy Problems

Vineyard Condition	*Cause*	*Corrective Measure*
A. Too Shaded	1. Shoots too closely spaced because vines not pruned or trellised properly.	• Prune to 10–15 buds per pound of pruning weight. • If necessary, divide canopy vertically or horizontally using the appropriate training-trellis system; e.g., Scott Henry, GDC.
	2. Excessive growth because of too much nitrogen or soil moisture.	• Reduce nitrogen fertilization. Increase vine size by dividing canopy. • Plant grass cover crop in row middles and mow. • Hedge shoots to 14–20 nodes. • Remove secondary shoots and position shoots to reduce crowding. • Remove leaves in fruit zone.
	3. Within-row vine spacing too close, or canopy height/distance between canopy greater than 1.	• Remove alternate vines in the row. • Remove leaves in fruiting zone. • Top or hedge canopy to lower level.
	Rootstock too vigorous for the site.	• Use divided canopy trellis system. • Plant competitive cover crop in row middles. • Root pruning is another possibility.
B. Too Exposed	1. Lack of growth because of shallow soils, insufficient water or fertility; competition from weeds or cover crop; or too many buds left at pruning.	• Install drip irrigation system. • Fertilize with appropriate plant nutrients (N, P, K, etc.) or organic matter (compost/manure). • Control weeds; clean cultivate row middles. • Prune more severely or thin clusters after fruit set. • Control insect and fungal diseases.

the flower clusters have emerged. This fine tuning of shoot number down the wire is called "shoot thinning." At the "count" nodes, secondary and tertiary shoots are removed, along with some of the primaries if necessary. In some vineyards, weaker and unfruitful shoots are especially targeted in the shoot thinning process. A shoot spacing of about 2.5–3 inches along the wire is normal. Caution is advised to avoid excessive shoot thinning, particularly in highly vigorous vineyards. Removing too many shoots under such circumstances stimulates rapid growth of the remaining shoots during the flowering period, which can reduce the fruitfulness of these shoots.

The removal of shoots on the arms and trunk near the head is commonly called "suckering," although the shoots themselves are technically watersprouts; true suckers arise from the crown of the plant. The goal of shoot thinning and suckering is to produce an open canopy in which the leaves are well illuminated and well ventilated. In Burgundy, this pass through the vineyard to sucker and shoot-thin is usually also used as the opportunity to manage cropping level.

True suckers are also removed during thinning because, if retained, they grow up into the canopy, contributing to crowding. In most vineyards, a second trip to remove suckers is often advisable later in Summer; otherwise, it must be done at dormant pruning. Sucker removal under dry conditions may provide less favorable conditions for infection by the crown gall pathogen. In less vigorous vineyards, or those where the cultural costs need to be low, suckering only once at midseason may be adequate. Above all, conscientious suckering early in the vine's life saves a lot of money and effort as the plant matures.

Shoot Positioning. As the emerging shoots become long enough, it is common for them to be positioned with the aid of catch wires along the trellis. Each trellis system has its own particular positioning requirements, but the goal is always the same—to array the shoots along the trellis in such a way that each shoot receives good sun exposure and ventilation. Shoot positioning, which usually requires several passes through the vineyard, is a time-sensitive operation. Catching the shoots at the right time is much more efficient than trying to tidy up a vineyard that is already sprawling all over the ground. The timing of the first pass is especially important on most clones of Pinot noir, which have a strongly trailing growth habit.

Hedging. Hedging (summer pruning) is commonly practiced on vertically shoot positioned vines when shoot growth becomes too long to be supported by the trellis wires. It can be accomplished either with a tractor-mounted machine or manually with long knives like those used in the Christmas tree industry. The purpose of hedging is to prevent unsupported shoot growth from drooping over and shading the rest of the canopy. Hedging is usually done when the shoots have grown well beyond the top wire and are just beginning to lean over. Shoots are commonly trimmed back to about 6–8 inches above the top wire.

Hedging may also be necessary on non–shoot-positioned training systems, to shorten stray shoots or long laterals. These may extend into the alleyways, shading the fruit zone and making equipment passage difficult. An early, light hedging done at bloom can increase fruit set in some situations (Smart and Robinson, 1991).

Hedging removes the shoot tip, which stimulates growth of lateral shoots from the nodes immediately below the cut position. This is particularly true for vigorous shoots, and the regrowth of laterals often necessitates repeated hedging.

The amount of hedging required in any particular block depends on the inherent vigor of the vine and the fertility of the soil. At Bethel Heights Vineyard, a couple of blocks rarely require hedging, but most need a few passes with the hedger before shoot growth slows. In tightly spaced or overly vigorous blocks, hedging may be required just to keep the aisles open for the passage of workers and equipment. In most Burgundian vineyards, for example, five hedgings are usually necessary. The goal of hedging is to maintain an open and well-illuminated canopy while retaining as many healthy leaves as the trellis can contain. Too much hedging or hedging late in the season delays ripening. Proper timing is critical.

Leaf Removal. The pulling of leaves in the fruit zone has become common practice in most Oregon vineyards. The goal is to allow the fruit to dry as quickly as possible after dew or rain, and to expose the clusters to sunlight. The result is a lower incidence of fungal diseases and improved wine phenolics. Most growers pull leaves in July, after fruit set but early enough to allow the clusters to acclimate to the sun before they begin to soften. Early leaf pulling is very important for controlling disease, reducing humidity in the fruit zone, and facilitating spray penetration onto the grapes.

The amount of leaf pulling done by Oregon growers varies greatly. Some pull only on the morning sun side of the plant; others pull all leaves in the fruit zone. Growers with an east–west row orientation should be especially cautious about pulling on the south side of the plant because of the risk of sun damage and excessive phenolics. Growers with a training system that positions the fruit at the top of the trellis should also be cautious with leaf pulling. The midday sun can burn fruit that has not been adequately acclimated to its effects.

Because of our relatively abundant labor supply, Oregon growers usually pull leaves by hand. Mechanical leaf removal is more common in other parts of the New World. A tidy canopy is required for most of these technologies. The machines are expensive, but so is manual removal. Leaf pulling by hand is more precise, but the hurricane force airstream used in some mechanical technologies can clear disease-related debris out of the clusters as it removes the leaves.

Change Training System. Much of the preceding discussion concerns the annual canopy management practices that Richard Smart would call "temporary solutions" to canopy management problems. Sometimes the best option is a more permanent solution—conversion to a divided canopy training system, if the spacing permits.

Divided canopy systems are designed to display a large canopy efficiently and can enable doubling the number of buds retained at pruning. Previously "overvigorous" vines are given additional trellis space for more shoots, which can improve the balance between leaf area and crop load. The Scott Henry training system and the related Smart-Dyson Ballerina system achieve this by dividing the canopy vertically; the Geneva Double Curtain system is horizontally divided.

At Bethel Heights Vineyard, we have employed four different systems to address excessive vigor problems. Probably the easiest retrofit in most vineyards is the Scott Henry system or the Smart-Dyson Ballerina system. In blocks where vigor is excessive, both yield and quality are often improved with a divided canopy thanks to a more appropriate vine balance.

For growers who are considering a change in training system, it can be very useful to visit vineyards where these systems are in use. There are excellent examples of all of these systems scattered through our industry. The growers who have built and managed these systems can be your best resources for information on how to bring your vineyard into balance.

Insufficient Canopy

Vineyards in Oregon that are too exposed by insufficient canopy do exist but are not common. Identifying and eliminating the causal factors can generally correct this condition. Investigate the possibility of disease- or insect-related devigoration. Excessive competition from floor vegetation can be reduced by various cover crop and floor management strategies. Installation of a drip irrigation system can eliminate water stress as a factor. Nutritional inadequacies can be identified and corrected with fertilizer applications. Pruning should be adjusted accordingly, perhaps retaining eight to ten buds per pound of pruning weight.

References and Additional Resources

Jackson, D. I., and P. B. Lombard. 1993. Environmental and management practices affecting grape composition and wine quality: a review. *American Journal of Enology and Viticulture* 44:409-430.

Kliewer, W. M. 1982. Vineyard canopy management. In A. D. Webb (ed.), *University of California Davis Grape and Wine Centennial Symposium Proceedings.* Department of Viticulture and Enology, University of California, Davis.

Kliewer, W. M., and R. E. Smart. 1989. Canopy manipulation for optimizing vine microclimate, crop yield and composition of grapes. In C. Wright (ed)., *Manipulation of Fruiting,* Proceedings of the 47th Easter School in Agric. Sci. Symp, April, 1988, Alneu, Nottingham. Butterworths, London.

Smart, R. E. 1985. Principles of grapevine canopy microclimate manipulation with implications for yield and quality: a review. *American Journal of Enology and Viticulture* 35:230-239.

Smart, R. E. 1992. Canopy management. In B. G. Coombe and P. R. Dry (eds.), *Viticulture,* Vol. 2: *Practices.* Winetitles, Adelaide, Australia.

Smart, R. E., and M. Robinson 1991. *Sunlight into Wine.* Winetitles, Adelaide, Australia.

Watson, B. T., E. Hellman, A. Specht, and H. P. Chen. 2000. Evaluation of nitrogen deficiencies in Oregon grapevines and musts. *Oregon Wine Advisory Board Progress Reports 2000–2001.* Agricultural Experiment Station, Oregon State University, Corvallis.

23

Management of Diseases

Jay W. Pscheidt, Walter Mahaffee, Robert R. Martin, and John N. Pinkerton

This chapter provides an introduction to the management of the principal grape diseases in Oregon. Symptoms of the diseases are important for diagnosing the cause of vineyard problems, and effective control strategies often require some knowledge of the pathogen's life cycle. Both of these topics are extensively covered in several standard references (Flaherty et al., 1992; Pearson and Goheen, 1988) so are not covered here. The emphasis instead will be on disease management strategies. Specific chemical disease control products and useage practices frequently change, so the reader is referred to the *Pest Management Guide for Wine Grapes in Oregon,* produced by Oregon State University Extension and revised annually. A comprehensive website, An Online Guide to Plant Disease Control, is produced by Oregon State University Extension and provides disease symptom descriptions and photos, pathogen descriptions, and the latest control recommendations.

There are a huge number of diseases that can afflict grapevines (Table 1). Oregon's relatively young grape industry has only a few of these problems, but a few can cause significant losses. Diseases were consistently responsible for a loss of 1–7% of Oregon's grape production during the 1990s.

Perennial disease problems that require annual attention in Oregon include powdery mildew, botrytis bunch rot and blight, and crown gall. Eutypa dieback is a problem on the increase around the state as vines age. In certain situations, armillaria root rot and phomopsis cane and leaf spot have been nuisances in Oregon vineyards. Several viruses such as corky bark, Kober stem grooving, leafroll, and rupestris stem pitting could become problems if current quarantines are ignored. Although several nematodes are present, including dagger, ring, and root-knot nematodes, their relative importance has not been fully assessed. Diseases to be on guard against since they pose a threat to vine growth and yield include black rot, downy mildew, and tomato ringspot virus decline, as well as damage from the root-lesion nematode. Some diseases that Oregon does not currently have a problem with could become additional economic burdens when they arrive. Also, as grafted vines become more commonplace, some diseases may take on new importance.

Principal Diseases Caused by Fungi and Bacteria

Powdery Mildew, *Uncinula necator*

This disease can be found in most grape-growing areas of the world. Uncontrolled, powdery mildew reduces vine growth and yield and negatively affects fruit and wine quality. As little as 3% diseased berries can be detected as off flavors in the wine. The costs associated with this disease are tremendous considering the frequent fungicide applications as well as potential yield loss. Direct yield losses in Oregon have been as high as 13% for some varieties such as Chardonnay. All vineyard managers must have a good understanding of this disease and a flexible management plan for its control.

Leaf blades, petioles, flowers, berries, and cluster stems are susceptible to infection. Cluster stems become brittle and may break as the season progresses. Cluster infection before or shortly after bloom may result in poor fruit set and considerable crop loss. Berries are susceptible to infection until their sugar content reaches about 8%, although established infections continue to produce spores until the berries contain 15% sugar. If berries are infected before they attain full size, the berry may split, dry up, or rot.

Combinations of cultural and chemical management tactics are needed to prevent losses from powdery mildew. Good viticultural practices and tactics used for other disease and pest problems also aid in the management of powdery mildew.

Table 1. Grape Diseases in Oregon

Common Name	*Causative Agent*	*Presence in Oregon (or Pacific Northwest)*[a]	*Relative Importance to Grape Production in Oregon*[b]
Fungi			
Angular Leaf Spot	*Mycosphaerella angulata*	NO	Low–None
Anthracnose	*Elsinoe ampelina*	NO	Low
Armillaria Root Rot	*Armillaria mellea*	YES	Low
Bitter Rot	*Greeneria uvicola*	NO	Low
Black Dead Arm	*Botryosphaeria stevensii*	NO	?–Low
Black Rot	*Guignardia bidwellii*	NO (Wash., Yes)	Low–Medium
Botrytis Bunch Rot and Blight	*Botrytis cinerea*	YES	Medium–High
Dematophora Root Rot	*Rosellinia necatrix*	NO	Low–?
Diplodia Cane Dieback and Bunch Rot	*Diplodia natalensis*	YES	Low
Downy Mildew	*Plasmopara viticola*	NO	High
Esca and Black Measles	unknown	?	?
Eutypa Dieback	*Eutypa lata*	YES	High–Medium
Grape Root Rot	*Roesleria hypogaea*	NO	Low
Leaf Blight	*Pseudocercospora vitis*	NO	Low
Leaf Blotch	*Briosia ampelophaga*	NO	Low
Macrophoma Rot	*Botryosphaeria dothidea*	NO	Low–None
Misc. Berry Rots	Many	YES	Low–Medium
Phomopsis Cane and Leaf Spot	*Phomopsis viticola*	YES	Low
Phymatotrichum Root Rot	*Phymatotrichum omnivorum*	NO	None–Low
Phytophthora Crown and Root Rot	*Phytophthora spp.*	Not Confirmed	Low
Powdery Mildew	*Uncinula necator*	YES	High
Ripe Rot	*Colletotrichum gloeosporioides*	NO	Low–None
Rotbrenner	*Pseudopezicula tracheiphila*	?	?–Low
Rust	*Physopella ampelopsidis*	NO	Low
Septoria Leaf Spot	*Septoria ampelina*	NO	Low
Verticillium Wilt	*Verticillium dahliae*	NO	Low–?
White Rot	*Coniella diplodiella*	NO	Low
Zonate Leaf Spot	*Cristulariella moricola*	NO	Low
Bacteria			
Bacterial Blight	*Xylophilus ampelinus*	NO	?
Crown Gall	*Agrobacterium vitis*	YES	Own Root, Low–Med Grafted, High
Pierce's Disease	*Xylella fastidiosa*	NO	? (suspect Low)
Yellows			
Boir Noir and Vergilbungskrankheit	unknown	? (NO)	?
Flavescence Doree	Phytoplasma	NO	Low
Viruses			
Corky Bark	?	YES	Own Root, None Grafted, Medium
Fanleaf Degeneration	Grapevine fanleaf virus	NO	Very Low
Kober Stem Grooving	?	YES	Own Root, None Grafted, Low–Med
Leafroll	?	YES	Medium–Low
Peach Rosette Mosaic Virus Decline	Peach rosette mosaic virus	NO	Low

table continues

Common Name	Causative Agent	Presence in Oregon (or Pacific Northwest)[a]	Relative Importance to Grape Production in Oregon[b]
Rupestris Stem Pitting	?	YES	Own Root, None Grafted, Low–Med
Tobacco Ringspot Virus Decline	Tobacco ringspot virus	Grapes, No Other Crops, Yes	Low
Tomato Ringspot Virus Decline	Tomato ringspot virus	Grapes, No Other Crops, Yes	Medium
Nematodes			
Citrus Nematode	*Tylenchulus semipenetrans*	NO	None
Dagger Nematodes	*Xiphinema americanum*	YES	?
	Xiphinema index	NO	Medium
Pin Nematodes	*Paratylenchus* spp.	YES	?
Root-Lesion Nematode	*Pratylenchus vulnus*	NO	Medium
Ring Nematode	*Mesocriconema xenoplax*	YES	?
Root-Knot Nematodes	*Meloidogyne* spp.	YES	?
Other			
Young Vine Decline	various	YES	?

[a] YES = It has been reported or found in Oregon; NO = It has not been reported (or found) in Oregon; Not Confirmed = someone believes it is present but lab tests have not confirmed its presence or it has not been reported; ? = unknown presence.

[b] High = has the potential to affect vine growth or yield severely; Medium = may affect vine growth or yield; Low = not expected to become an industry-wide problem; None = not a threat to grape production; ? = unknown threat.

Cultural Management. Cultural practices can reduce the amount of disease and can increase the effectiveness of chemical control. Viticultural practices and vineyard sites that encourage lush, vigorous growth are very conducive to disease development. Preventing excessive vigor and growth helps manage powdery mildew. This can be accomplished through the proper selection of rootstocks, training systems, irrigation and fertilization practices, and canopy management practices. Training systems that allow good air movement through the canopy and prevent excessive shading are helpful. Some growers practice hedging to reduce canopy size and shading. Timely sucker control and topping canes on vertical trellises reduce the amount of susceptible tissue that is not normally sprayed with fungicide. Leaf removal to control bunch rot improves fungicide coverage of the clusters.

Chemical Management. The goal of chemical management is to reduce the number of powdery mildew colonies that form at the beginning of the growing season and keep the rate of disease increase to a minimum.

The use of chemicals to reduce the amount of overwintering inoculum has met with limited success in various viticultural areas. Although it has not been formally investigated in Oregon, many growers make applications that target this part of the lifecycle. These include a spray of lime sulfur during the dormant season to kill overwintered cleistothecia or micronized sulfur at 100% bud break to reduce primary infections. The lime sulfur spray should be directed to cover only the trunks and cordons, since this is the place cleistothecia overwinter. Even with these spray applications, a normal season-long spray program is still needed for adequate control of powdery mildew.

The cornerstone of most powdery mildew management programs is the regular use of foliar fungicides from prebloom through verasion. This practice reduces the rate of disease increase during the growing season. There is a wide choice of effective fungicides to choose from including synthetic and organically acceptable materials. These include benzimidazoles, demethylation inhibitors (DMI), strobilurins, sulfurs, oils, soaps, and bicarbonates.

Fungicide applications should begin when shoot growth reaches 6 inches and continue at regular intervals through the growing season. The strongest materials and shortest intervals should be used from prebloom through the end of bloom. Many growers stop applications around veraison, although powdery mildew can continue to increase on green tissues. A postharvest application may help control late-season infections in some years. Thorough, complete coverage of all actively growing tissue is essential for good control.

Prevent the development of resistant fungi by alternating use of fungicides with different modes of action. Limiting the number of applications during the same growing season of a fungicide with the same mode of action also helps. Use of many of these materials once powdery mildew is out of control can also lead to resistance.

Sulfur is one of the most commonly used materials in a powdery mildew management program. It is also useful to control several mite problems. Sulfur is commonly applied as a dust or wettable powder. Much of the fungicidal activity of sulfur is associated with its vapor phase. The optimal temperature range for sulfur activity is 74–86°F, and the fungicide may not be effective below 64°F. Sulfur is less active in humid air than in dry air.

There are several forecasting programs for scheduling fungicide applications. The standard Oregon phenology-based program begins applications at 6 to 8 inches shoot growth and continues at regular intervals based on grapevine development. The Gubler-Thomas U.C. Davis program uses leaf wetness and temperature early in the year to predict ascospore infection periods and only temperature during the summer to predict conidial infection periods. The New York (Gadoury) program is based on rainfall and temperature. The Kast (Oi Diag) program incorporates relative humidity along with temperature and rainfall. All programs have been effective at timing fungicides and controlling powdery mildew.

Biological Control. The fungus *Ampelomyces quisqualis* has been formulated and registered for use against powdery mildew. Although effective in Oregon tests, it should not be used as a standalone treatment, for it produces unacceptable results. Exactly where to position this material in a season long program remains unknown. The manufacturer advises its use with a mineral oil–based surfactant (which may be as effective alone). It should be applied during early evening or morning when humidity or leaf wetness are high. It also is not compatible with certain fungicides such as sulfur and soaps.

Botrytis Bunch Rot and Blight, *Botrytis cinerea*

The gray mold fungus, *Botrytis cinerea,* is not specific to grapevines and lives as a normal resident on necrotic, senescent, or dead tissue as well as infecting almost all dicotyledons, many monocotyledons, and some pteridophytes. Because of our relatively mild climate and abundance of alternative hosts, *B. cinerea* spore loads in the air during the wet months are extremely high, especially in the Willamette Valley. These high spore loads serve as a continuous source of inoculum during these periods. Botrytis bunch rot can be particularly damaging to Oregon vineyards in years when fall rains occur early.

Diseases caused by *B. cinerea* generally appear as blights (sudden severe necrosis of tissue) or rots (decomposition and putrefaction of tissue) of various plant tissues. In early spring, buds and young shoots may become infected. During flowering, infection of flower parts can appear as necrotic spots. Under humid or wet conditions the characteristic gray cottony sporulating mycelia can appear from infected tissues, thus the common name "gray mold." In late summer or fall, berries start showing symptoms from latent infections or become newly infected. Often, rotted berries are near the center of the bunch. The rot can spread quickly and may engulf the entire bunch. Additionally, other fungi and bacteria can invade the rotting berries, producing a large variety of colors, smells, and tastes. Occasionally, immature berries may develop a soft brown rot early in summer under favorable conditions.

Although *B. cinerea* is generally considered to be an extremely damaging pathogen, under certain environmental conditions and on specific white varieties infection by *B. cinerea* can result in a desirable condition called "noble rot." The wines produced from these grapes are exceptionally sweet.

No single measure gives effective control of *B. cinerea* during grape production. Disease management depends on the integration of cultural methods and pesticide (biological or chemical agents) applications.

Cultural Management. Cultural management of *B. cinerea* can be accomplished through canopy management and sanitation. Through canopy management, the microclimate around the berry clusters can be manipulated to reduce duration of berry wetness by increasing airflow, light penetration, and temperature. This alteration of the berry microclimate can be done through shoot positioning, hedging and leaf removal. The canopy should be maintained so that light and airflow can penetrate through the canopy.

Leaf removal increases the airflow and light penetration even more. This is particularly important in the growing areas prone to rain after fruit set and just before harvest. Leaf removal should be done at the berry shatter stage. Later removal may be ineffective to prevent infections. Generally, leaves are removed from the east or north side to avoid sunscald of berries.

Sanitation consists of removing the previous year's infected tissue (clusters and canes) from the vineyard. Removal of fruit dropped for yield adjustment may also be warranted. In wet falls the dropped fruit could serve as an extremely large source of *B. cinerea* inoculum. Plant material waste should be removed from the vineyard and burned or disposed of off site.

In years when dry weather prevails during harvest, cultural practices are often sufficient for bunch rot control.

Chemical Management. Effective use of pesticides depends on thorough coverage of plant tissue, particularly within the middle of the plant canopy. These tissues are probably more susceptible to infection because of the presence of debris on the leaf surface and senescing tissue. Proper coverage begins with a well-calibrated sprayer. It also depends on spray volume and droplet size.

Numerous fungicides are available for application to grapes, but resistant *B. cinerea* populations have developed to some of these compounds. To manage resistance development effectively, it is extremely important to use mixtures of or rotate with different classes of chemicals that do not already have resistant *B. cinerea* populations in your production system. Some laboratories can test *B. cinerea* isolates for resistance to pesticides (e.g., Plant Disease Clinic, Oregon State University).

Make applications at the end of bloom or beginning of fruit set, just before berry touch, start of veraison, and three weeks before harvest. Sprays later in the growing season are preferable to earlier ones. Applications right after a damaging hail event may help prevent some damage to young shoots and vines.

Biological Control. A product containing a protein from *Bacillus subtilis* (strain QST 713) as the active ingredient is available for use on grapes. It has shown variable efficacy in tests in western Oregon.

Crown Gall, *Agrobacterium vitis*

Crown gall is typically a problem after winter freezes that injure grapevines. Not all grape-growing regions experience cold injury each year, so the problem is sporadic at any one time throughout Oregon. Vines often can be recovered if grown on their own roots, but the disease takes on much greater importance in vineyards with vines grafted to rootstocks.

Crown gall disease is named after its principal symptom, fleshy galls produced in response to infection. Galls are composed of disorganized primary and secondary phloem tissues from the grape. Fleshy galls typically are formed at the crown and on the first 2 feet of the vine above the soil line, but they can also be found along damaged canes, the base of rooted cuttings, sucker removal sites, and at graft unions.

Large galls may develop rapidly and completely girdle and kill young vines in one season. Galled vines frequently produce poor shoot growth and few clusters. If fruit are formed they generally do not ripen. Portions of the vine above the galls may die. Galls are rarely seen on roots, but the bacteria may cause areas of dark necrotic lesions on the roots. Crown gall can also occur at the graft union and be confused with callus formation. Isolation of the bacteria from these areas is needed to confirm crown gall adequately.

The primary management of crown gall is through cultural methods. Chemical treatments can be made, but their efficacy has not been satisfactory.

Cultural Management. The planting of pathogen-free nursery stock on soils that are free of the crown gall bacteria may offer the best long-term management. Research vineyards planted using this goal in other areas have shown much less crown gall development.

The production of pathogen-free stock can be achieved through hot water treatments or shoot tip propagation. Dipping dormant cuttings in hot water (129°F or 54°C) for 30 minutes may help eliminate the bacteria. Primary buds are killed, but secondary buds grow out well. Some grape varieties may, however, be more sensitive to heat treatment than others. The bacteria are rarely found in the green shoot tips of vines. Propagating plants using this method has produced pathogen-free vines and vineyards free of crown gall.

Avoiding any injury near the base of the vine helps reduce disease problems in established vineyards. This is especially true when using mechanical weed control devices. Practice timely sucker removal and be careful not to get dirt into these wounds. Frequently disinfect all grafting equipment, especially between varieties or different sources of scion or rootstock.

Once crown gall has appeared in the vineyard, there are not many management options. It is advised to remove seriously diseased vines from the vineyard and burn them. Vines with less extensive disease may be recovered if the gall is high enough on the trunk to enable a sucker replacement to be brought up from well below the galled areas. This may be difficult or impossible on grafted plants. Sterilize pruning tools with 10% Clorox (bleach, which also oxidizes pruning equipment) or shellac thinner (70% ethyl alcohol) before working on healthy vines.

Chemical Management. Chemicals used to treat established galls have given variable results and are not worth the time and expense. For example, a mixture of 2,4-xylenol and metacresol (Gallex) is registered for use on grapes. This mixture is painted on very young galls to reduce further development. Galls may return the next year or, if treated late, may continue to develop. Tissue surrounding the gall may be injured especially on younger vines.

Biological Control. Unfortunately, the bacterium used to control crown gall in other crops, strain K84 of *Agrobacterium radiobacter*, is not active against the grape crown gall bacterium. It is, however, an active area of research that may result in effective agents in future years.

Eutypa Dieback, *Eutypa lata*

Eutypa dieback is a problem on the increase around Oregon as grapevines age, reflecting the long lifecycle for the disease, and, in some cases, vineyard conversion from cordon to cane pruned vines. It is rarely seen until vines are at least eight to ten years old. Early recognition and management may help avoid an industry-wide problem well into the twenty-first century.

This is a disease of the woody tissues of the vine, killing the vascular cambium and adjacent vascular tissues. These dead areas, known as cankers, are generally found centered on an old pruning wound. Symptoms may not appear on diseased vines for more than three years after infection. Cankers continue to expand each year; they expand length-wise in both directions from the wound and girdle and kill arms or trunks of infected vines in five to ten years. A study of infected Concord vineyards in eastern Washington found that moderately infected vineyards can lose 19–50% of yield, and severely affected vineyards can have 62–94% yield loss.

Cultural Management. Recognizing conditions that lead to infection and taking preventative action help in the long-term avoidance and management of this disease. Early detection and recognition of infected vines also help.

Avoid large pruning cuts when possible. The disease usually gets started when large cuts are made to remove or renew major vine structures such as arms or cordons. Chemicals that prevent spore germination and early colonization can be used on the cut surfaces and are discussed below. Avoid pruning during and before wet weather. This may not be practical in western Oregon, but it could be useful in more arid areas. One technique that may be effective is to leave long pruning stubs when making big cuts in the trunk or cordon during rainy weather. The stubs (which may become infected) are removed during the following summer when conditions are dry and the chance of reinfection low.

Once the disease is in the vine, it must be cut out to ensure future productivity. In spring, when symptoms are evident, mark diseased vines. Remove diseased wood 4–6 inches below the canker and train a new, healthy shoot into position. This may be difficult if the canker is low on the trunk and near the graft union. If the canker extends below ground, remove and replace the entire vine. This disease is so severe in the eastern U.S. that two trunks are routinely used per root system. With this method an infected trunk can be removed, and twice as many buds are retained on the other trunk to maintain fruit production.

Infected vine debris should be removed from the vineyard and destroyed by burning. The fungus can continue to sporulate on old trunks left in or near the vineyard for several years.

Chemical Management. The fungicide benomyl was previously labeled for eutypa control, but it is no longer available. Other chemicals have been investigated but have not been as effective. Dreft baby diaper detergent (30% weight/volume solution) applied to freshly cut surfaces has been shown to have wound protectant activity.

Armillaria Root Rot, *Armillaria* spp.

There are many fungi that attack the roots of grapes, but armillaria root rot is most likely to be encountered in Oregon. It is frequently a problem on land that has been recently cleared of natural vegetation or old orchards. *Armillaria mellea* is the species associated with conifer mortality in the Pacific Northwest. Although *A. mellea* has not yet been positively identified, it most likely is the one found on grapevines.

Aboveground symptoms are not diagnostic and are typical of any root problem where water and nutrients are not transported to the leaves. These may include rapid wilting, slow decline, lack of vigor, stunting, chlorosis, or sunburn.

Little can be done once grapevines are infected. Infected vines should be dug out and destroyed. Removal sites should be left fallow for a few years, or the soil can be spot-fumigated before replanting.

The best control measures are implemented well before planting. Before new land is cleared or an old orchard removed, girdle the larger trees (in the summer) to reduce the nutritional reserves of the roots and to begin the decay process. Remove the old and dead vegetation, including all roots greater than 1 inch in diameter. Burn the roots on location to prevent movement to other areas. If at all possible, leave the ground fallow for at least one year to ensure the decay of roots left behind.

Soil fumigation has been the best (but limited) means of control. Contrary to soil fumigation for nematodes, the soil conditions should be as warm and dry as possible. Fumigation should be done in the late summer prior to any rain, and the fumigant should be placed as deep as possible. Some problem spots may still appear because of infected, large roots deep in the soil from the previous vegetation.

Trenches lined with plastic sheeting may help prevent growth of rhizomorphs if they originate from an adjacent stand of native vegetation.

Phomopsis Cane and Leaf Spot, *Phomopsis viticola*

Phomopsis has been found in vineyards west of the Cascades but has not been associated with much damage. Several different forms of the fungus have been found in Australia. Only one form caused the typical spotting on new shoots. All forms, however, bleached canes. Historically, the disease known as "dead arm" was associated with this fungus. The dead arm symptoms, however, are now known to be those caused by eutypa dieback.

Susceptible varieties may show elongated lesions or tears on the basal internodes of canes. Small leaf spots also can appear early in leaf development, which can severely crinkle or misshape the leaf. Leaf petioles and cluster rachises can also develop lesions in severe years. Fruit rots can occur but have not been associated with this fungus in the Pacific Northwest.

The disease is easily managed through pruning practices and early-season spray applications. Remove infected canes during normal pruning operations during the dormant season. Debris from previous years such as old rachises and petioles should also be removed.

Several different chemicals are effective at protecting young tissues from infection. Some labels still carry the old "dead arm" name for this disease. Spray when shoots are 0.5–1 inch long and again when shoots are 5–6 inches long. Control is not needed most years in the Pacific Northwest.

Potential or Less Common Diseases Caused by Fungi

Black Rot, *Guignardia bidwellii*

This disease has been reported in Washington State but not in Oregon. It is another disease native to eastern North America that could become a problem in western Oregon if introduced. The disease causes spots on leaves, petioles, tendrils, shoots, and fruit. Fruit infections begin as small white spots that gradually enlarge. Within a few days the berry begins to shrivel and dry like a raisin, but becoming very hard. Only new, immature growth is susceptible to infection. No infections occur once leaves and berries mature. It is hard to guess how the disease might behave in western Oregon. Spring rains would be conducive to disease development, but arid summer conditions would inhibit further development. The disease might become a nuisance, causing leaf spot, but only rarely threaten fruit yield in years with rainy weather up to and past bloom.

The quarantines in place at this writing should help reduce the possibility of introducing the disease into the area. Once here, simple sanitation measures (removal and destruction of infected plant debris) and maybe a protective fungicide application or two would keep the disease to a minimum. Some fungicides already used for powdery mildew control are effective against this disease.

Downy Mildew, *Plasmopara viticola*

The fungus *Plasmopara viticola* has not been reported on grapevines from the Pacific Northwest. However, in 2001, the downy mildew fungus was detected on Boston ivy plants in the landscape or grown as nursery plants in Multnomah, Washington, and Marion counties of Oregon. This represents a serious threat to the grape industry. Weather west of the Cascade Mountains would be highly conducive to disease development if the pathogen ever established on grapevines. All growers should be on guard for this disease, for it has the potential to cause significant crop loss if left untreated. If it becomes established in Oregon, it could change production practices and increase production expenses.

Downy mildew can cause lesions on any green part of the vine. Severely infected leaves may drop and shoot tips can die back. Vines may experience premature defoliation. Immature berries can also be infected; infected fruit remains firm compared to healthy berries, which soften as they ripen.

Continued use of quarantines and disease scouting are the best ways to keep downy mildew out of the region. Do not bring infected plant debris or soil into the region. Also, do not transport infected stock into the region. Inquire about this disease before purchasing material from out of state.

Scout vineyards for any sign of the disease. Check newly planted stock originating from out of state. If downy mildew is suspected, get diseased plant samples to the Oregon State University Plant Disease Clinic as soon as possible to confirm the diagnosis. These will be processed free of charge. We would much rather see one hundred samples that are not downy mildew than to miss the one that introduces the disease into Oregon.

The Pacific Northwest does not have this disease, so there is no need for chemical applications specifically for its control. Many other viticultural areas use copper products at some time during the season, partially for control of downy mildew. Use of these materials is questionable under our conditions. Some products registered for powdery mildew control, such as the strobilurins, also have a side benefit of downy mildew control.

Young Vine Decline

Young vine decline (also referred to as black goo or black xylem disease) is actually a disease complex for which there are potentially multiple pathogens that can cause similar symptoms. An individual pathogen or a complex of any of them can infect young vines and result in a general decline. There are at least two distinct diseases associated with the problem: cylindrocarpon black-foot disease, caused by

Cylindrocarpon obtusisporum and C. destructans; and phaeoacremonium young vine decline, caused by *Phaeacremonium aleophilum, P. chlamydosporum, and P. inflatipes.* These fungi have also been shown to be associated with the disease measles, or esca. Young vine decline is generally associated with grafted vines in the first ten years of establishment, and is not specific to any scion-rootstock combinations.

Good vineyard management techniques, both before and after planting, are the best preventative control for this disease. These include proper planting practices, irrigation, fertilization, and weed control; reduce or avoid any sources of devigorating stress. In addition, proper selection of plants for planting is important. Cull any plants with a weak graft union, poor callus formation, or small, spindly root system. In California, young vine decline is commonly associated with vines that were fruited in their second or third year.

Nematodes

Nematodes are microscopic, unsegmented roundworms that are residents in all soil. Most soil-inhabiting nematodes are free-living, feeding on bacteria or fungi, but some nematodes are parasites of plants. These plant-parasitic nematodes have hypodermic needlelike mouthparts, which they use to puncture plant cells and suck out the cell contents. Plant-parasitic nematodes can damage plants by killing the cells at the site of feeding or by modifying root growth, that is, stopping root elongation or producing galls. Nematode feeding can disrupt root function by reducing water uptake and translocation and by modifying carbohydrate partitioning in the plant. Nematodes can interact with other plant pathogens by increasing the severity of some fungal and bacterial pathogens and by vectoring viruses.

A survey of seventy vineyards in the Willamette Valley and southern Oregon found five nematode genera that are reported to damage grapes (Pinkerton et al., 1999). Ring nematodes *(Mesocriconema),* dagger nematodes *(Xiphinema),* root-lesion nematodes *(Pratylenchus),* and pin nematodes *(Paratylenchus)* were recovered from more than 85% of the vineyards, and 10% of vineyards had detectable populations of root-knot nematodes *(Meloidogyne).* There is little experimental data on the economic loss caused by nematodes in established vineyards in Oregon. Damage thresholds for population densities of various nematodes have, however, been established for California and other regions. In the Oregon survey, *Mesocriconema xenoplax, Xiphinema americanum, and Meloidogyne hapla* were found in 20%, 8%, and 5% of vineyard blocks, respectively, at population densities reported to cause moderate yield loss in California (McKenry, 1992). Based on growers' observations of vine vigor and yields in nematode-infested vineyards, however, nematodes do not appear generally to cause economic damage to established vines in Oregon. Since most vineyards in Oregon are young compared to those in other regions where plant-parasitic nematodes are a major constraint in grape production, the long-term impact of nematodes in Oregon vineyards is unknown.

Nematodes Potentially Important in Oregon Vineyards

Ring nematode. *Mesocriconema xenoplax* is a migratory ectoparasite with a wide host range. This nematode prunes the roots, causing darkening and poor development of feeder roots. Vines in several Oregon vineyards infested with *M. xenoplax* were stunted and had low yields. In greenhouse trials conducted at USDA-ARS in Corvallis, two rootstocks, 420A and 101-14, demonstrated tolerance or resistance to *M. xenoplax* (Pinkerton, unpublished data). Diagnostic soil samples should be collected in early spring.

Dagger Nematode. *Xiphinema* is a migratory ectoparasite with a wide host range. These nematodes feed at the root tips, causing swollen root tips and root stunting. *Xiphinema americanum* can reduce vine vigor in other regions but has not been associated with damage in Oregon. *Xiphinema americanum* and *X. pachtaicum* have been associated with unthrifty vines in eastern Washington. Several dagger nematode species are vectors of nepoviruses. These viruses have not been detected in grapevines in Oregon, but tomato ringspot virus is present in other crops in the Willamette Valley. *Xiphinema index,* a major pathogen of grape and the vector of grape fanleaf virus, has not been found in the Pacific Northwest. Soil samples should be collected in the early spring when population densities of *Xiphinema* are highest.

Root-knot Nematode. These nematodes are a major pest of grapevines in most areas of the world. *Meloidogyne* is a sedentary endoparasite that causes root galls and can reduce root growth and vine vigor. The northern root-knot nematode, *M. hapla,* which is found in Oregon and Washington, is not as damaging a root-knot nematode species as found in warmer climates. Injury to vines is greater in sandy to sandy loam soils compared to clay soils, which explains why *M. hapla* has impacted vineyards in eastern Washington more than vineyards in Oregon. Soil and root samples should be collected in the early fall.

Root-lesion Nematode. *Pratylenchus* is a migratory endoparasite with a wide host range. These nematodes produce necrotic lesions on roots and can girdle feeder roots as these necrotic areas coalesce. The species found in Oregon vineyards do not appear to be causing economic damage. *Pratylenchus vulus*, an important parasitic nematode of grape worldwide, has not been found in Oregon or Washington. Soil and root samples should be collected in early fall.

Pin Nematode. *Paratylenchus* is a small, ectoparasitic nematode. These nematodes are not known to reduce yield or quality of grape in Oregon.

Management

Oregon vineyard managers have rarely observed unacceptable yields in nematode-infested vineyards. Vineyard trials with synthetic chemical and biological nematicides have been conducted in Oregon. Under conditions in the trials, these nematicides failed to reduce nematode densities, predominantly *M. xenoplax*, or increase vine vigor. Additional trials are in progress in Oregon and Washington vineyards.

In areas of vineyards affected by plant-parasitic nematodes, the management should be modified to reduce other stresses on the vines, for example, change irrigation practices or remove competing ground covers. Although plant-parasitic nematodes have been shown to reduce the establishment of new vines significantly in microplot experiments in Oregon, little is known about their impact when replanting commercial vineyards. This subject warrants further investigation before management recommendations can be developed.

Soil fumigation reduces nematode population densities prior to planting but has been used in Oregon infrequently. Some rotation or cover crops, which can suppress nematodes before planting grape, are being evaluated in research trials by the USDA-ARS in Corvallis. Soil solarization, the process of heating soil under a clear plastic film laid on the soil surface, can control soilborne pathogens. In a trial in southern Oregon, solarization significantly reduced densities of *M. xenoplax* and increased by 300% the growth of one-year-old vines. Rootstocks resistant to several of the nematodes found in Oregon vineyards are available and should be considered when planting vines on nematode-infested sites (McKenry, 1992; Nicol et al., 1999; Ramsdell et al., 1996).

The Oregon winegrape industry is young, as is our understanding of the role of plant-parasitic nematodes in grape production. The situations in which nematodes damage grapevines, as well as viable management strategies to meet these situations, will be defined through further research and feedback from the Oregon winegrape industry.

Viral Diseases

Forty-five viruses have been reported to infect grapevine (Martelli and Walter, 1998). For growers in the Pacific Northwest, five viral diseases are of importance: decline (nepoviruses), fanleaf (grapevine fanleaf virus), leafroll (grapevine leafroll associated viruses, at least eight viruses), rugose wood disease complex (corky bark, rupestris stem pitting, and two or three other viruses), and fleck (grapevine fleck virus). As can be seen from the list, for these five diseases there are about fifteen different viruses that should be of concern. In addition to these fifteen, there are at least thirty other viruses known to be present in grapevine in other parts of the world. These other thirty viruses are the reason all imported grapevine plant material must come through approved channels.

Plants cannot be cured of viruses in the field. The best way to control viruses of grapevine is to plant vineyards with virus-tested vines and, if possible, to isolate new plantings from virus-infected sources. To ensure that new viruses are not introduced into the country and the region, all imported plant material must come through one of the three federally approved locations in California, New York, and Missouri (see Chapter 11). When grafted vines are used, care must be taken to ensure that the rootstock as well as the scion is virus-free. In cases where the virus vector is known, care should be taken to reduce vector populations. Vineyard operations should be scheduled to work first in the vineyard block that has the least viral infection and to finish with the block that has the highest level.

Fanleaf Degeneration and Decline (nematode-transmitted viruses)

These diseases are caused by related viruses and so are discussed together. In Europe the disease caused by nepoviruses is referred to as "degeneration," whereas in North America this disease is referred to as "decline." There are thirteen nepoviruses reported to infect grapevine; of these, grapevine fanleaf virus (GFLV) is the most devastating. In North America there are five nepoviruses that infect grapevine: peach rosette mosaic, tomato ringspot (ToRSV), tobacco ringspot, blueberry leaf mottle and GFLV. Europe has GFLV plus five other nepoviruses that are not present in North America. Arabis mosaic virus (ArMV) also exists in Europe but is very rare in North America and has been reported only in grapevines from Canada.

These viruses generally cause a stunting of the vines and in some cases a severe chlorosis. The severity of symptoms depends on virus strain and grapevine variety. GFLV is the most serious virus in this group of pathogens. It is present in California and was found recently in Washington. GFLV can cause mild to severe chlorosis, leaf distortion, shoot or leaf fasciation, and a decline in plant vigor. There are many strains of this virus; mild strains cause nearly imperceptible symptoms, and severe strains cause a severe decline.

The Pacific Northwest currently has few if any problems with the viruses that cause this disease complex. In a survey done in Oregon in the mid 1990s, *Xiphinema americanum*, vector of TomRSV, was found in more than 70 vineyards, but *X. index*, the vector of GFLV, was not found. In a virus survey carried out in Oregon in 1998, no plants infected with ToRSV were found, even though all the vineyards sampled were those that had high populations of *X. americanum* in the previously mentioned nematode survey. GFLV was recently found in Washington, but at this time it is not clear if there is a vector present or if the virus came in on planting stock. Strict maintenance of our certification and quarantine restrictions for these viruses should prevent them from becoming a problem in Pacific Northwest vineyards.

The suspected presence of a nepovirus should be confirmed by ELISA (enzyme linked immunosorbent assay) or PCR (polymerase chain reaction) tests. With the presence of the nematode vector, *X. americanum*, in many vineyards in Oregon, growers should be alert to the potential threat of ToRSV infections, since this virus is present in other crops in the Willamette Valley. Planting stock should be free of these viruses. Each of these viruses is readily detected by ELISA (a quick laboratory test), so plant material from certified nurseries should be free of these pathogens.

ToRSV has been found in pastures and other agricultural crops in the Pacific Northwest, so sites should be tested for the presence of *Xiphinema* prior to planting. Ideally, sites should be selected that are free of these nematodes, but that may be difficult in many situations. If the site is selected when there is still vegetation, such as other crops or dandelions, plantain, or other broadleaf weeds, these plants should be tested for the presence of ToRSV prior to site preparation for a vineyard. If ToRSV is present, another site should be selected.

If one of these viruses is known to be present in a vineyard, management practices should be conducted last in the infected vineyard area to minimize the movement of soil to healthy areas. Field equipment should be washed down after use in an infected vineyard block to minimize the chance of moving viruliferous nematodes to a healthy block.

Grapevine Leafroll Complex

Eight closteroviruses are associated with the leafroll disease and this number will likely increase in the future; hence, the disease is referred to as grapevine leafroll complex. GLRaV-1, -2-, and -3 are known to be transmitted by mealybugs. In surveys done in Oregon and Washington in 1999 and 2000, GLRaV-1 and GLRaV-3 were the most common of the GLRaVs found. The vectors for the other GLRaVs have not been determined. The disease causes delayed fruit ripening, reduced sugar content, and smaller clusters of berries.

There are reports of leafroll virus spreading in vineyards in Washington and Oregon. This may be due to growers taking cuttings or budwood from their own vineyards rather than using certified planting stock, thus starting new vineyards with infected wood. Also, there has been an apparent increase in the number of mealybugs in vineyards in Washington since 1995. This mealybug increase may account for the reported increase in the spread of grapevine leafroll in vineyards in Washington.

Leafroll has also been observed in Oregon, with some evidence of spread in the vineyards. Mealybug populations have been increasing in orchards but have not become a problem in Oregon vineyards. Leafroll has the potential to have a serious impact in cool regions such as the Willamette Valley, since it delays ripening and reduces sugar levels. Other effects of leafroll infection on wine quality are unknown.

The foundation for control of leafroll and all other viruses is the use of virus-free planting stock when establishing a vineyard. Once leafroll is present in a vineyard, control is difficult if mealybug populations are high. Management practices that may result in movement of mealybugs on equipment, or that induce the mealybugs to move on their own, should be minimized. Control measures for mealybugs may be beneficial in reducing the spread of the virus as long as the control measure does not induce the mealybugs to migrate to other grapevines. There is little information on the effectiveness of mealybug control on reducing the spread of grapevine leafroll viruses.

Rugose Wood Disease Complex

Rugose wood is a disease complex characterized by pitting and grooving of the wood cylinder or trunk. This group of disorders includes corky bark, rupestris stem pitting, LN33 stem grooving, and kober stem grooving. Several viruses have been implicated as causal agents of the rugose wood disease complex in grapevine. These viruses belong to two newly recognized genera of plant viruses: the vitiviruses, which include grapevine virus A (GVA), grapevine virus B (GVB), grapevine virus C (GVC), and grapevine virus D (GVD); and the foveaviruses, grapevine rupestris stem pitting associated virus (GRSPaV).

Vines affected by rugose wood disease are usually less vigorous and may show delayed bud break in the spring. *Vitis vinifera* may carry the virus and be symptomless until grafted onto *V. rupestris* rootstocks.

The measures used to manage grapevine leafroll viruses also apply to rupestris stem pitting and corky bark. The vector for corky bark has been reported to be mealybugs, but the vector for rupestris stem pitting is unknown. The use of virus-free certified planting stock for establishing a vineyard is the best means of controlling these diseases. Although rupestris stem pitting has been removed from the quarantine list, it may be prudent to maintain it as a controlled virus in certification programs. With the increased use of grafted vines in Oregon because of the threat of phylloxera, there is much we do not know about the reaction of the newer rootstocks to rupestris stem pitting when grafted with various scions. One concern is that rupestris stem pitting, in combination with other viruses, could lead to graft incompatibility with certain rootstocks.

Grapevine Fleck Disease

All varieties of *V. vinifera* are susceptible to infection by grapevine fleck virus, but not all show symptoms. Fleck is often present in combination with other viruses, so its impact alone is not well understood, although it has been reported to be a damaging disease. Since fleck shows symptoms in *V. rupestris*, its effect on graft unions with *V. rupestris* rootstocks should be examined.

The use of virus-free certified planting stock for establishing a vineyard is the only means of controlling fleck. There is not a known vector for this virus. The virus is limited to the phloem tissue, and so transmission by pruning is unlikely. Grapevine fleck virus is readily detected by ELISA or PCR.

General References

An Online Guide to Plant Disease Control, Oregon State University Extension. http://plant-disease.orst.edu/

Flaherty, D. L., L. P. Christensen, W. T. Lanini, J. J. Marois, P. A. Phillips, and L. T. Wilson (eds.). 1992. *Grape Pest Management,* 2d ed.. Publication No. 3343, University of California, Division of Agriculture and Natural Resources, Davis.

Pearson, R. C., and A. C. Goheen. 1988. *Compendium of Grape Diseases.* APS Press, St. Paul, Minn.

Pest Management Guide for Wine Grapes in Oregon. EM 8413. Oregon State University Extension, Corvallis. Revised annually.

Pscheidt, J. W., and C. M. Ocamb. 2002. *Pacific Northwest Plant Disease Management Handbook.* Oregon State University Extension, Corvallis.

Powdery Mildew

Belanger, R. R., W. R. Bushnell, A. J. Dik, and T. L. W. Carver. 2002. *The Powdery Mildews. A Comprehensive Treatise.* APS Press, St. Paul, Minn.

Gubler, W. D., M. R. Rademacher, S. J. Vasquez, and C. S. Thomas. 1999. Control of powdery mildew using the UC Davis powdery mildew risk index. http://www.apsnet.org/online/feature/pmildew/

Botrytis Bunch Rot and Blight

Broome, J. C., J. T. English, J. J. Marois, B. A. Latorre, and J. C. Aviles. 1995. Development of an infection model for Botrytis bunch rot of grapes based on wetness duration and temperature. *Phytopathology* 85:97-102.

Johnson, K. B., T. L. Sawyer, and M. L. Powelson. 1994. Frequency of benzimidazole- and dicarboximide-resistant strains of *Botrytis cinerea* in western Oregon small fruit and snap bean plantings. *Plant Disease* 78:572-577.

Crown Gall

Burr, T. S., C. Bazzi, S. Sule, and L. Otten. 1998. Crown gall of grape: biology of *Agrobacterium vitis* and the development of disease control strategies. *Plant Disease* 82:1288-1297.

Eutypa Dieback

Munkvold, G. P., and J. J. Marois. 1993. The effects of fungicides on *Eutypa lata* germination, growth, and infection of grapevines. *Plant Disease* 77:50-55.

Armillaria Root Rot

Shaw, C. G., and G. A. Kile. 1991. *Armillaria Root Disease.* USDA Forest Service Agriculture Handbook No. 691.

Phomopsis Cane and Leaf Spot

Merrin, S. J., N .G. Nair, and J. Tarran. 1995. Variation in Phomopsis recorded on grapevine in Australia and its taxonomic and biological implications. *Australasian Plant Pathology* 24:44-56.

Pscheidt, J. W., and C. M. Ocamb. 2002. *Pacific Northwest Plant Disease Management Handbook.* Oregon State University Extension, Corvallis.

Black Rot

Pearson, R. C., and A. C. Goheen. 1988. *Compendium of Grape Diseases.* APS Press, St. Paul, Minn.

Downy Mildew

Pscheidt, J. W., and C. M. Ocamb. 2002. *Pacific Northwest Plant Disease Management Handbook.* Oregon State University Extension, Corvallis.

Young Vine Decline

Scheck, H., S. Vasquez, D. Fogle, and W. D. Gubler. 1998. Grape growers report losses to black-foot and grapevine decline. *California Agriculture* 52(4):19-23.

Nematodes

McKenry, M. V. 1992. Nematodes. In D. L. Flaherty, L. P. Christensen, W. T. Lanini, J. J. Marois, P. A. Phillips, and L. T. Wilson (eds.), *Grape Pest Management*,2d ed. Publication No. 3343, Division of Agriculture and Natural Resources, University of California.

Nicol, J. M., G. R. Stirling, B. J. Rose, P. May, and R. Van Heeswijck. 1999. Impact of nematodes on grapevine growth and productivity: current knowledge and future directions, with special reference to Australian viticulture. *Australian Journal of Grape and Wine Research* 5:109-127.

Pinkerton, J. N., T. A. Forge, and R. E. Ingham. 1999. Distribution of plant-parasitic nematodes in Oregon vineyards. *Journal of Nematology* 27:515.

Ramsdell, D. C., G. W. Bird, F. W. Warner, J. F. Davenport, C. J. Diamond, and J. M. Gillett. 1996. Field pathogenicity studies of four species of plant-parasitic nematodes on French-American hybrid grapevine cultivars in Michigan. *Plant Disease* 80:334-338.

Viruses

Bovey, R, W. Gartel, W. B. Hewitt, G. P. Martelli, and A.Vuittenez. 1980. *Virus and Virus-Like Diseases of Grapevines.* Verlag Eugen Ulmer, Stuttgart, Germany.

Frazier, N. W. (ed.). 1970. *Virus Diseases of Small Fruits and Grapevines.* University of California Division of Agricultural Sciences, Berkeley.

Krake, L. R., N. S. Scott, M. A. Rezaian, and R. H. Taylor. 1999. *Graft-transmitted Diseases of Grapevines.* CSIRO Publishing, Collingwood, Victoria, Australia.

Martelli, G. P. (ed.). 1993. *Graft-transmissible Diseases of Grapevines: Handbook for Detection and Diagnosis.* FAO, Rome.

Martelli, G. P., and B. Walter. 1998. Virus certification of grapevines. In A. Hadidi, R. K. Khetarpal, and H. Koganezawa (eds.), *Plant Virus Disease Control,* APS Press. St. Paul, Minn.

Martin, R. R. 1998. Advanced diagnostic tools as an aid to controlling plant virus diseases. In A. Hadidi, R. K. Khetarpal, and H. Koganezawa (eds.), *Plant Virus Disease Control,* APS Press. St. Paul, Minn.

Martin, R. R., D. James, and C. A. Levesque. 2000. Impacts of molecular diagnostic technologies on plant disease management. *Annual Review of Phytopathology* 38:207-239.

Meng, B., R. Johnson, S. Peressini, P. L. Forsline, and D. Gonsavles. 1999. Rupestris stem pitting associated virus-1 is consistently detected in grapevines that are infected with rupestris stem pitting. *European Journal of Plant Pathology.* 105:191-199.

Rowhani, A., J. K. Uyemoto, and D. A. Golino. 1997. A comparison between serological and biological assays in detecting grapevine leafroll associated viruses. *Plant Disease.* 81:799-801

Zhang, Y. P., J. K. Uyemoto, D. Golino, and A. Rowhani. 1998. Nucleotide sequence and RT-PCR detection of a virus associated with grapevine rupestris stem-pitting disease. *Phytopathology* 88:1231-1237.

24

Management of Insect and Mite Pests

James R. Fisher, Philip D. VanBuskirk, Richard J. Hilton, and Jack DeAngelis

Throughout the grape-growing regions of the west coast of North America, insects and mites have the potential to cause serious economic damage. Despite over 10,000 acres of winegrapes in Oregon, there have been few problems. A recent survey of Oregon vineyards found nearly every terrestrial order of insects with more than one hundred species represented. As grape plantings increase, vineyards will become more commonplace in the landscape of western Oregon. Inevitably, insects that are occasional pests today and even some undiscovered (for Oregon) pests might become a nemesis for the industry in the future.

Over the past decade the grape phylloxera, a small aphidlike creature that feeds on the roots, has come to the forefront as a formidable problem for growers with self-rooted winegrapes. In 1990 grape phylloxera was discovered to be present in three vineyards. A decade later, more than forty vineyards were recorded to be suffering from this pest. From time to time other pests have been noted to be problems isolated in individual vineyards and often confined to small areas within those vineyards. These are thrips, grape leafhopper and other leafhoppers, spider mites, cutworms and other lepidopterous pests, black vine weevil, branch and twig borers, mealybugs, and grasshoppers. Yellow jackets are primarily a health threat or nuisance to vineyard workers, but they may also damage grapes.

The intent of this chapter is to provide growers with information that assists in the recognition of key insects and damage and the development of informed management strategies for these pests. When encountering a possible insect pest problem, the grower should assess the extent of damage and have an estimate of the number of insects that are causing the damage on a per-leaf or per-cluster basis. In the case of belowground insects, estimates should be of the number of insects on a per-shovel or other measured basis. Often there is visible damage or symptoms but no culprit insects to be found; the grower should contact the local extension agent or extension entomologist or cooperating consultant.

There are many techniques to control insect pests. Sometimes it may be just sanitation processes such as pruning and destroying the wood, or it may be through the use of cover crops or other cultural techniques. In other situations a chemical may be needed, or a biological control. Methods are improving and changing from season to season. It is best to obtain the current year's recommendations or consult with an insect control advisor or crop consultant before taking action.

In this chapter we review the aforementioned insects in a manner for quick reference. For each insect group or species we list the common and scientific names, a description of damage and when and where the insect and damage may be found, economic damage thresholds, if known, and other interesting facts that may help provide an understanding of the insect and the problem. We also list some methods for control or damage amelioration that do not involve pesticidal chemicals.

Grape Phylloxera, *Daktulosphaira vitifoliae*

The grape phylloxera has been found in Oregon since the early 1960s, but damage to modern commercial *V. vinifera* vineyards was not found until 1990. Grape phylloxera is a native American insect that affects winegrapes worldwide. It has caused billions of dollars in replant costs in California since the late 1980s.

Damage. Aboveground damage is initially noticed as a loss of vigor and chlorosis of new leaves and early senescence. By the time this damage is noticed, the insects have probably been there at least three years. A look below ground will reveal clublike formations (nodocities) on the tips of new roots and maybe deformation of larger older roots (bumps and depressions called tuberosities). Examination of the nodocities or the root may reveal a goldlike dust, which if placed under a 15x hand lens will reveal golden aphidlike insects with red eyes (nymphs), large yellow to gray lumps of an insect-like organism (mothers or adults), and yellow glistening football-shaped eggs. These are, indeed, phylloxera.

Older damage results in short to very short shoots, lack of winter hardiness in shoots, reduced leaf size, a "bunch-head" appearance of the plant when fully leaved, later ripening, or in advanced cases early drying and falling off of the fruit and, eventually, dead plants. In the vineyard the area of serious damage progresses from a center plant (sometimes called the "center of origin," referring to the first plant to show damage) and progressively moves out concentrically or in a half-moon shape year after year. Digging in the soil within a foot or so of the trunk of the plants with the most damage will reveal that nearly all of the roots found are rotted, or there may be no roots except for major (greater than 5/8 inch) roots to be found.

Researchers at the University of California, Davis, are finding that it is actually the secondary invaders, rot fungi, that are primarily responsible for the damage that causes eventual death of the plant. Phylloxera are piercing-sucking feeders. Thus, they pierce their mouthparts or styli into the xylem of the root tissue. Something in the saliva of the insect causes swelling of the plant tissue surrounding the wound. Within this swelling there is a reduction of the starch granules usually present and an increase in sugar content. This may then make an excellent medium for growth of fungi, which can enter through the wound hole, or they may be already present as spores attached to the stylets of the insects.

The severity of injury within a vineyard varies from site to site. Even blocks within the same vineyard may develop symptoms at different rates. Some sites take fewer than five years to succumb to damage; others may take more than twenty years to succumb totally.

Monitoring. Sampling to find phylloxera in a suspected center of origin should begin away from the center, at the first one or two plants that appear to be without symptoms. Few phylloxera will be found on unhealthy plants. Dig close to the trunk until roots are found. If many roots and rootlets appear to be rotting or dead, keep digging. If no good roots are found in the first 1–2 feet, then go to the next healthy-looking plant and repeat the process. If nodosities or tuberosities are found, this is good evidence of phylloxera; most likely, phylloxera will be found on close inspection of the physical evidence. A 15x hand lens makes identification much easier.

Thresholds. There are no thresholds for phylloxera. A single nymph or egg introduced into a vineyard that has self-rooted *V. vinifera* plants can start an infestation that ultimately devastates the vineyard. These insects are parthenogenic (without sex); thus, all insects that survive to adulthood lay eggs.

Other Considerations. Speculatively, it appears that grapevine vigor and plant health have something to do with damage progression. Recent studies have shown that, when vigor has been increased, such as by using organic fertilizers or irrigation, symptoms may be retarded and damage attributed to phylloxera kept at a minimum. Eventually, however, the field succumbs to the damage.

Spread of phylloxera can be limited through sanitation measures such as washing machinery, boots, and tools and restricting travel into vineyards. But this is not a sure cure; vineyards that have followed sanitation practices strictly have contracted phylloxera. Eggs are durable and live in bleach water for more than 15 minutes. Phylloxera have also been reared in containers with soilless mixes at the USDA laboratory in Corvallis. Nymphs sometimes come above ground and may be blown on the wind to other places. They have also been found to float in irrigation water to other vineyards and to wash downhill in heavy rains. Eggs and nymphs may be transported by dogs (digging and carrying on muddy claws), gophers (burrowing among plants and carrying roots to other places), deer (on hooves), and any other animals that dig or roam the vineyard.

The most effective measure to prevent problems with grape phylloxera is to plant *V. vinifera* grapes that have been grafted to a resistant rootstock. Some rootstock varieties have been shown to have antibiotic effects against phylloxera. Most resistant rootstock varieties are actually tolerant to grape phylloxera and tend to support small numbers, if any, of the insects. Although phylloxera feed on their roots, they do not show any typical symptoms of phylloxera damage.

Black Vine Weevil, *Otiorhynchus sulcatus*

Black vine weevils have a worldwide distribution and are present throughout western Oregon. They are usually found in northern areas, typically, within 300 miles of a coast. These insects are known to feed on more than two hundred species of plants. Adults are about 1/3 to 1/2 inch long, with an oval body and a snoutlike head area. The insect is black with gold-colored flecks on the wing covers. The larvae are white with a brown head, C-shaped, and from 1/4 to 1/2 inch long. They live in the soil and feed on the roots of plants for nearly nine months of the year. Pupation takes place in the soil in the spring. Adults emerge from mid-May to June. Eggs are laid on the soil surface and in soil litter from July through September. There is one generation per year, with the larval stage being the primary overwintering stage. These insects are parthenogenic and have an egg capacity of 300–400 eggs each.

Damage. Damage to grapes is by the adults. Usually, adult feeding goes undetected or is of little concern. Typically one or two weevils are present per plant in the vineyard. Still, economic damage may occur to vines with only a small number of growing points: newly planted vines in spring, young plants with late budburst, or vines that have been recently field-grafted. Damage is detected as notches chewed in new leaves or chewed buds. Weevils are usually not seen because they are nocturnal. In California, adults have been known to feed on the flowers, causing poor set.

Monitoring. Monitoring for these insects is not necessary. If weevil damage is suspected, there are several ways to detect their existence. During the daylight, one can carefully dig near the base of the plant. An adult weevil may be found just below the surface. To count weevils or find them on the plants requires night surveillance. Activity usually takes place about two hours after sundown. A flashlight quickly shown on a plant near the buds may reveal one or two adults.

Thresholds. There are no established thresholds for this pest on grapes. One per plant on newly planted or grafted stock can cause severe retardation of the plant and even kill it.

Other Considerations. If a small block in the vineyard is affected, a simple cultural control is possible. About midnight, walk along a row, holding a 5-gallon bucket (smaller buckets if the vines are small, new plants) under each vine and gently tapping a majority of each plant. The weevils fall into the bucket and can be disposed into a plastic bag at the end of the row. The insects should then be destroyed, either by placing in fire, boiling water, or heating past 110°F for one or two hours.

Twospotted Spider Mite, *Tetranychus urticae*
Pacific Spider Mite, *Tetranychus pacificus*
Willamette Spider Mite, *Eotetranychus willamettei*

Spider mites are sporadic pests in Oregon vineyards. The Pacific spider mite, the Willamette spider mite, and the twospotted spider mite are all known to be present in Oregon. The twospotted spider mite has caused occasional problems in vineyards in southern Oregon. It usually becomes evident three to four weeks prior to harvest, causing concern as to whether the vines will maintain sufficient leaf area to mature the fruit. So far, treatment has rarely been necessary for control of this pest during the growing season, and no problems with fruit maturity have been reported.

Twospotted spider mites are small (about 0.007 inch); the adults have eight legs and are light tan or greenish in color with a dark spot on each side of the body. Overwintering adults, which turn orange in color, are found under the bark or in the debris around the base of the vine, moving up the vines as buds begin to grow in spring. Since the generation time for spider mites is very short, especially when temperatures are high, there are multiple overlapping generations of these species during the summer months.

Damage. These small mites use their mouthparts to penetrate the cells of a leaf, feeding on its contents. The extent of damage done by spider mites is often not realized until hot summer weather develops in mid-July and August. As temperatures warm up, vine transpiration rates increase dramatically and the damaged leaves cannot maintain their healthy appearance. Green leaves quickly turn yellow as mites begin to feed on them. On white grape varieties, yellow leaves quickly turn to brown or bronze, whereas on red varieties yellow leaves turn red and then purplish. If sufficient leaf damage occurs, vine vigor and berry size are reduced, clusters may sunburn and shrivel, and fruit quality is reduced. If vines are under other stresses, the effects of the mite feeding are exacerbated.

Monitoring. Monitoring for spider mites can begin as early as bud break and extends through the foliar season. During the dormant period, groups of overwintering mites are found under the bark and in the leaf litter around the base of the vine. Once the vines have leafed out, spider mites are primarily found on the underside of the leaves, along the midrib and veins. The mites, along with their spherical eggs and webbing, can easily be seen with the aid of a 10x hand lens. When the population level is high, the growing tip of the vine may be completely encased with webbing.

Be aware that some areas of the vineyard, particularly borders along dusty roads, may develop damaging mite populations distinct from the rest of the block. These areas should be sampled and, if needed, treated separately. Before sampling begins, the vineyard should be divided into smaller more manageable blocks, possibly by variety. In some crops it has been shown that the spider mite damage threshold varies with variety. Although it is likely that winegrape varieties also vary with respect to their tolerance of spider mite damage, precise damage thresholds for different winegrape varieties are not presently available.

Sampling protocols specific to twospotted spider mite on winegrapes have not been developed, but three methods of monitoring grapes for spider mites, primarily Pacific and Willamette mites, have been described: average number of mites per leaf over time, or mite-days; presence/absence sampling; and average number of mite-damaged leaves per shoot.

The method of sampling for average number of mites per leaf was described for Willamette mites on Chenin blanc and Zinfandel. A total of six leaves, one midcane leaf from each of six shoots randomly selected throughout the block, are collected and the average number of motile mite stages per leaf is determined for each block. This same information can be used to determine mite-days by taking the average number of mites from two sample dates and multiplying by the numbers of days between sample dates.

The presence/absence method has not been fully developed for winegrapes but is described in Flaherty et al. (1992) for use on the Thompson seedless variety. Three leaves are randomly collected from each of fifteen randomly selected vines within each block, for a total of forty-five leaves per block. If rows are oriented east–west, the leaves should be collected from the west side of the vine; if oriented north–south, collect from the south side of the vine. Leaves are examined with a 10x hand lens for the presence or absence of all stages of spider mite, and the total number of infested leaves is recorded. This sampling method requires that, as well as recording spider mites, you record the number of leaves that have predatory mites (i.e., the western predatory mite or other phytoseiids), because treatment thresholds are based on predator–prey ratios. Divide the total number of leaves infested with spider mites by the total number of leaves collected and multiply by 100 to determine the percentage of infested leaves. Then repeat the procedure for the predatory mites so that the predator–prey ratio can be determined.

The method of calculating the average number of mite-damaged leaves per shoot on the basis of observed injury is described in Winkler et al. (1974). Randomly select vines throughout the vineyard and count the number of leaves found to show mite damage (yellowing or bronzing of leaves) on each vine. Keep a running total and divide by the total number of vines sampled to obtain an average number of mite-damaged leaves per vine. Care should be taken to avoid confusing yellowing or bronzing caused by mites with water stress or leafhopper damage.

Thresholds. Thresholds specific to twospotted spider mite on winegrapes have not been developed. Thresholds for Pacific and Willamette mites on grapes have been developed for the sampling methods described above.

Based on a study conducted in California on Zinfandel, 50 mites per leaf have been shown to cause reduced levels of sugars or delayed harvest. Fewer than 30 mites per leaf had no effect on productivity or quality. When using mite-days, the same study indicated that damage occurred in the range 621–1,308 mite-days.

The presence/absence method of sampling uses percentage of infested leaves and factors in the abundance of predators. Table 1 lists the treatment threshold developed for Thompson seedless grapes.

An average of eight or more mite-damaged leaves per vine indicates the need for a control measure to be applied.

Many variables influence the treatment threshold for spider mites on winegrapes. Vine vigor, weather,

Table 1. Treatment Guidelines for Presence/Absence Monitoring

	Predator-prey distribution ratios			
Mite injury levels (percentage of leaves infested)	*Rare (<1:30)*	*Occasional (1:30 to 1:10)*	*Frequent (1:10 to 1:2)*	*Numerous (>1:2)*
Light (<50%)	Delay treatment to increase predators.	Delay treatment.	Treatment not necessary.	Treatment not necessary.
Moderate (50–65%)	Treat if population is increasing rapidly.	May delay treatment to increase predation.	Treatment may not be needed if the predator-prey distribution ratio is increasing rapidly.	Treatment not needed.
Heavy (65–75%)	Treat immediately.	May delay treatment a few days to take advantage of increasing predation.	Treatment may not be needed if predators are becoming numerous.	Treatment not needed if damage is not increasing.
Very heavy (>75%)	Treat immediately.	Treat immediately.	Treat immediately unless predator-prey distribution ratio increasing very rapidly;carefully evaluate damage.	Treatment may not be necessary if population is dropping because of very high (1:1) predator-prey distribution ratios; carefully evaluate damage.

Source: Flaherty et al. (1992).

cultural practices, natural enemy populations, and possibly variety may all play a part in determining the appropriate treatment threshold and optimal control timing. Vines that are stressed because of conditions such as heat or drought can tolerate less mite damage and need to be treated at a lower threshold. The use of herbicides to control weeds in the vineyard may force mite populations from the groundcover into the vines. It is a good practice to examine the groundcover prior to any herbicide treatment, as well as sampling the vines for mite populations. Field bindweed is a particularly good host for twospotted spider mites. Leaf samples for mites should be made five to seven days after an herbicide application. Likewise, any planned miticide treatments should be made after, not before, an herbicide application. The use of carbaryl (Sevin) in the vineyard for other pests, such as leafhoppers, has been shown to cause outbreaks of spider mites. Natural enemies, particularly predator mites and *Stethorus* lady beetles, can provide some control, but the full extent of their effect in Oregon vineyards is not adequately known.

Grape Leafhopper, *Erythroneura elegantula*
Variegated grape Leafhopper, *Erythroneura variabilis*
Potato Leafhopper, *Empoasca fabae*

Leafhoppers are found in all winegrape growing regions of Oregon. They occasionally reach economic thresholds in southern Oregon and in the Columbia Basin. The three most important leaf-hopper pests are the potato leafhopper, the grape leafhopper, and the variegated leafhopper. The potato leafhopper is a migratory species, so its damage occurs at the end of the summer and monitoring is difficult. The other two groups of leafhoppers rarely cause damage in Oregon.

The potato leafhopper is small (less than 1/8 inch) and green. The adults are winged and have six yellow legs with spines. The adults do not overwinter in the north and die after the onset of the first heavy frost. Because they are migratory, they are not detected in the early summer months. Leafhoppers appear as a sudden immigration into the vineyard in late summer.

The grape leafhopper and variegated leafhopper are small (less than 1/4 inch). The grape leafhopper is green, and the variegated leafhopper is white with reddish-orange marks over its wings and head. The nymphs look similar to the adults but are wingless.

Damage. The leafhoppers described above are all mesophyll feeders. They penetrate the leaf cells with their piercing mouthparts and suck the contents. Damage first appears as white speckling on the leaf surface and later develops into chlorosis and early senescence of the leaves. Leafhoppers produce sugary excreta called "honeydew," which has been shown to promote the growth of sooty mold. If extensive leaf damage occurs, vine vigor is reduced. Berry size decreases, shriveling and sunburn of clusters occur, and disease pressure increases.

Monitoring. There are several monitoring methods for leafhoppers in vineyards. Monitoring should commence in early spring and continue throughout the entire growing season. The most accurate way to monitor is with the use of yellow sticky traps. Sticky traps are an absolute method for sampling population densities. At least seven yellow sticky card traps should be placed at the border of the vineyard and checked at least every fourteen days. Another suitable method for sampling is the sweep net method; this is best for surveying the presence/absence of leafhoppers in the vineyard. Although this method does not give quantitative results, it is still a reliable method for determining what species are present in the vineyard. When sweeping, a person should move the net side to side in a steady manner, making 180° sweeps while walking slowly but steadily over the sample area. Ten such sweeps should be taken in at least three different areas in the vineyard. A third method for sampling is with the use of a D-vacuum, a suction device to pull insects directly off of the vine. This is primarily a research tool, but it has the ability and relatively low cost (not as low as a $20–$25 sweep net) to be adapted by growers. This method has the advantage of sampling directly from the vine without causing any damage to the vine.

Thresholds. Presently there are no specific economic thresholds developed for leafhoppers. Damage is visible in the vineyard and is directly proportional to the number of leafhoppers in the vineyard.

Blue-green Sharpshooter, *Graphocephala atropunctata*
Green Sharpshooter, *Draeculacephala minerva*

These two sharpshooters reach 1/2 inch in size and are both dark green during the summer months. They both have a winter phase that is light to dark brown. The green sharpshooter has a sharply pointed head, whereas the blue-green sharpshooter has a head that is rounded at the end. Both species have been found in all grape-growing regions of Oregon. They do not cause any appreciable damage, but they are known vectors of Pierce's disease in California, a disease that has not spread into Oregon at this time. Pierce's disease is caused by a xylem-limited bacteria, *Xylella fastidiosa.*

Three-Cornered Alfalfa Hopper, *Spissistilus festinus*

The three-cornered alfalfa hopper has been known to damage grapevines occasionally. It is usually a pest of leguminous crops and also feeds on fruit trees, shrubs, grasses, and herbs. The adult is about 1/4 inch long and green to green-brown. From the front it appears to have a triangular shape; hence the common name. The nymphal stages have the same general shape without the sharp corners but are covered with spines and hairs. The over-wintering adults lay eggs singly in the stems of the host in the spring.

Damage. This insect feeds by puncturing the very young stems or leaf petioles of grapes. It tends to feed in a continuous line, thus leaving a group of in-line punctures. The wounds may cause girdling. Feeding causes a concentration of anthocyanins, and thus the leaves get a reddish appearance, much like leafroll virus. Insertion of eggs into stems may also cause injury.

Monitoring. There are no monitoring methods available. If sticky traps are in place early in the spring, they may detect the presence of these insects.

Thresholds. There are no thresholds, for these insects are seldom economically damaging.

Branch and Twig Borer, *Melalqus confertus*

The branch and twig borer, sometimes called cane borer, is a beetle. The adult is dark brown and about 3/8 to 3/4 inch long. Eggs are laid singly in the early summer in crevices and cracks in the bark on the cordons or on the trunk. Larvae, about 1/2 inch long when mature, bore into the wood at a pruning wound or injured area of the vine. Here they develop from midsummer until the next late spring, when they pupate for a few weeks and emerge as adults. There is one generation per year.

Damage. Both adults and larvae can cause injury to grape vines. Adults may burrow into the canes at the bud axils or into the crotch formed by the cane and the spur. This weakens the canes and may cause them to break in strong winds. Larvae burrow into the wood at dead and dying parts of the vines and feed on living and dead wood. Larvae burrow slowly and plug their tunnel with frass and chewed wood.

Monitoring. These insects are not a chronic problem in Oregon vineyards, but they have occasionally been found damaging small areas of some vineyards and sometimes in heavy infestations. If the vineyard is near a forested area or was recently planted on an old orchard site or is next to an orchard, examination of grapevines is necessary. Also, piles of old prunings or other wood attract the adults. Examine old pruning scars, dead parts, and crotches or bud axils. Look for entry holes or frass or wood chewings. If insect signs are present, action should be taken.

Thresholds. There are no established thresholds, but when damage is found the wood should be examined for presence or absence of the larval or adult stage. If either stage is found, the wood should be pruned and burned. Vineyards that have had infestations or are likely to should be kept as clean as possible of any prunings and wood. Dead wood on vines should be removed. Healthy vine vigor contributes to thwarting damage from this insect.

Grape Mealybug, *Pseudococcus maritimus*

In recent years grape mealybug populations have been increasing in Oregon orchards as orchardists have moved to softer insecticides, eliminating or reducing broad-spectrum materials such as organophosphates from their spray programs. To date, however, grape mealybug has not become a problem in commercial vineyards.

Damage. Nymphs and adults feed on shoots and leaves and within clusters, producing honeydew that can drip onto fruit. The droplets of sugary honeydew support the growth of a black sooty mold fungus that contaminates the fruit and affects juice quality.

Monitoring. During the dormant season look for eggs and crawlers (newly hatched nymphs) in the white cottony egg sacs found under the bark. Once the vines have leafed out, grape mealybug is detected primarily by the presence of honeydew or sooty mold. If grape mealybug is found, a more in-depth examination of the vineyard should be conducted. During harvest, the clusters should be examined for the presence of grape mealybugs and signs of infestation to determine the percentage of clusters infested.

Thresholds. Although a treatment threshold has not been well defined, it is recommended that controls be applied if 2% of the clusters are infested at harvest. Dormant treatment is preferable, with a summer treatment being applied only if necessary. Treatments applied in the spring have been found to be ineffective.

Variegated Cutworm, *Peridroma saucia,* and other cutworms

Cutworms are only intermittent pests in commercial vineyards. Damage generally occurs in the spring, in the period from bud swell until the shoots have grown about 6 inches. These caterpillars are primarily nocturnal, so they are rarely seen on the vine during the day. They have a wide host range and feed on many grasses and broadleaf weeds as well as grapes.

Damage. In the spring, as buds begin to swell, cutworm larvae emerge from their overwinting sites in the leaf litter and soil around the base of the vine and under the bark. During the night, the caterpillars climb the vines to chew into developing buds and feed on young shoots. Often, succulent shoots are eaten only partway through such that they are weakened and fall over, referred to as "shoot flagging." Vines respond to this damage with growth of the secondary bud. Secondary shoots are less fruitful than primary shoots, so some crop reduction may occur, depending on how widespread the injury is to primary buds.

Monitoring. There are two main methods to monitor for cutworms. If signs of possible cutworm damage are observed in the vineyard, note the location of damage and return after dark with a flashlight to look for the actively feeding larvae. Also search around the base of vines showing signs of cutworm feeding, looking for cutworm larvae hidden under the soil and leaf litter or in clumps of weeds. In areas where cutworm damage occurred the year before, randomly select and check three vines in each of twenty locations for bud damage caused by cutworms.

Thresholds. If the grape variety has highly fruitful secondary buds, no treatments may be necessary. If secondary buds are not fruitful, treatment should only be applied if 4% or more of the buds are found to be damaged; then, where possible, controls should be applied as a spot treatment.

Other Considerations. Vineyards with good weed control or clean cultivation generally have few cutworm problems. Cultivation in the late summer or fall may be of some help in suppressing cutworms by removing their weed hosts. Spring cultivation prior to bud swell may cause cutworms to feed more extensively on grape by eliminating favored weed hosts and is therefore not recommended. The use of furrow or flood irrigation can force cutworm larvae to the surface, thereby exposing them to birds and other predators.

Grape Leaffolder, *Desmia funeralis*
Omnivorus Leafroller, *Platynota stultana*
Orange Tortrix, *Argyrotaenia citrana*

These lepidopterous insects have rarely been noted as problems in Oregon vineyards. Each has small larvae that fold the grape leaves and sometimes eat them. In California these pests have been known to web fruit clusters and contaminate and damage fruit. They are common pests on raspberries and blackberries, which are major crops in the Willamette Valley.

Western Flower Thrips, *Frankliniella occidentalis*
Grape Thrips, *Drepanothrips reuteri*

Two species of thrips that are known to cause plant injury have been identified from Oregon vineyards: the western flower thrips, and the grape thrips. Thrips range in color from yellow to brown and are difficult to see because of their size; adult thrips are about 0.04 inch long and have feathery wings. The nymphs lack wings and are smaller. Thrips can move rapidly. Often when a leaf is turned over to be examined, thrips quickly move to the opposite side of the leaf. Multiple generations occur within a season; when temperatures are high, a generation can be completed in two to three weeks.

Damage. In spring, thrips cause shoot and foliar damage by stunting new shoots that are less than 12 inches long. In high populations, thrips may also attack flowers from prebloom to fruit formation. As fruit begins to develop, the female thrips may deposit an egg into a developing berry. Egg laying causes scarring that may crack as the berry develops, allowing decay organisms to enter. Early-season injury caused by thrips is more of a problem in cool wet springs when shoot growth is slowed, allowing the thrips to feed on the new shoot for an extended period of time.

Damage from thrips can also occur in mid to late summer when temperatures rise and surrounding vegetation dries out. Thrips are attracted to the succulent vegetation in vineyards, where they attack shoot tips and cause stunted vines and leaf growth. Leaves turn bronze, internodes are shortened, and vines may not have sufficient leaf surface to mature the crop and accumulate good reserves for the following year. By the time the damage becomes apparent, the population has often peaked and treatments are of limited benefit. Stressed or weak-growing vines are damaged more readily than healthy vines. Thrips have been reported to prefer white grape varieties, but red varieties can also be affected.

Monitoring. Thrips can be monitored early in the season by vigorously knocking the young grape cluster against a flat surface of cardboard (8.5 by 11 inches) that is dark-colored, preferably blue. Sample clusters from randomly selected vines throughout the vineyard; count all adults and nymphs found on the cardboard and obtain an average per cluster for the vineyard. As berry size increases, knocking the cluster to dislodge the thrips becomes a less efficient method of sampling. To simply determine the presence of thrips on the vines at any point in the season, knock the shoot tip onto cardboard or even your hand and see if thrips are visible.

A three-year study conducted in southern Oregon utilized blue sticky cards to monitor for thrips. Cards were 3 by 5 inches in size, placed randomly on the borders and interior of the vineyard. The sticky cards were hung with clothes-pins on the trellis wire closest to the upper growth of the vine. The cards were replaced every two weeks, and the collected cards were wrapped individually with clear plastic, the number of thrips was counted, and the average per card was determined.

Thresholds. California recommendations indicate that a thrips population level of 25 adults and 50 nymphs per cluster is considered normal, with 150 adults and 300 nymphs per cluster considered excessive. Treatments should therefore be applied when the average count of adults or nymphs per cluster exceeds the normal range but before the population reaches an excessive level. When sticky cards are used, damage in the form of stunted shoot growth, leaf cupping, and bronzing occurs when populations exceed 1,800–2,000 thrips per card during any two-week period. Controls should be applied as populations are increasing but before they reach damaging levels, about 1,200–1,500 thrips per card over a two-week period.

Grasshoppers, *Order Orthoptera, Family Acrididae*

Grasshoppers are present in and around vineyards nearly every year. Some of the species most economically damaging to grapes belong to the genera *Melanoplus* and *Schistocera.* These insects typically become damaging on a cyclical basis, about every seven to twelve years, but they have never reached economically damaging populations in Oregon vineyards. During abnormally dry years they could present a problem, for grapes may be the most succulent food material to be found. Nymphs are similar in appearance to adults but they do not have wings

Damage. Both nymphs and adults feed on the foliage and fruit. Even though nymphs are present early in the season, feeding is not noticed until the grasshoppers are in their last nymphal instars or adults, which is in late July and early August. These stages feed voraciously and damage is readily detectable, with large areas of leaves raggedly chewed and fruit injured. In Oregon, the presence of cover crops usually prevents problems by providing an alternate food source. Even so, some feeding on grape leaves may be found. In dry years and in areas next to rangeland and forage crops, the grapes and leaves should be examined closely to assess damage.

Monitoring. Sweep netting of cover crops is the best monitoring method for these pests. If a problem has been noticed in previous years, or if the season is predicted to be drier than normal and the vineyard is situated next to rangeland or forage crops, the following assessment (see thresholds) should be made about every two weeks from early June.

Thresholds. Since grasshoppers tend to rise from grasses and ground forbs, twenty sweeps should be made around each edge of the vineyard. If one nymph per sweep is found in early June, action may be warranted. If the count remains below one per sweep, no action may be needed. Unfortunately, most assessment is done when it is too late for effective action, when grasshoppers are flying and feeding as adults. At that point, damage has been incurred and harsh chemicals are needed to stop further feeding. If populations are recognized early, some biological control measures may be available for use. These change from year to year, so consult with your viticulture consultant, extension entomologist, or extension viticulturist.

Yellow Jacket Wasps, *Vespula* spp. and *Dolichovespula* spp.

Yellow jackets are mainly yellow and black (sometimes black with white markings), stout-bodied wasps about 5/8 to 3/4 inch long. Their nests are made of gray-brown papery material with a single, small opening. Nests may be as large as a soccer ball and are constructed above ground (aerial nests) or below ground (ground nests). Ground nests often are constructed in abandoned rodent burrows or similar cavities. The two most common ground-nesting species that cause concern each year are *Vespula pennsylvanica* (western yellow jacket) and *V. vulgaris* (common yellow jacket).

An egg-laying queen dominates yellow jacket nests. Queens start nest building in the spring. A few workers are reared, which then provision, build, and defend the nest. The queen may not leave the nest again; her role now is to lay eggs. By the end of summer, nests may contain hundreds or even thousands of female workers and a few males (drones). This is when the nests are most trouble-some and dangerous.

Yellow jacket "workers" search for prey, carrion, or rotting fruit. They are attracted to any meat or sugary item. Food is carried back to the nest where it is fed to nestmates. Stings usually occur through accidental contact with the nest or nest entrance. Workers vigorously defend the nest and queen against intruders.

Fall is when the nests produce a new batch of queens (female reproductives). The queens and drones mate, and by first frost most female workers and male wasps have died. Newly fertilized queens find a protected spot to spend the winter. They reemerge in spring to begin the cycle again.

Damage. Yellow jackets pose a threat to people working in vineyards, primarily in the fall. During years when the density of nests is high, loss of productive work time can be significant. Yellow jackets also damage grapes directly by feeding on and breaking the skin of thin-skinned varieties.

Controls. If problem nests can be located, usually by yellow jacket activity around nest entrances, nests can be treated directly. At dusk when wasps are inside and relatively calm, treat with an approved aerosol insecticide. Use one of the "wasp and hornet" aerosols that propel a stream of insecticide so that you can stand off a safe distance. Treat directly into the nest opening. For ground nests, seal the nest entrance with rock or soil. Do not pour flammable liquids such as gasoline or paint thinner into nests.

Poison baiting is the single most effective method for area-wide control of scavenger, ground-nesting species. Scavenger species, such as the common and western yellow jackets, are most likely to cause human injury because they build the largest nests and are the most difficult to control. Poison baits can be hazardous, but they are effective for severe yellow jacket infestations. The idea is to get the worker yellow jackets to carry a bit of poisoned food back to the nest, thereby effecting control. The method uses an encapsulated insecticide as the poison in the bait. The instructions accompanying the bait describe how to use it and must be followed exactly. Bait stations must be protected so that other animals cannot get to the poisoned bait. Poisoned baits should only be used after about July 15, when nests have begun to grow rapidly.

Non-toxic yellow jacket traps are also available in yard and garden stores. The most effective traps use a synthetic attractant called n-heptyl butyrate to lure worker yellow jackets into a trap from which they cannot escape. Fruit juice or various meats can also be used as attractants but are not as effective. Traps can provide some temporary relief by drawing workers away from people, but they are not effective for area-wide nest control even though many yellow jackets may be trapped.

Other Considerations. Some people (about 1% of the population) are highly allergic to yellow jacket venom. Even a single sting can evoke a life-threatening allergic reaction in sensitive individuals. Sensitivity can increase, or even decrease, with repeated exposure to the venom, so be prepared. Be cautious around nests, especially in late summer and early fall. Individuals that are stung should be observed for unusual reactions and immediately taken to medical care if adverse reactions develop (usually within 30–60 minutes of sting).

Remember, all wasps including yellow jackets are generalist predators that attack a wide range of potential vineyard pests. Therefore, care should be exercised so that area-wide control of yellow jackets is not so successful that it disrupts natural control of vineyard pests.

References and Additional Reading

Acorn, J., and I. Sheldon. 2001. *Bugs of Washington and Oregon.* Lone Pine, Edmonton, Alberta.

Akre, R. D., A. Greene, J. F. MacDonald, P. J. Landolt, and H. G. Davis. 1980. Yellowjackets of America north of Mexico. Agriculture Handbook 552, USDA.

Borror, D. J., and R. E. White. 1970. *A Field Guide to the Insects of America North of Mexico.* Houghton Mifflin, Boston.

Connelly, A. E. 1995. Biology and demography of grape phylloxera, *Daktulosphaira vitifoliae* (Fitch) (Homoptera: Phylloxeridae), in western Oregon. M. S. Thesis, Oregon State University, Corvallis.

Fisher, J. R., and E. Hellman. 2000. Status and progression of infestations and management of the grape phylloxera in the Pacific Northwest, USA. In K. S. Powell and J. Whiting (eds.), Proceedings of the First International Symposium on Grapevine Phylloxera Management, January 21, 2000, Melbourne, Australia.

Flaherty, D. L., L. P. Christensen, W. T. Lanini, J. J. Marois, P. A. Phillips, and L. T. Wilson. 1992 (eds,). *Grape Pest Management,* 2d ed. Publication 3343, University of California, Division of Agriculture and Natural Resources, Davis.

Granett, J., B. Bisabri-Ershadi, and J. Carey. 1983. Life tables of phylloxera on resistant and susceptible grape rootstocks. *Entomol. Exp. Appl.* 34:13-19.

Omer, A. D., J. Granett, J. A. De Benedictis, and M. A. Walker. 1995. Effects of fungal root infestations on the vigor of grapevines infested by root-feeding grape phylloxera. *Vitis* 343:165-170.

Weigle, T., and J. Kovach. 1995. *Grape IPM in the Northeast.* Number 211, New York State Integrated Pest Management Program, Ithaca.

Winkler, A. J., J. A. Cook, W. M. Kliewer, and L. L. Lider. 1974. *General Viticulture.* University of California Press, Berkeley.

25

Winter Cold Injury

Bernadine C. Strik, Donald Moore, and Porter Lombard

When looking at overall cold hardiness, it is important to consider midwinter, spring, and fall hardiness. The process whereby vines enter dormancy (winter rest) is called *acclimation;* the rate of acclimation is affected by short days and cold temperatures. Once the vine's chilling requirement is satisfied (generally considered to be the number of hours between 32° F and 45° F), deacclimation occurs. This period ends with bud break in the spring. The rates of acclimation and deacclimation differ among varieties. Cold hardiness is also affected by these physiological stages. For example, the cold hardiness of a variety varies throughout the winter, depending on the stage of dormancy at which the cold spell occurs. A period of warm weather in late winter can lead to a rapid loss of cold hardiness. This makes the vines relatively sensitive to subsequent cold temperatures.

The weather preceding a cold spell also impacts cold hardiness. Continuous periods of below-freezing, nonlethal temperatures lead to a slow movement of water out of the plant cells and increases cold hardiness (this has been documented in vine buds but not in canes). During acclimation, tyloses (blockages in the vine's vascular tissues) are formed that reduce vine water content and may be important in the maintenance of low water content throughout most of the dormant period. Toward the end of dormancy, cane and bud tissues rehydrate; vines are therefore very sensitive to subfreezing temperatures after this stage.

Overall, the degree of cold vines tolerate without damage depends on the variety, the maturity of the wood, the physiological stage of dormancy, and the weather pattern preceding the freeze. Site mesoclimate and topography are also important determinants of the risk for cold injury and should be considered when selecting a site for a new vineyard. These aspects are discussed in Chapter 3.

Varietal Differences

Comparison of varieties for cold hardiness should distinguish between the ability of different varieties and, perhaps, clones to tolerate low temperature (cold hardiness) and their ability to avoid damage by frost or freezing events by, for example, having later bud break. Varietal differences in timing of bud break, flowering, and particularly fruit maturation are important in the selection of grapes for a given vineyard; however, these are not discussed in this chapter. Here we focus on varietal differences in cold hardiness that growers should recognize and plan into their vineyard management practices.

The relative cold hardiness of varieties may change depending on the physiological stage of dormancy—acclimation, dormany, or deacclimation. For example, a research study in Virginia showed Cabernet Franc buds were 1.8–3.6°F more hardy throughout the winter than those of Cabernet Sauvignon (Wolf and Cook, 1991). However, Cabernet Franc deacclimated more rapidly in the spring and was therefore more susceptible to spring frost or late winter freeze conditions.

Relative bud hardiness of several wine grapes in Washington was compared in January 1992/93 and 1993/94 (Table 1) (Wample et al., 2000). The temperature required to kill 50% of the buds was recorded. The most cold hardy variety in midwinter was Maréchal Foch, a French-American hybrid (-21.1°F for 50% bud death). Meunier, Pinot noir, Cabernet Franc, Pinot gris, Riesling, and Chardonnay ranked next in hardiness in this study (-13°F to -13.9°F). Zinfandel, Merlot, Petite Sirah, Chenin blanc, Sémillon, and Cabernet Sauvignon were moderate in cold hardiness (-11.2°F to -12.1°F), and Sangiovese had the least hardiness (-7.6°F). The temperature reported to kill buds in most of these studies is not necessarily reflective of what would be required in the field to do damage, but it does provide a good guide of relative hardiness.

In western Oregon, we have categorized varieties as to susceptibility to freeze damage based on our experience of damage after winter freezes in 1972 (lows of -8°F to -12°F), 1989 (-1°F to 3°F), 1990 (-3°F to -13°F), and 1998 (0°F to 4°F) (Table 2). Varieties may differ in susceptibility to cold injury because of conditions that affect acclimation. Chardonnay and Pinot noir vines were damaged more in 1989 than in 1972 because warm temperatures preceding the cold

Table 1. Relative cold hardiness of buds of winegrapes in Washington, as observed by controlled freezing in 1992/93 and 1993/94 (Wample et al., 2000; Wample, unpublished). Varieties are ranked from most to least hardy under each category based on temperature required to kill 50% of buds. Relative hardiness across categories is also presented.

	Winter	*Winter after Warm Spell*	*Spring (3/21)*
Most Cold Hardy			
	Maréchal Foch		
	Riesling		
	Pinot noir		
	Meunier		
	Cabernet Franc		
	Pinot gris		
	Chardonnay		
	Merlot		
	Petite Sirah		
	Zinfandel		
	Cabernet Sauvignon		
	Chenin blanc		
	Sémillon		
		Maréchal Foch	
	Sangiovese		
		Riesling	
		Pinot noir	
		Cabernet Sauvignon	
		Sémillon	
		Meunier	Pinot noir
			Riesling
			Meunier
			Cabernet Sauvignon
			Merlot
		Zinfandel	Sémillon
		Chenin blanc	Chenin blanc
		Chardonnay	Chardonnay
			Petite Sirah
			Pinot gris
		Cabernet Franc	Cabernet Franc
		Merlot	
		Petite Sirah	Zinfandel
		Pinot gris	
		Sangiovese	Sangiovese
			Maréchal Foch
Least Cold Hardy			

spell in 1989 led to deacclimation. The loss in cold hardiness that occurs after a warm spell was also shown in controlled freezing studies (Table 1). Müller-Thurgau, which had a heavy crop in 1988, was damaged considerably in 1989. On the other hand, in Cabernet Sauvignon and Sauvignon blanc, the 1989 freeze had less effect on bearing vines; these varieties had more extensive damage in 1972, because vines were not sufficiently acclimated before the cold spell.

Differences in cold hardiness of varieties grown in Washington were observed in December 1990 when a 36-hour period of temperatures as high as 50°F preceded temperatures as low as -19°F (Wample et al., 2000). After this cold spell, Sémillon and Chenin blanc had almost 100% primary bud death, Chardonnay and Cabernet Sauvignon about 96%, Pinot noir about 71%, Gamay Beaujolais and Meunier about 59%, and Concord 10% in an irrigated vineyard. During the winters of 1992/93 and 1993/94, additional evaluations were made regarding the response of different grape varieties to midwinter warm temperatures and spring deacclimating conditions (Wample et al., 2000). Controlled freezing laboratory methods were used for this work. After a warm spell, the ranking of varieties for cold hardiness did not change much, although hardiness dropped considerably as the warm spell led to deacclimation. Although Cabernet Sauvignon ranked relatively low for cold hardiness in winter, it maintained its hardiness well after a warm spell compared to most other varieties (Table 2). After the deacclimation period (spring), Maréchal Foch ranked lowest for cold hardiness, whereas Pinot noir ranked the highest.

Most of the research done on grapevine cold hardiness has involved evaluating dormant buds. There may also be changes in cane and trunk cold hardiness. Preliminary work by Wample (Wample et al., 2000) and observations by the authors in Oregon (Table 2) indicate that varieties do differ in cane cold hardiness. Cane damage can have more influence on subsequent crop than bud damage. For example, with primary bud damage only, vines can compensate by growth of secondary buds, or more buds can be left at pruning. With cane damage, however, shoots may grow in spring yet subsequently collapse because of insufficient capabilities of xylem and phloem tissues to keep up with demands.

Rootstocks and Grafted Vines

Research conducted in Washington indicates that rootstocks (nongrafted) are generally more cold hardy than European winegrape varieties (Wample et al., 2000). Rootstocks differed slightly in cold hardiness, with 3309C buds having more cold tolerance than those of 110R, 5C, and SO4.

Table 2. Observed bud and cane damage to winegrape varieties in western Oregon following freezing events in 1972, 1989, 1990 (-3 to -13° F), and 1998 (0 to 4° F). Varieties are ranked from least to most damage.

Bud	*Cane*
Least Damage	
Maréchal Foch (1989, 1990)	Pinot noir (1989, 1990)
Riesling (1972, 1989, 1990)	Chardonnay (1989, 1990)
	Cabernet Sauvignon (1998)
Cabernet Franc (1990, 1998)	Cabernet Franc (1998)
Muscat Ottonel (1990)	
Sauvignon blanc (1990)	
Auxerrois (1990)	
Cabernet Sauvignon (1989, 1998)	Gewürztraminer (1972, 1989, 1990)
Sémillon (1990)	
Pinot noir (1972, 1989, 1990)	
Pinot gris (1990)	Malbec (1972)
Chardonnay (1972, 1990)	
Gamay noir (1990)	
Cabernet Sauvignon (1990)	Sauvignon Blanc (1972)
Pinot blanc (1990)	
Cabernet Sauvignon (1972)	
Lemberger (1990)	Muscat blanc (1972, 1989, 1990)
Nebbiolo (1990)	Syrah (1998)
	Viognier (1998)
Chardonnay (1989)	
Muscat blanc (1989)	
Tempranillo (1998)	Müller-Thurgau (1989, 1990)
Viognier (1998)	
Gewürztraminer (1972, 1989, 1990)	Sémillon (1972)
Syrah (1990)	
Muscat blanc (1972)	Chenin blanc (1989)
Chenin blanc (1989)	
French Colombard (1972)	French Colombard (1972)
Merlot (1989, 1990)	
Syrah (1998)	
	Merlot (1989, 1990)
Malbec (1990)	Tempranillo (1998)
Most Damage	

It is not known whether rootstocks have a direct effect on cold hardiness of the scion in grafted vines. They do, however, have an indirect effect through their control of vine vigor and maturity, for example. Rootstocks that produce excessive vine growth may increase the risk of freeze damage because of late maturing wood. For example, two vineyards in Western Oregon, grafted onto St. George, suffered severe trunk and bud damage in 1990.

Conclusive work on the hardiness of the actual graft union is not available. In grafted vines, trunk damage below the graft union requires that the entire vine be replaced or regrafted. In southern Oregon, some growers have direct experience with the loss of grafted vines to cold damage. In the winter of 1998/99, many acres of two-year-old Merlot were lost due to trunk damage below the graft union. Self-rooted vines were also damaged, but suckers originating below the cold damage could be used to retrain a trunk. Growers in this region recommend delaying sucker removal on grafted vines after a cold spell, because the buds just above the union frequently remain viable and can be used to retrain the trunk. One grower even leaves a two-bud spur just above the graft union for this purpose in young vines.

Another suggestion is to have the graft union as close to the ground as possible, where it is warmer, to minimize the risk of loss from cold injury. Graft unions that are close to the ground may be physically protected from cold injury as well (see below).

Vine Management to Influence Hardiness

Crop Load. Carbohydrates accumulated by ripening berries can, alternatively, be used to develop mature shoots or canes. Overcropping vines thus reduces the amount of carbohydrates available for shoot maturation. Thus, maintaining an adequate crop load balanced with sufficient vegetative growth produces well-matured shoots that have maximal hardiness for that variety. Studies in Washington have found very little consistent effect of crop load on bud hardiness, but cane hardiness tended to be reduced at the high crop load treatment (Wample et al., 2000). Wolf (Wample and Wolf, 1997) also found that higher crop levels reduced the acclimation rate of canes, and buds were less hardy during acclimation.

Some vineyards in southern Oregon have direct experience with the impact of overcropping on cold hardiness. In one vineyard, half of a two-year-old, 9-acre block of Merlot/101-14 was cropped and the adjacent half was not. In the block that was cropped the previous season, 60–70% of the vines were killed in a cold spell in the winter. Much less than 5% of the vines from the noncropped block was killed. The severe winter damage in the early-cropped vines was likely due to low levels of carbohydrate reserves in these vines.

Canopy Management and Sunlight Exposure. The amount of canopy sunlight exposure provided by the training system and canopy management practices influences carbohydrate production and hardiness. Vine cold acclimation requires energy, which is derived from the carbohydrates produced by photosynthesis. In general, training systems and practices that enhance photosynthesis also enhance tissue cold hardiness, and the inverse is also true. Maximizing photosynthesis and carbohydrate production requires the maintenance of healthy foliage that is evenly distributed to prevent the development of heavily shaded canopies. Effective disease and insect control and adequate nutrient supply are also necessary for maintenance of healthy foliage.

Pruning. Pruning of vineyards in susceptible winegrowing regions should be delayed as late as possible to lessen the chance of cold damage. Pruning toward the end of the dormant period leads to deacclimation; thus, pruning prior to a freeze (within two weeks) can increase damage to cold. Late pruning also gives the opportunity to assess the damage after a freeze, thus allowing adjustment of the pruning level if necessary. Light pruning (leaving two or three times the necessary node number) increases the likelihood that enough live buds will be available after a freeze. Final bud assessment and adjustment can be made after bud break.

If some cold damage to buds has occurred and the basal nodes have less bud damage (wood is most mature in this area), spur pruning may be more successful in retraining live buds in heavily damaged vines than cane pruning. Thus, assessing the amount and location of bud damage after a freeze is important.

Irrigation. Where irrigation is available and cold hardiness is a concern, vines should not be drought stressed. Excessive irrigation near harvest can, however, be damaging, particularly if it results in late growth and delayed acclimation. Research in Washington has found very little consistent effect of irrigation on bud hardiness (Wample et al., 2000). Low irrigation levels have tended to increase cane hardiness. Also, field observations in Washington in 1990 showed a trend for greater bud hardiness of European winegrapes in stressed vines (only one irrigation) as compared to vines receiving four or more irrigations (R. L. Wample, personal communication).

Fertilization. Nitrogen nutrition and its relationship to grapevine performance, including cold hardiness, was the subject of an international symposium in 1991. Wample (Wample et al., 1991) reviewed the literature and found little evidence to support the contention that high nitrogen nutrition resulted in direct loss of grapevine cold hardiness. This was based on good management practices and precluded the application of nitrogen late in the growing season, which would maintain active vegetative growth and therefore decrease frost and cold tolerance.

There are, however, numerous reports demonstrating the dynamic nature of nitrogen metabolism in grapevines during the acclimation and dormant periods. Total and protein nitrogen levels have been found to rise at the outset of acclimation and into the second phase of hardening. Higher levels of nitrogenous substances have been found in the more cold hardy varieties. It is possible that these higher concentrations were the result of a slower growth rate found in these varieties and were thus indirectly related to cold hardiness. Wample's review cited Russian research that showed that grafting European grapevines onto winter hardy American rootstocks led to higher midwinter nitrogen levels and better cold hardiness. A high ratio of protein to total nitrogen has also been related to greater cold hardiness across several varieties. On the other hand, two studies (Wample et al., 1993; Wolf and Pool, 1988) have found little or no effect of different nitrogen fertilization levels on the cold hardiness and survival of Chardonnay.

Multiple Trunks. A severe winter freeze does not necessarily cause equal damage to both trunks of a multiple-trunked vine, because trunks often vary in hardiness.

Physical Protection. In cold grape-growing regions, the graft unions of vines are commonly covered, or "hilled-up," with soil in the fall. The ridge of soil provides many degrees of protection. For hilling-up to be effective, vines must be set at planting with the graft union within 2–3 inches of the soil line. Soil is graded toward the trunks from either side of the row in the fall with a tractor-mounted, hydraulically controlled blade, or the operation can be done with a hand hoe in smaller plantings. The soil ridge must be removed at least every other year to prevent permanent scion rooting (which eliminates the effect of the rootstock). Hilling-up vines is not easy. Carelessness causes injury to trunks, and the disturbance of soil complicates chemical weed control. But in the event of severe injury to aboveground portions of the vine, basal buds protected by the soil are the source of shoots that can be used to reestablish trunks and vines much more rapidly than replanting a new grafted vine. In heavy-textured soils that are wet and sticky in spring, hilling-up may be difficult and impractical.

Reflective trunk paint may reduce winter injury to trunks when temperatures are low and days/nights are clear with snow cover present. Apply 50% white latex paint to the south side of the trunk from ground level to 4–5 inches above the graft union.

Assessing the Amount of Cold Injury

When winter cold injury is suspected, it is important to determine the extent of the damage and what you can do about it. When cold temperatures happen before bud break, injury may occur to one or more of the following tissues: the primary, secondary, or tertiary buds; the phloem of the trunk, canes, or roots; the xylem of the trunk, canes, or roots; pith (soft tissue central to the cane or trunk); and the vascular cambium of the permanent structures. Depending on the weather conditions, there is potential for damage to all of these tissues. Research on cane hardiness has shown that the least hardy tissue appears to be the phloem and cambium (Wample et al., 2000). Early in the acclimation stage, buds are hardier than canes. From November through spring, when deacclimation begins, cane hardiness exceeds that of buds. During the deacclimation stage through bud break, canes may be more cold tender than buds.

The simplest and perhaps the most frequent case, low-temperature injury of the primary bud, results in few changes in overall growth of the vine but may lead to a significant loss of crop for that season. Some varieties that have fruitful secondary buds may still produce nearly a full crop when most primaries are damaged. Loss of more than the primary bud is often associated with damage to other vine parts and frequently results in the loss of permanent structures and necessitates major retraining of the vine if it survives. The entire vine may be lost if cold injury is below the graft union in a grafted vine.

In Washington in 1978/79, cold damage occurred to the vine root system while the majority of the shoot system was undamaged (Wample et al., 2000). There had been very little rainfall after harvest, and there was a shortage of irrigation water. The soil was dry and frozen to a depth of about 1 foot during a prolonged cold period. During this time, the temperature stayed below 32°F but did not generally get below -4°F. Under these conditions, minimal bud damage but significant root damage occurred. The following spring, bud break occurred and shoot growth began normally. However, after a few weeks of shoot growth and during a very warm period, cold-injured roots were unable to meet the water requirements of the shoot system, and the shoots collapsed. Weakened root systems may also be more sensitive to attack by soil pathogens.

Low-temperature injury to the vine's permanent structure may also increase the chances of crown gall (*Agrobacterium vitis*). This is often the cause of a more severe vine decline than the low-temperature injury itself.

It is important to assess the extent of bud and wood damage accurately to determine whether pruning methods need to be adjusted.

Bud Injury. Assessment of the extent of winter injury involves sampling as many as one hundred buds per acre. This can be done in one of two ways: cut through one hundred randomly selected buds on canes with a razor blade, or collect ten ten-node canes randomly through the vineyard. These should be as similar as possible to the canes you would retain for fruiting wood when pruning; terminal canes can be sampled so as not to damage basal canes, which are usually kept for the coming season's crop. Sample buds at the same height or position as you would leave for fruiting wood. If the first thirty buds sampled indicate 10–100% damage, then sample a total of one hundred buds to get an accurate assessment of damage. The hundred buds collected should be evaluated visually for freeze damage to arrive at a percentage of live buds. Sample varieties separately and keep track of buds that come from frost pockets or low areas of the vineyard, for these are likely to be more damaged.

Buds can be sliced lengthwise or crosswise for evaluation. Each compound bud contains three buds or growing points, the primary (1°), secondary (2°), and tertiary (3°). The 1° growing point is the largest and is central to the 2° and 3°. For lengthwise sections, slice each bud at its midpoint with a razor blade. In crosswise sections, make the initial cut at a point far enough out on the bud to bisect the tip of the 1° growing point within. This cut normally should not expose any of the 2° or 3° growing points. You may wish to take a series of very thin slices off the tip of the bud until the 1° is reached, at least until you become familiar with the procedure.

Dead buds appear brown or black, whereas live buds are light green. In bright light, you can see bud damage without the aid of a hand lens or dissecting scope, but a hand lens helps. A blackening or browning of the entire bud indicates bud kill by cold temperatures. Sometimes the bud is partially brown or black, indicating that one or two of the 1°, 2°, or 3° growing points is dead. Often, if the l° bud is dead, the 2° and 3° buds will grow, thus leading to a partial or near full crop. In some cases the 2° or 3° growing points are killed by cold, leaving the 1° healthy; this happens about 10% of the time in partial bud death. A bud can also have a thin layer of black tissue just under the bud scales (the outer protective tissue); in this case, the bud may grow, leading to normal shoot growth.

Another way to assess bud and cane damage is to collect some canes (again sample the entire block) and force these in a bucket of water indoors. This is a time-consuming technique (it may take up to four weeks at 80°F), and you may not have results in time to adjust pruning levels. The subsequent bud growth indicates the extent of bud damage. Sometimes buds break, produce a short shoot, and then grow no farther; this is often an indication of damage to the conducting tissue at the base of the bud or of cane damage.

If a vineyard has some winter damage and has not yet been pruned, pruning can be adjusted to compensate for the loss. If bud mortality is less than 20%, do not adjust pruning severity, for the vines will compensate by producing larger clusters. Bud damage in the range of 20–50% requires more judgment; leave more buds to make up for the loss due to damage. At the lower end of the range, leave an extra cane or two. When mortality is 50%, leave twice as many buds as you would normally. A formula for adjusting pruning levels is [number of buds usually left per vine divided by (1 - percentage of bud damage)]. For example, assume that you normally leave thirty buds per vine and that you have determined that there is 30% bud damage. Using the formula, the number of buds to leave to compensate for freeze damage is 30/(1 - 0.30) = 43 buds per vine. Be prepared to make adjustments after bud break, if necessary. If you appear to have 100% bud damage, we recommend that you still prune the vines. Leave five or six times as many buds as you would otherwise. For vertical shoot positioned vines, these extra canes can be trained as in a Scott Henry system, or leave them on the ground. Reevaluate the vines after bud break.

Cane and Trunk Injury. Cold injury to the vascular tissues of canes and trunks is more serious than bud injury to the long-term health of the vine, and it is more difficult to assess in the winter. Cane tissues, like buds, brown after being killed and allowed to warm several days. Cane and trunk cold injury is diagnosed by making shallow longitudinal cuts into the wood (or scraping away the bark tissue) and examining the phloem and cambium regions for browning. If browning of the phloem and cambium is not yet apparent, try again in March when the vines begin to deacclimate.

When the cambium is killed, vascular tissues of the cane or trunk are not regenerated during the subsequent growing season. The sudden collapse of shoots after the first hot days of summer is a delayed manifestation of cambium injury. The xylem of the cane and trunk supports shoot growth during the early part of the season. With hot weather, the transpiration rate of new shoots exceeds the water-conducting capacity of the damaged vascular system and the vine (or shoot) wilts and dies.

If wood injury has occurred, extra canes can be retained at pruning. It is possible that injury will not be uniform and that some canes may be unaffected.

Trunk injury is also diagnosed by making shallow, longitudinal cuts into the wood. Trunk injury is more common on young vines, and on the southwest side of the trunk in areas that received some snow cover. The snow reflects sunlight, causing the trunk to thaw during the day and refreeze at night, leading to bark splitting. Injured trunks may split or be affected by crown gall one or two years after the injury. Often, however, trunk injury is difficult to assess. On a trunk-damaged vine multiple suckers often emerge at the base of the plant (this may be above the graft union in a grafted plant). Retain a good sucker to replace the trunk.

References and Additional Resources

Howell, G. S. 1988. Cultural manipulation of vine cold hardiness. Proceedings of the Second International Symposium for Cool Climate Viticulture and Oenology, Jan. 11-15. Auckland, New Zealand.

Howell, G. S. 2000. Grapevine cold hardiness: mechanisms of cold acclimation, mid-winter hardiness maintenance, and spring deacclimation. Proceedings of the ASEV Fiftieth Anniversary Meeting, Seattle, Washington. American Society for Enology and Viticulture, Davis, Calif.

Wample, R. L., and A. Bary. 1992. Harvest date as a factor in carbohydrate storage and cold hardiness of Cabernet Sauvignon grapevines. *Journal of the American Society for Horticultural Science* 117:32-36.

Wample, R. L., S. Hartley, and L. Mills. 2000. Dynamics of grapevine cold hardiness. Proceedings of the ASEV Fiftieth Anniversary Meeting, Seattle, Washington. American Society for Enology and Viticulture, Davis, Calif.

Wample, R. L., S. E. Spayd, R. G. Evans, and R. G. Stevens. 1991. Nitrogen fertilization and factors influencing grapevine cold hardiness. In J. M. Rantz (ed.), *Proceedings of the International Symposium on Nitrogen in Grapes and Wines.* American Society for Enology and Viticulture, Davis, Calif.

Wample, R. L., S. E. Spayd, R. G. Evans, and R. G. Stevens. 1993. Nitrogen fertilization of White Riesling grapes in Washington: nitrogen and seasonal effects on bud cold hardiness and carbohydrate reserves. *American Journal of Enology and Viticulture* 44:159-167.

Wample, R. L., and T. K. Wolf. 1997. Practical considerations that impact vine cold hardiness. In T. Henick-Kling, T. E. Wolf, and E. M. Harkness (eds.), *Proceedings for the Fourth International Symposium on Cool Climate Enology and Viticulture,* Rochester, N.Y., 16-20 July, 1996. New York State Agricultural Experiment Station, Geneva, New York.

Wolf, T. K., and M. K. Cook. 1991. Comparison of 'Cabernet Sauvignon' and 'Cabernet Franc' grapevine dormant bud cold hardiness. *Fruit Varieties Journal* 45:17-21.

Wolf, T. K., and R. M. Pool. 1988. Nitrogen fertilization and rootstock effects on wood maturation and dormant bud cold hardiness of cv. Chardonnay grapevines. *American Journal of Enology and Viticulture* 39:308-312.

26

Strategies for Frost Protection

David Sugar, Randy Gold, Porter Lombard, and Alfonso Gardea

Vineyards in regions of Oregon where freezing temperatures occur during the spring months or early in the fall have suffered reduced production and quality of grapes. Spring frost damage in Oregon has occurred as late as May and mid-June. Losses from spring frost are principally due to primary shoot damage. Primary buds usually develop first in the spring and have the greatest crop potential. Because of their early development, primary buds are also more susceptible to frost damage than are secondary and tertiary buds. In most vinifera cultivars secondary shoots may develop and set fruit after damage to the primary shoots, but the resulting clusters are fewer and immature at normal harvest time. First-year vineyards, where tissue is close to the ground, may be injured significantly by a frost that does not damage older vines in the same vineyard.

Ripe fruit is rarely injured by fall frosts, but cluster stems can be; frost-injured clusters should be used immediately to avoid berry abscission. Leaf damage from fall frost slows or stops the fruit-ripening process, depending on the extent of damage. Frost shortly after harvest may delay cane hardening and render the canes more susceptible to winter damage, and consequently affect the following year's crop. Spring frost injury to buds and leaves may be observed within a few hours of damage as darkening and collapsing of the affected tissues, and it is irreversible.

Protecting a grape crop from frost damage requires knowledge of the vineyard site, the weather conditions that are conducive to frost, and advance preparations to apply practices that prevent or reduce frost injury. The circumstances of each vineyard are unique, so vineyard managers must make informed decisions on the most appropriate practices and methods. Like most other vineyard operations, timely application of frost protection measures are the key to success.

Frost Basics

Critical Temperatures for Frost Damage

As grape buds go through deacclimation to the green shoot stage, they are increasingly sensitive to low temperatures. The critical temperatures for frost damage at various stages of bud and shoot development were studied for Pinot noir in Oregon (Gardea, 1987). The number of "flat" leaves, leaves that had expanded enough to have an orientation that is nearly perpendicular to the ground, defined shoot stages. The critical temperature data are presented in Table 1. Beyond bud break, damage may occur when developing shoots experience temperatures of 31°F or lower for one-half hour or longer. During bud break, shoots cannot tolerate lower temperatures, becoming increasingly sensitive as bud development proceeds.

Conditions Under Which Frost Occurs

During the day, the earth and objects on it receive short-wave radiation from the sun. This radiation can pass through clouds and glass but not through smoke. Some of the radiation is absorbed by the earth as heat, and some is reflected back. After being once absorbed by the earth, some heat is also radiated back to space, but this re-radiation is in the long-wave form, which cannot pass through clouds or glass but can pass through smoke. At night, under a clear sky, the earth loses heat through re-radiation and cools the air closest to it. The cool air, being more dense than the warmer air above it, tends to stay close to the ground and is drawn by gravity to seek the lowest elevations. Cool air forming on sloped ground tends to flow downhill to the valley floor.

Inversion Layers. In the absence of wind, the cooling process described above creates an inversion layer. Whereas during the day there is a continuous decline in temperature with increase in elevation, under an inversion the coldest air is near the ground, and there is an increase in temperature with increased elevation. Commonly, with an inversion, the temperature very near the ground can be 1°F to 2°F lower than at 5–6 feet above the ground. The area of temperature inversion varies under different conditions but is generally within 50 feet above the ground.

In most cases, the temperature at the inversion layer is well above the critical temperature for damage to grapevines. As discussed later, it is this warmer air above the vineyard that is used by wind machines to warm the air around the vines; a good breeze can

Table 1. Critical Temperatures for Pinot Noir Buds and Young Shoots

Growth Stage	None Killed		50% Killed	
	°F	°C	°F	°C
Dormant bud enlarged	-	-	6.8	-14.0
Green swollen	-	-	25.9	-3.4
Budburst	30.2	-1.0	28.0	-2.2
First flat leaf[a]	30.2	-1.0	28.4	-2.0
Second flat leaf	30.2	-1.0	28.9	-1.7
Fourth flat leaf	31.0	-0.6	29.8	-1.2

[a] Shoot stages defined by the number of flat leaves, those that had expanded enough to have an orientation nearly perpendicular to the ground.

cause temperatures to rise by mixing warm air from above with the cold air around the vines. Where fuel is burned to provide heat, it is not the "whole outdoors" that is warmed but the area from the ground to the inversion layer, since the warmed air rises only to that level (weather forecasters call it the *ceiling*).

Clouds, Moisture, and Dewpoint. Since heat loss from the ground at night is in the form of long waves that do not pass through clouds, clouds act like blankets over the earth, and in most cases cloudy nights are free from frost. Most of the radiation from the earth bounces off water molecules in the clouds and back down to the ground, slowing the loss of stored heat. Radiated heat travels in straight lines, so a cloudy area stays warmer than an adjacent clear area. Lone clouds passing overhead during the night temporarily warm the areas they pass over.

Moisture in the air can reflect heat loss back to the ground, even if the moisture is not dense enough to create clouds. The more dense the moisture, the more effective the blanket effect, but even on clear nights moisture in the air slows the rate of temperature fall. Very dry air can allow temperatures to drop rapidly, so planning for frost protection must include knowledge of the moisture content of the air; most protective measures must be initiated at higher temperatures if the air is very dry. As the air near the ground cools at night, the moisture in the air concentrates (relative humidity increases). With continued cooling, the air may reach a point where it can no longer hold the moisture as water vapor, and the water condenses out of the air as dew. The temperature at which water condenses out of the air is called the *dewpoint* and is predictable from the difference between wet- and dry-bulb thermometer readings.

It is valuable to know the dewpoint because the condensing water vapor releases heat into the air, slowing or temporarily stopping temperature fall. This process can be understood by considering that it takes heat to vaporize water (make steam), and that this heat remains latent in the vapor. When the vapor condenses back to water, heat is released to the surrounding air. Weather Service frost forecasts always include the expected dewpoint.

If the dewpoint is higher than the critical temperature, the heat released at the dewpoint may provide sufficient heat to avoid reaching damaging temperatures, or at least delay the temperature fall and postpone the need for control measures. The higher the dewpoint, the more moisture in the air, and the more heat available for release. If the dewpoint is below the critical temperature, however, temperature will fall rapidly, since the low dewpoint indicates the lack of moisture in the air to reflect heat lost from the ground, and there will be no condensation to help warm the air.

Advective Freezing. On rare occasions, a very cold air mass moves into a grape-growing region during the time when new tissue is unfolding and susceptibility is high. *Advective* refers to the transfer of heat by the horizontal movement of an air mass. Under this condition, there is no formation of an inversion layer as in a radiation frost, dewpoints are very low, and there may be wind. Consequently, controlling the temperature may be beyond the capacity of available methods. Fortunately, although this situation is common in winter, it is unusual during the susceptible period of April to June in most fruit-growing areas. Advective freezing can be more frequent in fall, late in the harvest period before canes have hardened off.

Controlling Frost and Frost Damage

Cultural Methods

Site Selection and Management. Necessity for frost control may be reduced by locating the vineyard on a site that is not prone to frost. In general, these areas are on slopes and hillsides, from which cold air drains toward a valley floor. Since south-facing slopes offer the advantage of receiving the most concentrated solar radiation (more heat units), these slopes are the most desirable for vineyards in Oregon. If the slope is gentle, orienting the rows up and down the slope facilitates air drainage. If, on the other hand, the slope is sufficiently steep so that soil erosion and accessibility of the vineyard would be compromised by this orientation, those considerations must take precedence. In either case, dense shrubbery, windbreaks, or buildings at the lower end of the vineyard should be avoided. No site is immune to frost, but location on a slope makes it easier to live with. Locations 50 feet above a valley floor are usually the best situated.

Floor Management. Clean-cultivated vineyards are usually slightly warmer (1°F to 2°F) than vineyards covered by sod or ground cover. Vegetation reduces the amount of heat absorbed by the ground during the day and inhibits release of heat at night. Sod mowed close to the ground with a flail mower is nearly equivalent to a clean-cultivated vineyard floor for these purposes. Mulches other than black plastic interfere with heat absorption and radiation.

Maximum heat absorption and slow radiation are achieved by a moderately moist, smooth, firmly packed soil free of ground cover. Water in the spaces between soil particles is much more effective than air in absorbing and storing heat. The moist layer does not have to be deep; only the top foot of soil needs to be moist to aid in heat absorption. In general, sandy soils do not store heat as well as loam or clay soils. Also, darker colored soils may absorb more solar radiation and store more heat than lighter colored soils. Consequently, if all other factors are the same, a vineyard in sandy soil would have greater risk of frost damage than a vineyard on clay or loam soil.

Varietal Differences in Budburst. Budburst develops in response to ambient temperatures. Temperatures over 4°C (39°F) affect development and swelling of the bud, but budburst occurs when the mean temperature reaches 10°C (50°F). The timing of budburst generally coincides with bloom in apple varieties such as Red Delicious and Golden Delicious.

Grape varieties begin budburst at different times; varieties that have early budburst and development have a greater chance of encountering spring frosts than varieties with late budburst and development. For example, budburst of Chardonnay vines may occur two weeks earlier than Cabernet Sauvignon vines. Varieties with early bud break, such as Chardonnay, Müller-Thurgau, Gewürztraminer, and Pinot noir, may not be suitable for frost-prone sites unless active means of frost protection are available.

Delayed and Double Pruning. The difference in time of bud break among buds on an unpruned cane can be used to advantage to escape frost damage. Buds at the ends of canes develop shoots earlier than more basal ones, regardless of the length of the cane. Delaying pruning until terminal buds have produced shoots 2–4 inches long can postpone bud break at the basal nodes by one to two weeks, without detrimental effects or delay in fruit maturity. The length of the delay in bud break varies with temperature, being shorter if temperatures are warm. Delaying the entire pruning operation to this point, however, is not convenient and does not take advantage of the time available for pruning during the long dormant season. The same amount of delay in bud break may be achieved by pruning vines during the dormant season but leaving those canes that are ultimately to become bearing canes or spurs at their full unpruned length, returning to cut them to the desired number of buds when the terminal buds have sprouted 2–4 inches.

Promoting Acclimation. Early fall frosts can injure canes that are actively growing or otherwise lack hardiness. Good canopy management can help promote acclimation (hardening off); canes that are well exposed to sunlight have good periderm development and are the hardiest in the fall. Delay in cane hardiness is primarily due to excessive vigor and shading of the canopy in late summer and early fall. Excessive irrigation should be avoided after mid-August in Oregon to help vines harden off promptly. Similarly, excessive soil moisture from rainfall can be reduced by establishing a winter cover crop.

Mechanical Methods

Heaters. For many years, growers have protected fruit crops from frost by burning fuel to create heat. Diesel oil in individual heaters (smudge pots) is the most common fuel used for frost protection, although propane delivered through underground pipes has also been used. Burning these fuels at today's prices as the sole means of frost protection is prohibitively expensive for most grape growers. But vineyards that use wind machines may also have heaters to add extra heat during the infrequent nights when temperatures fall below the capacity of the wind machines.

When heaters are the sole source of protection, forty to fifty heaters burning at a rate of 0.5 gallons per hour per acre is recommended. Heaters should be evenly distributed through the vineyard, except on the borders, where they should be twice as frequent. Heaters may be lit within one degree of the critical temperature, unless temperatures are falling rapidly due to very dry air or there is inadequate personnel to light the heaters within one-half hour.

Not all heaters need to be lit at once, however. For example, if minimum temperatures are expected to drop to slightly below 31°F, only one-third of the heaters may need to be lit. Lighting one-half of the heaters can provide approximately 3°F protection on a typical night. These are not absolute formulas: actual needs vary by site and conditions. The cost of burning forty heaters per acre with a diesel price of $1.50 per gallon would be $30.00 per hour. There is additional cost for the labor to light heaters and put them out at dawn, as well as labor to refill the heaters with oil to prepare for the next night.

Over-vine Sprinklers. Over-vine sprinkling systems offer a high degree of frost protection at relatively low operating costs, although initial investment costs can be high for materials, installation, and development of an adequate water source. The ability of over-vine sprinkling to provide frost protection is based on the fact that, when water freezes, some of the energy present in the water is released as heat. One can easily imagine that heat is required to melt ice; physicists tell us that this heat remains "latent" in the water and is available for release when the water freezes again.

Water delivered to tender buds and shoots during a frost freezes on the plant, and most of the latent heat travels through the ice to keep the tissue safely at 31.5°F to 32°F. Water must be delivered continuously during the freezing period, so that there is a constant film of water turning to ice. Systems delivering 0.19 inches of water per hour can protect buds even when air temperatures approach 22°F. Protection to 26°F requires an application rate of 0.12 inches per hour. If the system stops functioning during the night, or if inadequate delivery rates allow all water on the plant surface to become ice before new water arrives, the process reverses; the bud tissue loses heat through the ice, and evaporative cooling can reduce bud temperatures to below ambient air temperatures.

Protection of the entire vineyard by sprinklers requires pumping capacity and water availability at a much greater rate than is normally used for delivering irrigation water. To apply 0.15 inches per hour requires 67.3 gallons per acre per minute; 0.10 inches requires 44.9 gallons per acre per minute. For a large vineyard, this may require construction of a pond or reservoir if continuous water sources cannot provide the volume needed. Pond capacity should be sufficient to provide for at least three consecutive nights of sprinkling, based on an average number of hours of sprinkling needed per night in the area of the vineyard.

Impact sprinklers should be spaced evenly through the vineyard and should rotate at least once per minute in order to keep a continuous film of water over the plants. Sprinklers should be arranged in a triangular or rectangular pattern with adequate overlap of water application to provide complete coverage of the vineyard. Lateral lines bringing water to the sprinklers should run perpendicular to the direction of the rows, rather than beneath the vine rows, and valves should be installed at every lateral to facilitate repair or cleaning while the system is in use. Underground pipe must be buried deep enough to be safe from damage by cultivation equipment and winter freezes; 2 feet is sufficient depth in most western Oregon locations. Finally, it is much more convenient and less damaging to install the sprinkler system before the vineyard is planted than to add it to an existing vineyard.

There is little experience with the use of over-vine minisprinklers for frost protection in Oregon. Studies in other areas with other crops have, however, demonstrated that systems using a single minisprinkler suspended over each plant can provide targeted frost protection with greatly reduced water use compared to impact sprinklers. The flow rate of the system must be sufficient to prevent freezing of the water in the tubing or before arriving on the plant surface. A study in a California vineyard found that a pulsing minisprinkler system provided frost protection equivalent to over-vine conventional sprinklers with 80% less water consumption.

Under-vine sprinklers have not been widely explored for frost protection. There is potential benefit to humidifying the air around vines through under-vine sprinklers, thus slowing the rate of radiation heat loss.

Wind Machines. Wind machines can provide a means of mixing warmer air above the vineyard with cooler air around the vines, thereby warming the air around the vines. Their effectiveness depends on the presence of a temperature inversion so that there is a source of warmer air available for mixing. If there is no temperature inversion, wind machines are of little benefit. Thus, wind machines can provide little or no warmth to the vines during advective freezing conditions. Likewise, if the ceiling or inversion point is very high, the effectiveness of wind machines may be reduced.

With a typical inversion layer at 40–50 feet, wind machines can be expected to increase the temperature around the vines by one-fourth to one-half of the difference in temperatures between air around the vines and the warmest air in the range of the wind machines. Typically, this means 1°F to 3°F of warming. In general, sites with air temperatures at 30–50 feet that are frequently at least 3°F warmer than at vine level on frosty nights are good candidates for protection by wind machines.

Oil heaters may be lit to supplement the wind machines on very cold nights. Supplemental heaters should be preferentially arrayed near the perimeter of the vineyard and in portions of the vineyard farthest from the wind machines.

The effective range of a wind machine may be 6–10 acres, depending on the site contour, the power of the wind machine, and the proximity of other wind machines. Two or more machines tend to reinforce the effectiveness of each other. Machines should be located where they can best take advantage of natural patterns of air movement; rather than placing a wind machine at the center of its allotted area, locate it slightly updrift with respect to patterns of cold air movement during frosty nights. Wind machine vendors generally have a great deal of experience in appropriate placement of machines.

In addition to the warming of air provided by mixing warmer air from above the vineyard with air around the vines, there is additional benefit provided by generating air movement around the plant surfaces.

On still, clear nights, plant tissues and a thin layer of air around them may become 2°F to 3°F colder than the surrounding air. In this respect, the plant is acting as a "radiating body," just as the soil does as it loses temperature during the night and gradually cools the adjacent air. By stirring up the air, wind machines can slow radiation heat loss from the plant and also interfere with ice crystal formation, both of which processes can help the plant endure cold temperatures.

Grow tubes. In a recent study in Washington, open-topped grow tubes were found to have no effect on temperatures around the vine, inside the tube. Covering the top of the tube at night could add some protection, but excessive heat could build up in closed grow tubes during the day. A preliminary assessment indicated that tubes with higher light transmission increased damage, since the higher light transmission increased the daytime temperature and therefore reduced the cold hardiness of the vines.

Ice Nucleation Bacteria Inhibitors. In trials conducted in Oregon, little or no frost protection has been obtained from treating vines with substances designed to depress the freezing point or inhibit bacteria that can serve as nucleators for ice formation.

Preparation for Frost

Thermometers. All information about critical temperatures and recommendations for start-up temperatures for frost protection are based on readings from standard orchard thermometers located in shelters designed by the National Weather Service. The standard thermometer is a straight-tub alcohol thermometer mounted with its bulb end just below horizontal; it contains a float that remains at the line of the minimum temperature experienced since the previous resetting. The thermometer is reset during the day after each frost by swinging its upper end down to vertical, so that the float sinks to the current temperature.

A written record of minimum temperatures should be kept to help develop an understanding of what can be expected from each vineyard site relative to the National Weather Service forecast for the area. The thermometer should be installed in a standard thermometer shelter; specifications are available from the National Weather Service and other sources. The objectives are to get the thermometer 5 feet above the ground, facing north, and shaded to protect it from direct sunlight. Thermometers should be calibrated each season by submerging them in a mixture of ice and water, which should register 32°F. Store thermometers between frost seasons in an upright position with the bulb end down to prevent separation of the alcohol column.

Frost Alarms. Temperature-sensitive alarms are available from several manufacturers, usually with a range of temperature options. The wake-up temperature should be selected based on the amount of time needed to get the frost protection system in operation. The sensor should be mounted in a standard thermometer shelter in order to correspond to the temperature system described above and placed in the typically coldest location in the vineyard. Like thermometers, alarm sensors can be calibrated by immersion in an ice-water mixture, which should trigger an alarm when set at 32°F.

Telemetry. Several manufacturers now produce equipment for monitoring temperature, humidity, and other parameters in the vineyard, and sending the information via radio frequencies to a receiver attached to a computer. A typical system setup might update the information every fifteen minutes. Computer software can display the information in a variety of formats and maintain archives of environmental information. Systems can also be set up to trigger alarms or dial telephone numbers when critical conditions are reached.

System Checks. Confidence in the frost protection system requires that the system be thoroughly tested at least one month in advance of anticipated use, to allow adequate time for acquiring replacement parts or making modifications.

References and Additional Resources

Gardea, A. A. 1987. Freeze damage of 'Pinot noir' (*Vitis vinifera* L.) as affected by bud development, INA bacteria, and a bacterial inhibitor. M.S. Thesis, Oregon State University, Corvallis.

Perry, K. H. 1998. Basics of frost and freeze protection for horticultural crops. *HortTechnology* 8:10-15.

27

Labor Management

Andy Humphrey and Edward W. Hellman

Vineyard Zen, part I

You have a vision for your vineyard. Look out your window onto your fields. Look out and in your mind's eye, you see each vine as a perfect individual entity. You imagine the cycle of the growing season, budbreak, bloom, verasion, and harvest. Each vine has been skillfully cared for. Each shoot has been carefully positioned and trained to yield the best quality fruit possible. You can feel the health and balance of the vines in your very bones. You can feel them thriving, growing, maturing and ripening the fruit. The vines talk to you in whispers all season. They tell you what they need. "Prune me, feed me, water me, spray me, pick me, I'm happy, I'm sad." You've responded to them as if they were your children, with care and love and individual attention. You can feel the finished wine on your pallet—the most elegant, beautiful, seductive wine imaginable. You are one with the vines, you have become the vines and they have become you. You have reached a state of Zen with your vineyard.

Now open the door and step outside. Uh oh, reality! It's mid-January in western Oregon. There is a constant drizzle of rain that has lasted for the past month, interspersed with the occasional downpour or sleet storm to liven things up. The forecast for the coming month looks the same as the previous month. The jet stream has decided Oregon is a fine place to be. The temperature is 38 degrees. The wind is blowing at 25 knots. Your reverie is somewhat interrupted by these facts. The soil is so saturated that after a half-dozen steps in the vineyard you've gained three inches of height from the mud stuck to the bottom of your boots. It smells like winter and you know that smell is here to stay for a while.

Next to weather, labor management is one of the greatest challenges a grower will face on a continuing day-to-day basis. In some cases, even when you have a critical mass of acres large enough to warrant some full-time help, the majority of the workforce can be transient. The crew that begins a season pruning may completely turn over by harvest. Although individual tasks change throughout the season, they can quickly become monotonous. Some tasks seem tedious and silly to the point of being torturous when it is cold, windy, raining, and muddy—or, conversely, unbearably hot and humid. Although the work may seem on the surface to require only unskilled laborers, the fact is that unskilled and inexperienced laborers who are unsupervised can quickly and easily cost a lot of money. Attentive labor management can help you attain your vineyard goals.

There are probably as many approaches to managing labor in the vineyard as there are vineyards, but most could be placed into one of two categories: the Blitzkrieg approach and the Delegation approach. Depending on the size and immediate needs of the vineyard, one or the other, or both, will be more appropriate or cost effective. From the smallest vineyard of only a few acres to the largest wineries with multiple sites involving hundreds of acres, both approaches are used to one extent or another. No matter how you approach completing the task at hand, qualified supervision is the key to success.

The Blitzkrieg Approach

The Blitzkrieg approach is simple: a task needs to be completed, send in a large crew of workers, move through the field like army ants, and move on to the next task. With a good crew and a good supervisor this can be very cost effective. With a less than motivated crew or an inexperienced supervisor, however, mistakes can be made repeatedly in a short period of time. Sometime corrections can be made in the same season, which increases costs for that season. But sometimes corrections cannot be made until the following season, or perhaps never, resulting in less than desired conditions or higher production costs. So, the quality of the work can have a direct effect on the quality of the fruit or the health of the plant, and ultimately on the profitability of the vineyard or winery. Blitzkrieg has its place and sometimes it is the smarter approach. But remember—the bigger the crew, the harder it is to really know what's going on.

Vineyard Zen, part II

You've spent hundreds of thousands of dollars to purchase, clear, prepare, and lay out your vineyard site. The bare-rooted dormant plants you ordered last year have arrived and the weather is right. You've hired a large group of people or a labor contractor to help you with the planting. This group claims to have "much experience" planting grapevines. In fact, they are so confident that you've negotiated a per-plant price for the job. You know exactly how much the cost will be, and it ties right in to what you've been told to expect. They swarm over your field in no time, and your field of dirt has suddenly, almost magically, transformed into a real vineyard. "Wow, those guys really know what they're doing," you think. Life is good. A warm glow overwhelms you as you reflect upon the fact that you now own a vineyard. You are in harmony with the land. You have attained the second level of Vineyard Zen.

Later that year, fewer plants survive than you expected. You inspect the dead vines and find that most have been planted with their graft union eight inches above the soil surface. The roots are only a few inches below the surface. No water, dried up, dead. Ouch! In year two, replacements for the dead vines are planted. But, only half of the second-leaf vines make it to the wire. Hmmm.

Year three, all the vines are up to the wire, but those weaker ones seem to remain weak. You attribute this to shallow soil or nutrient problems, or bad graft unions, disease …something. You inspect, test, find nothing wrong and finally decide that they will catch up eventually. A few years later when those weak vines are still weak and producing only half what the stronger vines are producing, you finally take a shovel and start digging to see what is really going on down there. That's when you discover that all of those weak vines were "J-rooted" when they were planted. Ouch!

A few years later, some strong vigorous vines become less vigorous, and sick-looking. You prune them hard but each season they become weaker and weaker. Again you take out the shovel and, with a magnifying glass, the verdict comes in: phylloxera! "No way Bubba! Those are grafted plants!" Unfortunately, the graft unions were below the soil surface when they were planted. The scion grew adventitious roots that became the dominant root system, and along came those hungry little guys. "Oh no! How did the planting crew make so many mistakes? Why did this happen to me"?

The Delegation Approach

The delegation approach divides not just the tasks, but the responsibility and accountability for the results among several people. It involves clearly defining who is responsible for what and training the right individuals into independent, creative thinking supervisors. Although the cost of farming per acre may increase over that of a pure Blitzkrieg approach, the benefits result in better work quality, consistency, attention to detail, and recognition of problems or potential problems with the vines. Simply put, you have more pairs of eyes watching what's going on, and you can accomplish multiple tasks simultaneously.

The following outline describes a chain of command the first author has found to be very successful for large acreages or multiple vineyard sites.

Chain of Command

Owner

- Creates the vision
- Sets standards of quality
- Holds Vineyard Manager responsible for implementing the vision within budget and on time. Sometimes the owner is the vineyard manager, supervisor, and crew.
- Answers usually to the Spouse or the Banker or both

Vineyard Manager

- Implements the vision
- Organizes, schedules
- Instructs and trains supervisors and tractor drivers
- Oversees quality control, purchasing, and hiring/firing
- Holds Crew Supervisor responsible for overall quality and timeliness
- Schedules tractor drivers, creates and implements spray plans
- Answers to the Owner

Tractor Driver

- Understands the safe operation of equipment
- Willing to work any day regardless of weekends or holidays
- Can recognize equipment problems before they become equipment disasters
- Responsible for minor repairs and regular maintenance of the equipment

- Can calculate tractor groundspeed and calibrate sprayers
- Answers to the Vineyard Manager

Crew Supervisor

- Has experience to do the task
- Cognizant of the time it takes to do a given task
- Able to relay the skill it takes to completely inexperienced crews
- Works with the crew, watches them, times them, follows up on them and holds them accountable for their work.
- Can recognize supervisory qualities in other crew members and utilizes them to spread the supervision over a larger crew
- Answers to the Vineyard Manager

Crew

- Does the hands-on, monotonous, repetitive tasks in the vineyard
- Answers to the Crew Supervisor

Supervision and good communication are essential factors in successfully managing the labor in your vineyard. For a given task to be completed correctly and efficiently, both communication and supervision must mesh. One without the other will not yield the desired results.

A work crew in the vineyard is only as good as the supervisor who is leading them. A good supervisor has a broad base of vineyard work experience. He (or it could be a she) has good communication skills and is able to train completely inexperienced workers. He is able to recognize workers' talents or lack thereof and able to adjust the crew accordingly. Possession of rudimentary math skills is a necessity. Perhaps most important, he must be a natural leader, someone to whom the crew will listen to and follow instructions. All crews who show up at your door will have a spokesperson, but that does not necessarily mean that they have a qualified supervisor. An unsupervised crew is a costly disaster waiting to happen.

A crew supervisor is only as good as the vineyard manager who is leading him. If the vineyard manager does not explain her (or it could be his) expectations, no one knows what the plan is. The vineyard manager must explain the task, not just to the supervisor, but to the entire crew. The reason it is being done should be explained, and why it must be done in a particular way. Both correct and incorrect techniques should be demonstrated, and of course with emphasis on the correct method. Generally, even inexperienced crews are intelligent, curious, and eager to please.

It is important for the vineyard manager to keep in mind the perspective of the crew members. Remember that boring, repetitious, unrewarding, minimum wage job you did as a teenager? The only thing you were really concerned with was payday. That is the mentality of the majority of the crew. It is easy for the vineyard manager to overexplain a task and confuse not just the laborers but the supervisor as well, especially with inexperienced help. Here is where the "Three-Thing Rule" is applied.

Once the manager has explained the task and all the reasons why, it should be broken down into no more than three things, either to remember or to do. For example, to explain leaf pulling: (1) pull only leaves that are within the fruit zone; (2) pull only big, dark leaves; (3) reach into the canopy and pull leaves that are congesting and shading fruit; (4) reach down and strip the suckers off of the trunk. *Warning! Danger!* You have exceeded the Three-Thing Rule. Do not try to accomplish too much in one pass through the vineyard. It seems that it would be a simple thing to reach down and strip the suckers off the trunk while pulling leaves. What usually happens, though, is that one of the four things is forgotten, or remembered inconsistently. Some plants will be perfect, some will have one aspect or another undone, or done incorrectly.

Supervise the supervisor. The manager should work with the crew for a long enough time to estimate when the task will be completed. An excellent supervisor, when asked to do the same estimate, should be able to count the plants in a row, count the plants that are finished, look at his watch, do the math, and give a similar answer.

Table 1. Example Timeline of Vineyard Labor Requirements

Workforce Requirements					*Number of Workers*			
Activity	*Jan*	*Feb*	*Mar*	*Apr*	*May*	*Jun*	*Jul*	*Aug*
Prune	3	3	3					
Pull brush	2	2	2					
Tie cane	1	1	1					
Sucker trunks					2		2	
Train/raise wires					2	2	2	2
Total workers	6	6	6	0	4	2	4	2

Table 2. Example Timeline of Labor Cashflow Requirements[1]

Activity	*Jan*	*Feb*	*Mar*	*Apr*	*May*	*Jun*	*Jul*	*Aug*
Prune	3,680	3,680	3,680					
Pull brush	2,891	2,891	2,891					
Tie canes	638	638	638					
Sucker trunks					789		789	
Train/raise wires					2,535	2,535	2,535	2,535
Total cost	$7,210	$7,209	$7,209	-	$3,324	$2,535	$3,324	$2,535

[1]Dollars, pre-payroll taxes

The vineyard manager must also know when to back off. No one likes someone looking over their shoulder. The crew should be allowed to work for a while without the manager. Later, the manager should check their work with the crew supervisor. When mistakes are found, individuals should be called back and their errors explained to them. Crew members should fix their own mistakes and go on. When the supervisor is seeing things that the manager is not and calling individuals back on his own, good communication has been accomplished.

The vineyard manager should come back and check on the work with the crew supervisor at least once or twice per week. It does not take long, especially with a large crew, to make a small mistake that is repeated into a very expensive mistake. A trip back through the vineyard to fix every plant, or worse, a repeated mistake that cannot be fixed and affects the quality of the fruit, will make everyone concerned unhappy.

Determining Labor Needs

Numerous management activities take place in a vineyard each season, but the type, frequency, and timing of activities typically vary among vineyards, and even from year-to-year in the same vineyard depending on weather and other conditions. Thus it is important for each vineyard to make predictions about the activities expected throughout the season to help plan the workforce needs and manage vineyard cash flow. Good planning can lower costs and enable incentives to be given to the crew.

Identify the tasks and timing and estimate the number of workers needed to complete each individual task in the desired timeframe. Use the formula below to calculate labor needs. Pruning, for example, is one of those activities that, regardless of other factors, takes place once each year, usually January through March. The average time, in minutes, needed to prune a single vine should be estimated by pruning enough vines to determine a reasonable average. Multiply the average pruning time by the total number of vines to calculate the total time (in minutes) required to complete the task. This figure is put in terms of workforce by dividing by the product of 60 minutes per hour, the number of hours per week the crew will work, and the desired timeframe to complete the task, in weeks. The result is the number of workers needed January through March to complete pruning. For a cost prediction, simply multiply the total hours (total minutes divided by 60) by the average labor rate. Labor requirements and costs for other activities can be calculated in the same manner (see equation below).

Follow this process for all the activities planned for the vineyard, and map out the season to give estimated timelines of labor (Table 1) and cashflow needs (Table 2). This procedure also enables planning of crew goals for the day, week, or season. Crews should be rewarded if they exceed time expectations and there is no loss in work quality.

Assembling a Crew

Many times you can find local, year-round, residents that work as a group. More often than not, a group of workers is a family base with extended friends and relatives who join or leave the group depending on the availability of work to be found in the area. Talk to friends and neighbors in the industry for advice on crew availability.

Some neighboring vineyard owners create a crew-sharing situation. When there is not enough work in a single vineyard to employ a full-time crew for an entire season, two or more vineyards share a crew and thus keep them in the area. This allows a group of vineyards to have a dependable crew that has become accustomed to their employers' ways of doing things. There can be pitfalls to sharing a crew, such as disagreements over which vineyard gets what task done first. More

$$\text{No. workers} = \frac{\text{Average task time (minutes)} \times \text{Total no. vines} \times \text{No. repetitions of task/year}}{\dfrac{\text{60 minutes}}{\text{hour}} \times \dfrac{\text{hours/week}}{\text{person}} \times \text{No. weeks to complete task}}$$

than one friendship has been strained or broken by crew sharing, but there are also many instances where it has worked smoothly for years.

Occasionally, situations arise that require the short-term services of a licensed, bonded, labor contractor. There are advantages and disadvantages to this approach. Labor contractors are usually able to draw from a larger pool of workers. If you need a large number of workers in a hurry (the Blitzkrieg approach), a labor contractor can provide them for you. The payroll is the contractor's responsibility. You write only one check, no matter the number of workers.

This sounds like a simple solution, but there are several possible drawbacks. A labor contractor obviously has to make a profit to stay in business. This means that, depending on the size and nature of the work to be done, there will be a charge of 25–50% per hour above the average direct labor rate a vineyard pays when it does its own payroll. It is sometimes possible to negotiate a piecework contract ($ per plant or acre) and actually get the job done for less than it would cost with the vineyard's own crew. The faster the contractor can push the crew, the lower the cost he can charge you. The contractor's motivation is that, the faster she (or he) can move on to another job in a time-limited season, the more money she can make overall. A word of caution: fast piecework many times results in a lower standard of quality. The expectations of the vineyard must be made clear to the contractor. Check the work often and hold the contractor accountable.

Hiring, Firing, and Layoffs

Employers have numerous regulations to follow and plenty of forms to fill out to fulfill government requirements. Many of these are discussed in Chapter 28. One such form is the U.S. Department of Labor WH-516, Worker Information, Terms and Conditions of Employment. The law requires you to inform newly hired employees of certain terms of their employment. The employer must fill out and keep on file this form, which includes basic information to show that the vineyard has informed the person that he or she has been hired as a temporary, seasonal, agricultural worker and stated the wage and estimated period of employment.

Although compliance forms seem innumerable and redundant, they do serve a good purpose. One of the main purposes of the WH-516 form is to make sure employees are aware that the job is temporary and their employment could be terminated at any time. The vineyard manager then should not feel uncomfortable or guilty when making a layoff. The workers understand that their employment is temporary. Thank them and tell them you will contact them when more work becomes available.

There is a difference between firing and laying off. If you are faced with a situation in which you feel an employee must be fired, make sure you document the circumstances surrounding the firing. It can save you time, trouble, and money when answering questions posed by the Oregon Unemployment Division or the U.S. Department of Labor. Compliance with labor laws and regulations is covered in more detail in Chapter 28.

Handling Problems and Disputes

The best way to avoid conflict is communication, from the top down and from the bottom up. The better everyone understands the desired end result beforehand, the less chance there will be of conflict along the way. Invariably, situations arise in which personalities or goals clash among the crew members, or somewhere along the chain of command. More often than not, personality conflicts or jealousies among crew members are better left to the crew to work out among themselves. As you move up the chain of command and disputes threaten the quality or timeliness of the desired end result, the vineyard manager must step in to solve the problem.

Although it does not happen often, an extreme example of labor conflict occurs when the harvest crew strikes during harvest. Direct conflict occurs between the owner and the crew; the owner needs the fruit picked according to the winery's schedule, and the crew simply stops, demanding more money. It is usually the vineyard manager's responsibility to solve this problem quickly in the midst of the anger and panic created by the situation. If you are the manager, your options are these:

1. Send the crew home, and let the fruit rot on the vine.
2. Send the crew home, try to find another crew, and risk the fruit rotting on the vine.
3. Recognize that you have underestimated the value of the crew, amicably agree to their terms and continue harvest with a pleasant smile.
4. Recognize that you are being held for ransom and grudgingly pay it.
5. Recognize that you are being held for ransom, grudgingly pay it, and spend the rest of the day and all evening lining up a different crew for the next day.

The point is this: Take a deep breath—and consider the options and outcomes before you act.

Establishing and Communicating Company Rules

Most vineyard owners and managers have certain expectations of their employees when they are on vineyard property. For example, employers expect their crew to show up on time for work, properly dispose of lunch trash, and put away tools when they are finished for the day. These expectations are more likely to be met if the vineyard manager develops some guidelines for the workers.

Large operations often end up developing an employee handbook. This may be necessary and justifiable, but without some restraint handbooks can easily develop into a quagmire of pages of pedantic rules, regulations, and legalese that few people will read all the way through and none will understand completely.

The simpler, and often more effective, alternative is a short document, one page or less, that simply lists the do's and don'ts in short sentences, in English and Spanish. Each worker reads and signs that he or she understands the rules at the time of hire. For example:

- Don't leave garbage on the ground.
- All vineyard tools and equipment are for vineyard use only. Do not borrow them for personal use.
- Always put tools back in the shed at the end of the day.
- You are expected to show up for work each day at the time determined by your supervisor. Call your supervisor if you will be late, can't make it, or need to leave early.
- Lunch break is 12:00 to 12:30.
- Do not hang your raingear to dry on my back porch railing. Use the hooks in the equipment shed.

Team Building

Developing an effective and efficient vineyard team requires an understanding of the reality of cultural differences. The majority of Oregon's vineyard workforce is Hispanic, mostly from Mexico, and some from Central or South America. Most workers, even if they have been in Oregon for years, understandably carry with them the habits of their culture. Sometimes these habits do not fit with the vineyard owner's or manager's outlook on things. Respect their culture and, when necessary, inform them that a particular habit is not appropriate while on vineyard property. Remember that most, if they were lucky, had the advantage of completing elementary school before they had to begin working. Each will have unique traits, just like any person you have worked with before. Smart or not so smart, brilliant or dull, hardworking or lazy, honest or not, gracious or selfish: Treat your workers with the respect they deserve. Teach them why and how. Show them with rewards that there are opportunities available beyond their paychecks.

It can take many years to establish a thinking, cohesive team that meshes together, is adaptable to the situation, and can take the initiative and run with it. Three keys to successfully establishing an "A" Team are communication, training, and rewards.

When farming multiple sites, the vineyard manager can optimize available time and solve problems simultaneously by holding a regularly scheduled meeting with the crew supervisors and tractor drivers. It can last ten minutes or three hours. The time involved will depend on the size of the operation and the immediate needs of the vineyard.

Discuss the current state of the vineyard; are you where you wanted to be? No? Why? Discuss what needs to be accomplished in the coming week. Use the combined past experience of your team to set goals, troubleshoot, solve logistical conflicts, and help determine manpower, equipment, supplies, and materials needs. Discuss options or workable scenarios that can be used if it rains, if the sun shines, if the cultivator is still not repaired, or if your lead supervisor has to leave for a family emergency. It is truly amazing how productive this can be—not only in terms of planning the work, but building a sense of confidence and cohesiveness among your team. Get them involved in the decision making. You will find that soon they will begin thinking, not just of the task at hand, but two or three steps ahead and how the tasks they perform on a given day will affect what happens next.

Although it can be frustrating at times, it is well worth the investment to spend some time cross-training your essential people. Build upon the strengths of the more experienced to teach the inexperienced. This can be something as simple as changing the oil in the tractor, or as complicated as teaching the algebra involved to calculate tractor ground speed and sprayer calibration. The better trained your supervisors and tractor drivers are, the less supervision they will require. You will find that they can solve logistical and planning problems among themselves. This can make an enormous difference in those times when the tasks are many and the time is short.

Reward people for their accomplishments; it works wonders. Give bonuses for under-budget costs through good management of work crews, or award incremental raises for mastering a new skill. Throw a party at the end of the season; let the crew plan and execute the arrangements. It pulls them together in a common goal, and it's fun! It will also show them that their efforts are appreciated. They are not just showing up to work for a wage; they are an important part of the process.

As your key people master new skills, give them new challenges. If you have chosen people with "the right stuff," they will rise to the challenge every time. This approach develops a team that enjoys coming to work each day, is proud of what they do, and has a sense of ownership and accomplishment about the vineyards.

Keys to Success

- Supervision is paramount to achieving the results you desire. Teach, train, and follow up. Even if your crew has prior experience elsewhere, you may want things done your own way. Old habits are sometimes hard to break. Better results follow if you maintain a certain amount of vigilance.
- Good and frequent communication is essential. Often, very few vineyard workers are able to speak fluent English. Learning some Spanish greatly facilitates communication.
- Make clear to your crew who is in charge, and to whom you have delegated the responsibility of supervision.
- Reward for positive accomplishments. Recognize the value of your crew.
- Always strive for the next plateau of Vineyard Zen. Be Happy. It's fun.

Additional Resources

Billikopf, G. E. 1994. *Labor Managment in Agriculture: Cultivating Personnel Productivity.* University of California Agricultural Extension, Davis.

Agricultural Labor Management. 2002. University of California Agricultural Extension, Davis. http://www.cnr.berkeley.edu/ucce50/ag-labor/

28

Compliance with Government Regulations

Mark L. Chien and Roberta Gruber

Farmers have the reputation for being highly independent-minded. They like to run their businesses with as little outside "interference" as possible. Until the early 1970s, grape growers led a rather unencumbered existence in terms of government regulation. Ironically, farmers now operate in one of the most regulated environments in the United States. Today, if you are careless or lack sufficient knowledge of current regulations, your vineyard could be shut down and put out of business almost overnight. More than ever, farmers must be mindful of the agencies and organizations that regulate the agricultural industry and learn how to mitigate the impact these agencies have on the farming enterprise.

Growers face a daunting array of compliance concerns. The dynamic nature of government regulations and forms further complicates compliance by making it a constantly moving target. This chapter provides an introduction to the broad and complex issues of government compliance, but it will quickly be seen that your vineyard requires someone on site to become your compliance specialist. The specialist must stay current on applicable laws and regulations and carry out all of the required activities and documentation. Use this chapter as an overview of the principal government regulatory issues rather than a recipe for complete, successful compliance with all regulations. A very useful reference of compliance concerns for all agriculture is the *Oregon Farmer's Handbook*, published (hard copy and on the Internet) and revised each year by the Oregon Department of Agriculture, (e.g., Oregon Department of Agriculture, 2001). Consider too, that many of the laws and regulations are complex; it is advisable to obtain professional advice from an agricultural labor attorney or consultant if you have questions. The primary emphasis of this chapter is on regulations related to labor and employee issues; other areas of regulatory concern are outlined.

Labor

All but the smallest vineyards have the need to hire additional people, at least on a seasonal basis, to help with the labor-intensive work of growing winegrapes. Many of the management practices regarding employees are highly regulated, and several different government agencies, both federal and state, are involved in enforcement of regulations. Each agency focuses on one or more specific aspects of labor issues, and compliance usually requires recordkeeping and reporting on official forms. The principal agencies involved with enforcing labor regulations are outlined below, and their major areas of concern and required forms are described. Additional requirements are also introduced.

Bureau of Citizenship and Immigration Services

The major concerns of the Bureau of Citizenship and Immigration Services (BCIS, formerly INS) are that workers are in the United States legally and are permitted to work here. Every newly hired employee is required to complete an BCIS Form I-9, Employment Eligibility Verification. The I-9 Form verifies that the employee has the right to work in the United States and verifies his or her identity. An employee may produce a variety of documents for this purpose as long as they are among those listed on the back of the I-9 Form. Be careful not to ask for any specific document; employees must choose what they present to you. In the winegrape industry, the most common document shown is the I-551 card (Alien Registration Receipt Card) – the "Green Card." The I-9 Form may be completed only after the person is hired and must be filled in completely. You may not use an I-9 Form as a screening tool for job applicants. BCIS can audit I-9 Forms and impose fines for missing forms, using outdated forms, and for some technical violations.

The employee must complete Section 1 of the I-9 Form on the date of hire, and you have up to three days from the start of work to complete Section 2. The I-9 Form must be kept for the duration of employment and three years after termination if the employee worked a year or less. Keep it one additional year for those employees who worked at least three years. To simplify the process, keep all I-9 Forms for a minimum of three years after termination. If an employee returns to work within a year or so from the date of layoff or termination and the documents he or she previously provided on the I-9 have not expired, you may use Section 3 to recertify the employee, or you may complete another I-9 form.

Although it is not the employer's responsibility to verify the authenticity of any work authorization or identification document, caution is advised about accepting documents with obvious errors, such as typographical errors. There is no requirement to photocopy any document given to you. However, if you do make a copy, it becomes part of the I-9 Form and must be kept with it. You also must be consistent with your practices; if you decide to copy documents, you must do so for all newly hired employees all the time. You may not copy some documents some of the time and not copy others when you are too busy—this is an all or nothing situation.

Keep in mind that the penalties are expensive if you are caught knowingly hiring an illegal alien: a maximum of $2,000 per worker for the first violation and up to $10,000 per worker for the third violation. A pattern of violations can land you in jail.

Internal Revenue Service (IRS)

As we all know, the IRS is concerned with the collection of taxes. Each new employee must complete an IRS Form W-4. This form gives a record of a Social Security Number and identifies the number of exemptions for tax purposes. Again, the form must be filled in completely and accurately. Be sure to use information from the W-4 form, not the I-9 form, to enter the payroll information. Due to the maternal-paternal name order, it is very likely that the order of the employee's name on the I-9 Form will be different from what is shown on his or her Social Security card. For payroll purposes (not to be confused with immigration purposes), it is a good idea to make a copy of the Social Security card and attach it to the W-4. This may also help you answer future questions from the Social Security Administration.

In order to comply with the New Hire Reporting Act, which went into effect in October 1998, employers are required to report all newly hired employees to the Department of Justice. This allows the government to track those employees who are ordered to pay child-support. The Department of Justice has a form you may use, or you may simply photocopy the W-4 form and mail it. You must report all newly hired employees within twenty days of the date of hire.

Social Security Administration (SSA)

The SSA has been active in verifying that Social Security numbers and names match when it receives tax reports from employers. If an employer's report contains too many mismatched numbers, the SSA either rejects the report and returns it to the employer or sends a letter to the employer listing the numbers that do not match with SSA records and instructs the employer to investigate. Such a notice from the SSA must be taken seriously, and a response is expected. Separate the list of numbers into two categories: former employees and current employees.

For former employees, if you have not kept a copy of their Social Security cards, you probably have no way to verify that the numbers in your records match the employees' Social Security cards. In this case, you must write to the SSA and inform them of those numbers that belong to former employees you are unable to verify. Those numbers that belong to current employees should be rechecked against the employee's Social Security card. Ask employees to bring you their Social Security cards so you can check that the name order and the number match with your records. Correct all mistakes. If no mistakes are found, write to the SSA indicating this. The key is to respond. It is an employer's responsibility to take "reasonable care" to report valid numbers.

U.S. Department of Labor (DOL)

The DOL is the enforcing agency for the Fair Labor Standards Act (FLSA) and the Migrant and Seasonal Worker Protection Act (MSPA). The FLSA regulates hours and wages, overtime, and child labor. The MSPA regulates the conduct of a grower, farm labor contractor, or agricultural association in relation to a farm worker. These laws are not to be confused with any State of Oregon law enforced by the Bureau of Labor and Industries (BOLI). It is important to understand that both federal and state laws exist, and sometimes they overlap. When this overlap occurs, the employer is required to comply with the law that most benefits the employee.

First, let's examine the Fair Labor Standards Act (FLSA). The FLSA is the law that sets rules for work hours and wages, child labor provisions, overtime, and recordkeeping requirements. It is also this law that establishes the definition of agriculture and the accompanying exemption from overtime for agricultural work. In most cases, all work a grower performs on his or her own vineyard is considered agriculture. If, however, the grower were to pay his employees to work in a neighbor's vineyard, the work is not agricultural and overtime must be paid for hours exceeding forty in a work week, even if the employees worked only one hour on the other farm. Any time that employees work with or on a product that is not grown on the employer's own farm, the work is not agricultural work. Furthermore, work in a winery is not considered to be agricultural work, and overtime must be paid even if the winery is processing grapes grown in the winery's vineyard. In situations such as this,

Oregon overtime laws are more strict than federal overtime laws.

The MSPA, on the other hand, regulates the activities of employers of migrant or seasonal agricultural workers, farm labor contractors, and agricultural employers or associations. There are many complexities to this law, so professional advice is recommended if you have questions.

Basically, the MSPA requires growers to provide specific information to migrant or seasonal field workers. It is critical to understand the definition of a "seasonal" or "migrant" worker. These definitions are solely for the purpose of this law, so do not assume that whom you consider to be a seasonal worker is the same as those defined by the DOL. A "migrant" worker is any field worker who has had to leave his or her permanent place of residence overnight to work for you; this could include parts of Oregon. A "seasonal" worker is any field worker who works for an employer on a crop of a seasonal nature. This could include year-round employees. Vineyard workers are seasonal workers since a grapevine is a crop of a seasonal nature.

Employers of migrant agricultural workers must provide specific information (in English and Spanish) about the terms and conditions of employment to the migrant worker at the time of recruitment. Completing DOL Form WH-516 satisfies compliance, or you may use your own form if it contains all of the information in Form WH-516.

There are also specific rules concerning the transportation and housing of migrant or seasonal workers. If you transport any migrant or seasonal worker in a company-owned vehicle, the vehicle driver must be properly licensed, the vehicle must have a safety inspection, and insurance liability limits must meet or exceed $100,000 per seat per vehicle.

Migrant and seasonal workers must be paid at least every two weeks, or semimonthly. Paycheck stubs (the part employees keep) must identify the deductions taken from the check and purpose, basis for wages earned, pay period dates, total hours worked or piece rate units earned, net pay, employee's Social Security number, and employer's name, address, and IRS identification number.

The activities of farm labor contractors (FLC) are regulated by the DOL under the MSPA. In essence, if you use an FLC, it is important to understand that this law has a joint-liability clause, which means that you, the grower, are equally liable for the violations made by the FLC. Therefore, make absolutely certain that the FLC is properly licensed with both the DOL and BOLI. The FLC's license should be reviewed to see whether or not this contractor is authorized to provide transportation or housing. If the FLC is not authorized to provide transportation, make sure none of the workers are transported by their supervisors or in FLC-owned vehicles. If transportation is authorized, make sure that the proper driver's license, insurance liability limits, and vehicle inspection requirements are met. The FLC is required to provide to the contracting vineyard copies of payroll records for the crew who works in the vineyard. The employer is required to have these copies, and it is also a good way to ensure that the workers are paid minimum wage. It should also be verified that the FLC provided all MSPA-required information to employees.

Oregon Bureau of Labor and Industries (BOLI)

Oregon's BOLI is responsible for enforcing some of the state's laws that relate to labor. In this respect, many BOLI rules mirror those of the DOL, in particular the FLSA. There are, however, subtle differences between federal and state regulations, and, as stated earlier, whichever is stricter provides the baseline for your compliance. For example, Oregon's minimum wage is higher than the federal minimum wage; therefore, Oregon's minimum wage supersedes the federal wage requirement. In addition to wage and hour regulations, BOLI oversees farm labor contracting, child labor laws, and to some extent working conditions. The agency also enforces drug use laws, mediates in employment disputes, and oversees record keeping such as payroll. BOLI operates primarily on a response to complaint basis.

Occupational Safety and Health Administration (OSHA)

All Oregon employers are required to provide a safe and healthy workplace. Oregon OSHA is responsible for enforcing health and safety regulations in the workplace, including the federal Environmental Protection Agency's (EPA) Worker Protection Standard. The extensive set of rules and laws, collectively called Division 4 regulations, constitute a document of more than eight hundred pages. A copy of the Division 4 regulations should be available in the vineyard manager's office for study and quick reference use. It is also very informative to request a voluntary audit from OSHA. The OSHA consultants come to a vineyard and review all safety-related paperwork and the physical aspects of the vineyard operation from a safety and health perspective. OSHA also conducts regular workshops and produces numerous educational hand-books and videos. Some of the OSHA requirements that commonly apply to vineyards are reviewed below:

1. The Worker Protection Standard (WPS) requires employers who use pesticides to train field workers and pesticide handlers about how to work safely

around chemicals. Training requirements are specific and must be done by a certified trainer. Simply showing an EPA-approved video and walking away cannot fulfill them. Employees must have an opportunity to ask questions during the training session. Other requirements include specific postings, decontamination facilities, pesticide application information, and more. Contact Oregon OSHA for information about compliance with this regulation. Other companies offer training services, including Americorp, the Oregon Farm Bureau's Farm Employer Education and Legal Defense Service (FEELDS), and the Farm Employers Labor Service (FELS).

2. There are several required written programs that agricultural employers must develop and maintain. These include a written Hazard Communication Program that teaches employees about the hazards they may be exposed to on the job as well as how to read product labels and material safety data sheets (MSDS). A written Respiratory Protection Program is necessary for those employees who are required to use a respirator. An Emergency Action Plan must be prepared that details the procedures to be followed in the event of an emergency, such as with an act of violence or an earthquake. Employers must also have written procedures for lockout/tagout of powered equipment such as forklifts and tractors.

3. OSHA requires annual submission of Form 300, the Log and Summary of Occupational Injuries and Illnesses. The summary portion must be posted where employees can readily see it throughout the month of February each year. It lists the recordable injuries and illnesses at your farm for the previous year. Retain this form for five years.

4. Forklift drivers and tractor drivers must receive training on safe equipment operation. Forklift drivers must be retrained annually. Keep documentation on file that this training has been done.

Other Requirements

Workers' Compensation. If you have employees, you are required by law to carry a workers' compensation policy. If an injury happens on your farm, the workers' compensation insurance covers the medical expenses. This insurance covers all work-related injuries and illnesses and provides for disability insurance in case an employee cannot return to work either temporarily or permanently. It is prudent coverage to have. Many Oregon vineyards turn to the State Accident Insurance Fund (SAIF) for their workers' compensation policy. But there are numerous other private companies that offer coverage, even at a potentially lower rate. It is a good idea to shop around. Some companies have an option that allows you to pay for medical expenses below $500, thus avoiding a claim and an increase in your premium.

If an injury happens, each insurance company has specific reporting guidelines that you must follow. Common to all is Form 801, the Worker's and Employee's Report of Occupational Injury or Disease form, which must be filled out immediately. Work closely with your compensation carrier to understand the services it provides and what you must do when you require these services.

Fraudulent claims are always a possibility. Protect yourself from exposure to fraud by insisting that your employees report each and every injury, regardless how slight. Be knowledgeable about how your workers' compensation company handles claims and what might happen if it considers a claim to be fraudulent. Keep accurate records. Throughout this chapter, a recurring theme is that legal action may be a consequence, so it is imperative that you keep current and accurate records.

Most workers' compensation carriers require employers to submit quarterly reports. Premiums are determined by the state and broken down into various categories. For vineyards, the primary classification is "vineyard workers and drivers." Be sure to understand how this system works. Workers' compensation is expensive, so if you can save some money, it is worth it. If you do clerical work as part of your job, then place those work hours into category 8810A. There can be as much as a tenfold difference in rates between the two job classifications. Remember that a contracted, subcontracted, or professional service is not required to be covered by your worker's compensation umbrella policy. And do not pay workers' compensation on your vacation time.

Safety will save you money. Your experience rating is determined by your claims record. You want to have the best experience rating possible in order to pay the lowest rate.

Payroll. If your total quarterly gross payroll exceeds $20,000, you are required to pay unemployment tax. Check with the Oregon Employment Department if you think you are near this threshold, because the penalty for exceeding the limit is very stiff. The relevant forms include the Oregon Department of Revenue Form OQ, the Oregon Quarterly Tax Report, and Form 132, the Unemployment Insurance Annual Wage Detail Report. You may also be subject to federal unemployment taxes (FUTA) and must file a federal Form 940, Employer's Annual Federal Unemployment Tax Return. Check with an accountant about these requirements. As with all of the prior requirements, you must keep absolutely accurate payroll records to be successful at complying with all of the tax laws. If you cannot do this, hire a professional bookkeeper or a payroll service provider. As with workers' compensation, your tax rate for unemployment insurance is based on your experience rating. Try to minimize the number of claims against your farm. A stable and loyal workforce is the key to controlling unemployment claims.

The IRS and its Oregon counterpart, the Oregon Department of Revenue (ODR), have specific guidelines that apply to businesses of various sizes. Be aware of when your tax payments are due. Most growers have to file an IRS Form 943, the Employer's Annual Tax Return for Agricultural Employees, and a Form WA, the Oregon Agricultural Annual Withholding Tax Return, for the ODR at the end of the year for agricultural businesses. You also have to distribute W-2 forms to employees and 1099s for nonemployee service expenses. You must have a W-4 for each employee. Aside from federal tax, other deductions include Medicare and Medicaid, workers' compensation insurance, and FICA. Professional payroll companies are available to handle your payroll needs if you elect not to do it yourself.

Employee Handbook. Although it is not legally required, an employee handbook is an important tool. An employee handbook effectively communicates the rules employees are expected to follow, and it forces you to apply those rules uniformly and consistently to all employees. This is the greatest defense to a discrimination claim. On the other hand, a poorly written employee handbook can cause as many problems as you hoped to solve, so it is important that you consult with your attorney before you write one.

A handbook should contain several important policies, such as at-will employment status, attendance, leaves of absences, workers' compensation and early return to work, sexual harassment and violence in the workplace, drug and alcohol use, conduct standards, disciplinary procedures, payroll, and pay dates. Commercial services are available to help you develop an employee handbook.

Poster Requirements. Many regulations require notification of employees through the prominent display of official posters. Labor compliance service providers commonly have appropriate posters for sale. BOLI has developed an inexpensive, laminated poster for agriculture that contains eleven of the posters commonly required in the industry. The poster is in English on one side and Spanish on the other. In addition to this poster, you must post your workers' compensation certificate; the Unemployment Insurance Notice (Form 11), if applicable; any nondiscretionary bonus information; and emergency numbers. The WPS requires pesticide application warning signs to be posted at all usual points of entry to a field for a period extending from twenty-four hours prior to treatment until the product reentry interval has expired.

Other Regulatory Concerns

Pesticide Use Reporting. The 1999 Oregon Legislature passed legislation requiring a state pesticide reporting system. The system, developed and maintained by the Oregon Department of Agriculture (ODA), began operating in January 2002. The law requires collection of information from all categories of pesticide use. Reports must include the following: date of application, crop, location, pesticide product name, EPA product registration number, amount of product used, and the purpose (e.g., disease control) of the application. Annual reports must be made, and they can only be submitted electronically via the Internet.

Water Rights. If you plan to use water for irrigation purposes, you must hold a water right from the Oregon Water Resources Department. This is true regardless of the water source—the ground, lakes, or streams. A water right is attached to the land where it was established, so if the land is sold, the water right goes with the land to the new owner. The presence of water flowing through or past a property does not automatically enable the landowner to divert the water without state permission. Application for a new water right can be a lengthy process, so plan well in advance, especially if the intent is to provide water to young vines. Be aware that most surface water in the state is no longer available for use during summer months. Additionally, many areas of the state have restrictions on further ground water appropriation. Construction of a pond or reservoir may be the only available option for developing a water source at some sites. Separate permits are required to store water in a pond or reservoir, and to divert it for use.

Agricultural Water Quality. The 1993 Oregon Legislature passed Senate Bill 1010, as the Agricultural Water Quality Management (AgWQM) Act, to formally organize agricultural efforts to address water pollution in watersheds across the state. The AgWQM Act directed the ODA to develop watershed-based plans that outline strategies to prevent and control water pollution from agricultural activities and soil erosion. For purposes of this act, Oregon is divided into more than forty AgWQM areas. Each area has developed, or is in the process of developing, a set of rules that outlines conditions or measures to implement the AgWQM Area Plan.

The ODA policy is to encourage the voluntary adoption of land management practices through education programs, on-farm demonstration projects, and technical assistance (Oregon Department of Agriculture, 1999). Enforcement action to achieve compliance with the Oregon administrative rules is pursued only when reasonable attempts at voluntary solutions have failed. Verification of noncompliance causes the ODA to issue a Notice of Noncompliance, a Plan of Correction, or both. Civil penalties may be assessed by ODA for failure to comply with area rules.

Wetlands. Wetland protection is a relatively recent regulatory concern. Wetlands are monitored by the Natural Resources Conservation Service (NRCS). It is best to work with the local county NRCS office to develop plans to protect your property's natural assets. They offer numerous cost share programs that may be useful.

Compliance Strategies

This chapter provided a quick introduction to a fairly large number of government compliance issues relevant to vineyards, but this is just the tip of the iceberg. Many more laws and regulations apply to vineyard operations. Unfortunately, many of these laws, though intended to protect people, were developed to address worst-case scenarios. The result is an avalanche of regulations that makes total compliance a practical impossibility. It is no exaggeration that a grower with just 10 acres of grapes and two or three employees could spend 100% of their time addressing regulatory issues and never have a chance to work in the field.

An effective strategy is to prioritize your compliance in a manner similar to how daily vineyard tasks are prioritized. Examine those areas where you have the greatest exposure to the high penalties, lawsuits, or audits. Remember that some laws provide an unhappy employee the right to sue you directly for your failure to comply (as with MSPA) rather than simply filing a complaint with a state or federal agency. In the case of MSPA noncompliance, the cost of every violation represents $500 per employee, per violation. So, if you have ten employees and you failed to post two posters, this could easily translate into a $10,000 penalty, plus the cost of an attorney to defend yourself.

A critical component of a successful strategy is documentation. Anyone who has pled a case in front of a judge or jury can tell you the importance of documentation. Documentation is the only way for an employer to prove that he or she has done what is required under the law in the employment situation. Employers are *always* required to prove the negative—to prove that they did not do whatever the claim is. The only way a grower can prove that he or she did not do something is to document what was done.

Keep accurate records. Document conversations with an injured employee's doctor. Document conversations with an upset employee and disciplinary discussions. Be sure to also document the actions of employees who have gone above and beyond your expectations. It is not important for documentation to be neatly recorded, but it must be filed away where it can be readily accessed when necessary. Keep a record of the date, the name of whom you talked with, and the facts of the conversation. Leave out extraneous opinion and commentary; document just the facts. Make certain that documentation is done immediately after a relevant event; many important points could be lost to memory if it is delayed for several days.

An established company policy is the other important component of the strategy. Overwhelmingly, courts tend to side with employees in cases where the employer has not been able to provide documentation and written policies. Although employee handbooks are not legally required, regulatory agencies and the courts both rely heavily on what is written in them.

Finally, it is highly advantageous to learn how to prove your case before your case is ever made. Actively protect your assets with documentation and written policies. Establish work practices that protect you and reduce your exposure to liability. The cost/benefit analysis greatly favors these proactive efforts.

References

Oregon Department of Agriculture. 1999. *Agricultural Water Quality Management Program—Enforcement and Compliance Process and Procedures.* Oregon Department of Agriculture, Salem.

Oregon Department of Agriculture. 2001. *Oregon Farmer's Handbook.* Oregon Department of Agriculture, Salem.

29

Crop Estimation and Thinning

Edward W. Hellman and Ted Casteel

Crop estimation is an important tool for the grape grower that enables appraisal of potential yield before harvest. Prior knowledge of potential crop levels is critical to managing the vineyard to produce a target yield that is often stipulated in winery contracts. It is also important for making plans for harvesting the fruit. Excessive crop levels are reduced by selective thinning practices that remove flower or fruit clusters.

Methods of Crop Estimation

Crop yield is a function of three factors—the number of bearing vines, the number of clusters per vine, and the cluster weight. Therefore, crop estimation methods use these factors at various times during the season to predict yield prior to harvest. Crop estimates must be done on an individual block basis because of the variability typically existing among vineyard blocks.

Pre-bloom Estimate

A simple method is used to provide a crop estimate early in the season. The number of bearing vines in a block is counted and multiplied by a count of the average number of clusters per vine (see below for advice on making these estimates). Then this product is multiplied by the average historical cluster weight at harvest to give a yield estimate for the block (Equation 1). This method provides an early rough estimate of yield potential that can be obtained prior to bloom. But it must rely on the historical average cluster weight at harvest, and it assumes that final cluster weight in the current season will be close to the average historical cluster weight at harvest. Because it is conducted before bloom, the effects of weather and growing conditions on fruit set and berry development are not accounted for by this method. Therefore, it is most suitable for early-season rough predictions of crop potential, but it is not accurate enough for final, pre-thinning estimates in regions or for grape varieties that commonly experience significant annual variation in cluster size.

Lag-phase Estimate

Many Oregon growers use the lag-phase crop prediction method developed by Price (1992). The following discussion is adapted from that description. The method uses the average cluster weight at lag phase multiplied by an "increase factor" as a better predictor of final cluster weight than the historical average cluster weight at harvest. Conducting the estimate at lag phase also allows enough time before harvest to reduce yields by thinning, if necessary, to the desired level.

The lag-phase method requires four factors to be measured or estimated each season for a block: the number of bearing vines; number of clusters per vine; cluster weight at lag phase; and cluster weight at harvest.

Vine Count. An accurate count of bearing vines per block is the starting point of a good crop estimate. Do not assume that the present number of bearing vines is the same as the number of vines originally planted, or even the number of vines in the previous season. Rate of establishment varies among vines, and the number of bearing vines is commonly the main variable affecting total yield in a young vineyard. In an established vineyard, the count of bearing vines can change from year to year because of temporary or permanent loss from disease-, insect-, or weather-related damage. Vine counts can be made by first determining the maximum number of vines possible in the block, then counting and subtracting the number of nonbearing vines. In a rectangular block with uniform spacing, multiply the number of vines per row by the number of rows. Rows of irregular length require individual row counts. Once an accurate count is made of an established vineyard, only nonbearing vines need to be counted each year. Go down every other row with a hand counter and count missing vines. This should be done after flower clusters are visible to ensure identification of bearing vines.

Equation 1. Pre-bloom yield estimate

Estimated Yield = No. of Bearing Vines per Block X No. of Clusters per Vine X Historical Avg. Cluster Weight

Clusters per Vine. The average number of clusters per vine can be determined easily prior to bloom, when flower clusters have emerged but are not yet obscured by foliage. Cluster counts are most accurate if conducted after shoot thinning is completed, since that procedure removes clusters along with shoots. A critical aspect of determining average cluster number is the collection of a sample that accurately represents the entire block. The number of vines necessary for a representative sample depends on vineyard uniformity. For small blocks of one to three acres, where all the vines are the same size, same age, and pruned to the same bud number, it is recommended that 4% of the vines be counted for the sample. Larger uniform blocks can sample a smaller percentage of vines. Methods to determine the required sample size for crop estimates are discussed by Wolpert and Vilas (1992).

Nonuniformity in the block makes collection of a representative sample more difficult, and the only solution is to sample a larger percentage of the vines. Price reported that, for one trial, a non-uniform vineyard required 30% of the vines to be counted to give an accurate, representative sample. Yield prediction in nonuniform blocks thus requires more time and greater cost compared to that in uniform vineyards.

Sample vines can be selected either randomly throughout the block or by a grid system (e.g., every tenth vine in every other row). It is best to determine the sampling system before you enter the block, and do not vary from the prescribed sampling routine. All the clusters on the sample vines are counted; a hand-held counter and notepad facilitate collection of the data.

As mentioned above, an early-season estimate of yield potential can be calculated with Equation 1 after vine and cluster counts are completed.

Cluster Weight at Lag-Phase. Cluster weight at harvest is difficult to predict because it is sensitive to growing conditions throughout the cluster development period. However, at about the halfway point between bloom and harvest, the effects of many of the variable factors that influence final cluster weight are apparent. This halfway point corresponds to the "lag phase" of berry development. It also corresponds to the time when seeds begin to harden, so this stage is sometimes referred to as "seed-hardening."

Grape berry development goes through three phases. The first and third stages are periods of rapid growth; the middle stage, the lag phase, is a period of relatively slow growth. Generally, at lag phase the berry has attained approximately 50% of its final weight. Thus, a measurement of average cluster weight at lag phase can be multiplied by an "increase factor" of about 2 to give an approximate prediction of the average cluster weight at harvest. It is recommended that new vineyards with no production history use an increase factor of 2.2 as a starting point for yield estimates. Lag-phase cluster weights should be collected and recorded each year to develop a long-term lag weight for each vineyard block.

Perhaps the most difficult aspect of this method is determining when the lag phase occurs each season. Price reported that the midpoint of lag phase for Pinot noir averaged 55 days after first bloom in a three-year trial. This corresponded to the time seeds within the berries were hardening. Since seed hardening is usually associated with the lag phase of growth, he recommends standardizing the timing of lag-phase cluster sampling to correspond with the occurrence of 75% hard seed tips. Seed tips are considered hard when they cannot be cut easily with a sharp knife or razor blade.

Just as with cluster counts, representative sampling of clusters is critical to a good cluster weight estimate. Sampling strategies are varied, but there is thought to be more variability within clusters on a single vine than variability between vines. Therefore, most recommendations are to collect all of the clusters on randomly selected vines representing all parts of the vineyard block. Bethel Heights Vineyard applies this method by identifying a set number of average vines per block and harvesting all of the fruit from these vines for lag-phase cluster weights. This method has been pretty accurate, and if you flag and reuse the same vines year after year the information can be even more useful.

A random sample of 200–400 clusters per block is usually adequate for Pinot noir and related varieties. This is weighed and divided by the number of clusters in the sample (keep an accurate count) to calculate an average cluster weight. Sample size should be increased for circumstances or varieties, such as Chardonnay or Merlot, with a wide range in cluster weights.

Cluster Weight at Harvest. Measuring and recording the average cluster weight at harvest every year enables development of a long-term average for each vineyard block. As described earlier, this figure can be used with Equation 1 for an early-season rough estimate of yield. But it can also be used to calculate an average increase factor by obtaining the ratio of average cluster weight at harvest to average lag-phase cluster weight. In his three-year study with Pinot noir, Price reported increase factors ranging from 1.9 to 2.5. Another vineyard in the Willamette Valley recorded increase factors ranging from 1.86 to 2.0 for Pinot noir (greater variability was seen in a young block) and from 1.90 to 2.54 for Chardonnay over a four-year period (MacDonald, 1993). The precision of your increase factor estimates is improved by collecting annual lag-phase and harvest yield cluster weights for each block to develop historical averages.

Harvest cluster weights should be collected by the same procedure used for lag-phase cluster weights. Although it is simpler and easier to collect cluster samples from picking bins, random samples from vines correlate better with lag-phase samples. Bin sampling may not be as random, and additional sampling error can result from difficulty in distinguishing whole clusters from partial clusters.

The four measured factors for each block are used at lag phase to predict the final harvest yield with a simple modification of Equation 1. Instead of using the historical average cluster weight at harvest, the current season's lag-phase estimate of cluster weight is multiplied by the increase factor calculated from historical data (Equation 2). This estimate of harvest cluster weight is then multiplied by the number of clusters per vine and the number of vines per block to estimate total yield.

The lag-phase cluster weight method is not foolproof. Reasonable accuracy requires that variability within vineyards be accounted for by adequate size and randomization of samples. Also, dependence on the increase factor means that the prediction can overestimate final yield if the vines experience water stress or other unfavorable conditions that reduce berry size and thus cluster weight from the predicted value. Similarly, yield can be underestimated if unusual factors contribute to greater than average berry growth in the final phase of development. Annual records of lag-phase and harvest cluster weights, accompanied by seasonal weather notes, will enable an experienced grower to adjust the increase factor for seasonal conditions.

Thinning to Adjust Crop Level

A variety of circumstances may lead to the situation of having more than the desired amount of fruit per vine. Correction of this problem by the selective removal of fruit clusters is called "cluster thinning" (or simply "thinning") and is sometimes referred to as "green harvest" since it is often conducted prior to color development of the fruit. Except in seasons with exceptionally poor fruit set, crop reduction of Pinot noir by cluster thinning is a routine practice in Oregon; thinning is commonly practiced on other varieties as well. Vineyard managers typically have a target range of yield that often is stipulated in the winery contract. The Oregon experience has been that fruit thinning can hasten ripening in late years or on cooler sites, especially on Pinot noir where thinning to levels of 2 tons per acre or less is commonplace for ultra-premium wines. Enhanced ripening can enable production of wines with greater intensity of aroma and flavor. The crop prediction methods described above can help determine the extent of thinning required to achieve the target yield desired for the vineyard.

Target Yields. Target yields are commonly established during contractual discussions between vineyard and winery. Many factors enter into cropping-level decisions and, ideally, vines are cropped at a level that is appropriate for their capacity and enables fruit to ripen to the desired quality level. Thus target yield levels are highly vineyard-specific and dependent on the inherent capacity of the variety and site and the weather conditions of the current season. Target yields must take into consideration the desired fruit quality and the market value of the grapes. These factors are directly related to the winery's price-point intentions for the wine produced from the grapes and are influenced by other market factors such as supply and demand. Vineyard management practices, including crop thinning, are often selectively applied to specific blocks depending on the market value of their fruit.

One large vineyard uses the lag-phase estimation method slightly differently—to estimate the number of clusters per vine that should be retained to achieve a target yield (personal communication, Betty O'Brien, Elton Vineyards). This is a very practical modification, since pruning crews cannot be told how many clusters to remove per vine because of vine-to-vine variability, but they can be told the number of clusters to retain per vine that will produce the target yield. This modified method is also simpler because it does not require cluster counts.

Calculations are "worked backward" from the target yield (Equation 3). Target yield (in pounds) is divided by the number of bearing vines in the block. The result is the average number of pounds per vine necessary to produce the target yield (Step A). This is divided by the product of the average lag-phase cluster weight multiplied by the increase factor to determine the average number of clusters to retain per vine (Step B).

What to Remove. Standard practice is to remove all clusters on short shoots, since their smaller leaf area may not have the capacity to ripen the fruit fully. Third clusters, if present, are routinely removed. Clusters

Equation 2. Lag-phase yield estimate

Yield Estimate = No. of Bearing Vines per Block X No. of Clusters per Vine X Lag-Phase Avg. Cluster Weight X Increase Factor

Equation 3. Number of clusters per vine to achieve target yield

Step A

$$\frac{\text{Target Yield (pounds)}}{\text{\# Bearing Vines per Block}} = \text{Yield per Vine (pounds)}$$

Step B

$$\text{No. Clusters per Vine} = \frac{\text{Yield per Vine (pounds)}}{\text{Lag-phase Average Cluster Weight X Increase factor}}$$

that are lagging in development can be identified when veraison is 75–80% complete; clusters with green or pink fruit are selectively removed. Removal of the greener fruit on the vine improves the overall uniformity of fruit ripeness at harvest. For similar reasons, growers sometimes thin within a cluster, removing shoulders or wings and retaining only the main body of the cluster. The berries on shoulders and wings may be somewhat behind in development compared to those on the main cluster.

Some growers thin to one cluster per shoot; the question becomes which one to retain. Basal clusters are usually more mature than those farther out on the shoot (Wolpert et al., 1983), but some growers prefer to leave the second cluster, which is usually smaller. Thinning to one cluster usually leads to a crop reduction of about 1/3 and spreads the remaining load equally over the vine. Moreover, instructions to the crew are straightforward, so there is a greater likelihood of achieving the desired result. The disadvantage of thinning to one cluster per shoot is that it is considerably more expensive than most other approaches.

Timing. The timing of thinning practices varies among vineyards. Some Oregon growers thin clusters at the onset of veraison. By this time in the growing season, the grower is able to make a reliable crop estimate, which an informed thinning requires. The onset of coloring in the ripening fruit also enables easier identification of clusters that are behind in development. Other growers prefer to thin earlier (lag phase or before), when it is usually less time consuming because the canopy is smaller and the clusters are easier to find.

Thinning Strategy. A thinning strategy should be developed for each vineyard block prior to the beginning of the season based on the target yields and market value of the crop. Implementation of the strategy is adjusted as crop estimates become available and the seasonal weather characteristics (e.g., cooler than usual) become evident. It must be recognized that the cost of elaborate thinning practices cannot be recovered with grapes sold at low prices. Therefore, grapes intended for reserve wines may require extensive crop adjustment and should receive appropriate price compensation to justify the expense.

Recently, a fairly standard practice has evolved for Pinot noir vineyards where reserve wines are expected. Clusters are thinned to one per shoot at lag phase or earlier. In years of exceptionally good fruit set, additional crop reductions may be necessary to meet the targeted fruit quality.

References

MacDonald, A. 1993. Crop evaluation and adjustment in the vineyard. *Proceedings of the Oregon Horticultural Society* 84:236-237.

Price, S. 1992. Predicting yield in Oregon vineyards. In T. Casteel (ed.), *Oregon Winegrape Grower's Guide*, 4th ed. Oregon Winegrowers' Association. Portland.

Wolpert, J. A., G. S. Howell, and T. K. Mansfield. 1983. Sampling Vidal Blanc grapes: I. Effect of training system, pruning severity, shoot exposure, shoot origin, and cluster thinning on cluster weight and fruit quality. *American Journal of Enology and Viticulture* 34:72-76.

Wolpert, J. A., and E. P. Vilas. 1992. Estimating vineyard yields: introduction to a simple, two-step method. *American Journal of Enology and Viticulture* 43: 384-388.

30

Evaluation of Winegrape Maturity

Barney Watson

Optimal winegrape quality is intricately linked to factors such as site, soil, climate, vintage, yield, vineyard management practices, and fruit maturity. Harvesting high-quality winegrapes at their optimal level of maturity in sound physical condition is the first step in producing high-quality wines. Determining the best time to harvest requires both experience and a careful assessment of winegrape maturity. Important factors in assessing winegrape maturity include the sampling methods, chemical analysis, and sensory evaluation.[1]

Sampling methods should provide samples that are representative of the fruit to be harvested. Chemical analysis useful for determining winegrape maturity includes monitoring the sugar content (soluble solids or degrees Brix), the titratable acidity and pH, the malic acid content, and the nutritional status of the fruit. Based on chemical analysis at harvest, modification of juice and must (crushed grapes) is commonly practiced to fine tune the sugar content (degrees Brix), the acidity and pH, and the fermentable nitrogen content in order to optimize fermentation behavior, wine composition, and wine quality.

Winegrape maturity cannot be based solely on chemical analysis, however, because the development of optimal aroma, flavor, and color does not coincide with specific levels of sugar, acidity, or pH. The Oregon experience has shown, for example, that color and varietal intensity develop at lower sugar levels during cool ripening seasons compared to warmer vintages. In addition to chemical analysis, a careful sensory evaluation should be done during ripening to monitor the development of aroma, flavor, and color. Overall, the best harvest decisions are made by winegrowers and winemakers working together using a practical, integrated approach to maturity assessment while taking into account the weather conditions and the physical condition of the fruit.

Grape Composition and Maturation

The grape berry consists primarily of skin, pulp, and seeds. The numerous components that affect wine quality change dramatically during ripening and are not distributed uniformly in the berries. The pulp accounts for about 78% of the weight of the berries, and its primary constituents are sugars (predominantly glucose and fructose), organic acids (primarily tartaric and malic), mineral cations (especially potassium), nitrogenous compounds (soluble proteins, ammonia, and amino acids), pectic substances (cell wall structural material composed of polymers of galacturonic acids), and nonflavonoid phenolic compounds (primarily benzoic and cinnamic acid derivatives). The phenolic compounds in the pulp represent about 10% of the total phenolic content of the berries.

The skins typically represent about 15% of the berry weight and are the principle source of aromatic compounds and flavor precursors. They also contain flavonoid phenolic compounds (including flavonols, anthocyanins, and large polymeric flavonoid compounds known as tannins). Phenolic compounds in the skins represent about 30% of the total phenolic content of the berries.

The seeds, which represent about 4% of the berry weight, contain both nonflavonoid and flavonoid phenolic compounds including relatively large amounts of tannins. The seed phenols represent about 60% of the total phenolic content of the berries. They also contain significant levels of nitrogenous compounds, minerals, and oils (primarily oleic and linoleic acids).

White wines are produced by fermentation of juice in the absence of skins and seeds after pressing destemmed and crushed clusters, or whole clusters. The total phenolic content of white wines is approximately 10% of that of red wines and is primarily due to nonflavonoid compounds present in the juice and small amounts of flavonoid compounds extracted from the skins during maceration and pressing. The total phenolic content for white wines ranges from about 150 to 300 mg/l expressed as gallic acid equivalents (GAE).

Red wines are produced by fermentation of destemmed and crushed berries, whole berries, or whole clusters with the skins and seeds present. New wines are generally pressed "from the skins" after completion of fermentation. Approximately 50% of the total extractable phenols of the berries are extracted during red wine production, primarily from the skins and seeds of the berries. Berry size is proportional to the number of seeds per berry, which can vary from zero to four. Larger berries have more seeds and a lower skin to pulp ratio by weight, whereas smaller berries have fewer seeds and a greater ratio of skins to pulp. The total phenolic content of red wines ranges from 1,500 to 3,000 mg/l expressed as GAE, depending on the variety and the wine style. The composition of the fruit at harvest, skin to juice ratio, number of seeds per berry, and processing and fermentation practices can have a dramatic effect on wine composition and quality, especially for red wines.

A thorough review of the annual cycle of growth and fruit development is presented in "Grapevine Structure and Function" (see pp. 5-19). The onset of ripening (veraison) begins during the final stage of berry growth and typically occurs in mid to late August in Oregon. During veraison the berries begin to increase in size rapidly due to cell enlargement. The sugar content of the berries increases rapidly, acidity decreases, pH increases, and cations accumulate in the berry tissue, particularly potassium (Figure 1). As the berries approach full maturity, berry size reaches a maximum and sugar accumulation slows. The yeast assimilable nitrogen content also increases during ripening, with a net decrease in ammonia offset by a net increase in the amino acid content of the juice/pulp (Butzke, 1998; Watson et al., 2000).

During veraison, the phenolic composition of the skin changes as the berry loses chlorophyll and begins to synthesize and accumulate phenolic compounds that are responsible for development of characteristic colors: yellow-gold (flavonols) and pink and red colors (anthocyanins). The differences in color development become very noticeable among clusters on a vine, and this is a common time to make final adjustments to crop levels. Clusters that are behind in maturity can be selectively removed to increase the uniformity of maturity within the vineyard.

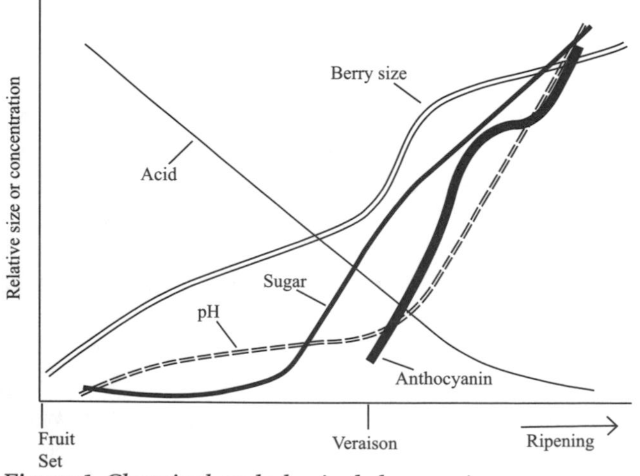

Figure 1. Chemical and physical changes in grape berries during development and ripening.

Canopy management practices that ensure adequate sunlight exposure on the clusters enhance ripening and the uniformity of fruit development. An average cluster exposure of about 50% is generally considered optimal (Smart and Robinson, 1991). Excessively shaded fruit tends to be lower in sugar content, higher in acidity (higher in malic acid), lower in pH, and lower in flavonols and anthocyanins, and they produce wines with lower color and lower total phenolic content (Price et al., 1995). As fruit reaches maturity, the berry stems (pedicels) turn from green to brown and decrease in green, "stemmy," herbaceous character. Also in the later stages of maturity, the phenolic composition of the seeds changes as they begin to turn from green to brown. The tannin content of seeds decreases at the same time as the degree of phenolic polymerization increases. The decrease in grape seed polyphenols (tannins) with ripening may be related to oxidative reactions as the tannins become fixed to the seed coat (Kennedy et al., 2000).

Reaction of tannins with other polyphenols, carbohydrates, or proteins may modify sensory properties such as bitterness and astringency. Skin tannins are also reported to increase in size during the later stages of ripening and to undergo reactions with pectins and anthocyanins, which may affect the mouthfeel and texture of red wines as well as color stability (Kennedy et al., 2001). Winemakers often describe the mouthfeel of red wines produced from fruit harvested at optimal maturity as having "ripe supple tannins," compared to "harder" more aggressive tannins in wines produced from less ripe fruit.

The aroma and flavor compounds present in wines are derived from the fruit and are also produced by yeast as fermentation by-products. Several hundred aroma-intensive compounds are present in wines including alcohols, aldehydes, ketones, fatty acids, esters, and a wide range of other aromatic compounds including terpenes and volatile phenols. Many compounds can have a sensory impact at concentrations of parts per million (ppm or mg/l), parts per billion (ppb or µg/l), or even lower concentrations.

With the onset of veraison, varietal aromas and flavors begin to develop, primarily in the skins of most varieties. Canopy management, yields, and degree of maturity all are known to have a dramatic effect on aroma and flavor development. Grape aroma compounds are present in the skins as both volatile aromatic compounds and nonvolatile aroma precursors (primarily glycosides) and are extracted during processing and fermentation. Nonvolatile aroma precursors can become volatile by nonenzymatic or enzymatic hydrolysis.

Some varieties have specific aroma compounds that tend to dominate their character. Riesling, Müller-Thurgau, Gewürztraminer, and Muscat varieties, for example, have a significant content of aromatic terpenes that gives them their characteristic floral, perfumey aromas and flavors. Cabernet Sauvignon, Sauvignon blanc, and Merlot have a distinctive herbaceous character because of significant levels of methoxy pyrazines. Other varieties such as Pinot noir and Chardonnay derive their characteristic aromas and flavors more from the overall complexity of the numerous aroma compounds present. Chardonnay is often described as having flavors of citrus, melons, and apples. Characteristic descriptors of Pinot noir include cherry, raspberry, plum, and rose petal.

Aroma and flavor complexity typically increases during the later stages of ripening. The number of aroma-intensive compounds detected in Pinot noir wine, for example, was twice as high in wines produced from fruit harvested during the later stages of maturity compared to wines produced from fruit harvested one to two weeks earlier (Miranda-Lopez et al., 1992). Wines made from the later harvest dates contained more floral, spicy, and fruity aroma compounds, and more than half of the aroma compounds present were not detectable in wines produced from fruit harvested at earlier dates.

After the berries have reached their maximum size and approach optimal maturity, they begin to soften noticeably. Average berry weights may decrease by 5–10% during this period due to dehydration, particularly during periods of dry weather. An increased concentration of color, aroma, and flavor are often observed in wines produced from fruit harvested at this stage of maturity. Both the rate of maturation and the physical condition of the fruit can be adversely affected by extremes of weather, especially during the later stages of ripening. Too much heat or drought can delay physiological maturation and cause excessive dehydration and shriveling, resulting in abnormal sugar increases due to a concentration effect rather than grape maturation. Fruit that remains on the vine for a prolonged time after reaching optimal maturity may shrivel excessively and drop off, a characteristic of "overripe" fruit. If significant rainfall occurs in the later stages of maturation, berry weight may increase through hydration, resulting in an apparent loss of sugar content from dilution. Late season rainfall also increases the risk of botrytis bunch rot development.

Winemakers decide when to harvest individual vineyard blocks on the basis of fruit sugar content, acidity, and pH for the intended type and style of wine. Of equal importance are the winemaker's observations on the development of varietal color, aroma, and flavor, their perceptions of skin, seed, and stem "maturity," and observations of the physical condition of the vines and the fruit. In Oregon, harvest of mature fruit generally occurs 100–125 days past full bloom: beginning mid to late September and continuing through October and occasionally into early November.

Sampling for Maturity Evaluation

Individual vineyards and distinct blocks within a vineyard should be sampled separately, and in a manner that is as representative as possible of the fruit to be harvested. An adequate sample is one that gives analytical results that fairly reflect the analyses of the juice/must at the time of harvest and processing. It is important to understand the considerable heterogeneity of fruit composition during winegrape maturation. Depending on the duration of the bloom period (flowering and fruit set), the range in maturity among berries on a cluster and among clusters on a vine may vary by up to two weeks.

An appropriate sample is representative of the heterogeneity within the vineyard. Therefore, proportional quantities of fruit should be collected from exposed and shaded locations in different parts of the canopy, at different heights on the vine, and on opposite sides of the row. Avoid sampling from vines at the end of rows, or from odd vines that are obviously different from the majority of vines in the vineyard block. Secondary clusters should be included in the field sample only if they will be picked at harvest. Underdeveloped secondary clusters can lower overall fruit maturity and wine quality if present in sufficient numbers, so they are best removed before or during veraison.

Samples may be taken as berries or whole clusters, but a sufficient quantity of fruit must be obtained to obtain a representative field sample. Typically, a random selection of 100–200 berries is taken from a large number of vines, or a selection of ten to twenty clusters is taken from a smaller number of vines. A larger sample may be required when a vineyard block does not have a high degree of observable uniformity of maturity among the clusters. Berry sampling can be flawed by the tendency to sample too few berries and to select riper, more mature ones. Exposed berries tend to be sampled more than the shaded, less visible berries, and berries on the inside of clusters tend to be less ripe than other berries on a cluster. This can result in sugar measurements as much as 1% higher in field berry samples than in the juice or must at harvest.

Cluster sampling has the advantage of providing a larger and more representative sample, giving compositional information that is typically closer to that of the fruit at harvest (a twenty-cluster sample consists of approximately 2,000 berries). Another advantage of cluster sampling is that the average cluster weight can be obtained and used to estimate the anticipated yield at harvest. A count of the average number of clusters per vine is multiplied by the average cluster weight to estimate average vine yield. A subsample of 200 berries from the clusters can be weighed to obtain average berry weight.

Samples should be taken weekly beginning about three weeks before harvest is anticipated. More frequent sampling should be done as the anticipated harvest date becomes closer, particularly if there are changes in the weather that could affect ripening or fruit condition.

Sample Preparation. Juice samples are prepared for chemical and sensory analysis by crushing the cluster or berry sample without breaking the seeds, squeezing it tightly through cheesecloth to obtain both the free run and the pressed juice. The different constituents of the juice are not evenly distributed in the pulp of the berry. A common mistake is to use only the free run juice for analysis, which tends to have higher Brix, higher titratable acidity, lower pH, and lower potassium than fully expressed juice. Berry samples can be crushed and pressed by hand, taking care to thoroughly crush each berry. Large cluster samples are more easily crushed with a small roller-crusher and pressed with a small bench-scale press. To ensure that the sample preparation best approximates commercial processing, a yield of approximately 65 ml per 100 grams (300 ml per pound) of fruit should be obtained. This corresponds to about 160 gallons per ton.

The juice should be settled to remove suspended solids and held in full containers to exclude air. Natural enzymatic oxidation (browning) occurs after the fruit is crushed and is more rapid when the fruit is very ripe, at warmer temperatures, or when rot is present. Refrigeration aids settling and delays enzymatic browning. To further help prevent browning and maintain sample freshness for sensory evaluation, 25 mg/l each of sulfur dioxide and ascorbic acid (vitamin C) can be added to the juice. Pectolytic enzymes may also be added to enhance juice clarity, if necessary. The settled, clarified juice can then be used for analysis. Analysis should include a sensory evaluation as well as chemical analysis for soluble solids (degrees Brix), titratable acidity, pH, malic acid content, and fermentable nitrogen content. Samples can be held refrigerated in full containers for one or two weeks for comparison with later samples.

Sensory Evaluation. With experience, winegrowers and winemakers learn to recognize varietal aromas and flavors as they develop during ripening. Most winegrape varieties have green, herbaceous flavors when underripe, which are replaced by fruity flavors that are characteristic of the variety when fruit is fully ripe. Sensory evaluation can be conducted by tasting fruit in the vineyard; however, too often only a few berries are tasted and the large heterogeneity in maturity among berries makes evaluation difficult. A more objective evaluation can be made when an adequate fruit sample size is processed to approximate commercial yields of juice.

The crushing and pressing process extracts aroma, flavor, and color from the grape skins. The processed juice can be evaluated for both intensity and quality of aroma and flavor, acidity and taste balance, and color. Unripe fruit has green juice because of high chlorophyll content and low pigment content in the skins. At full maturity, white winegrape varieties produce light yellow to golden juice; red grape varieties yield red to purple juice. The ease with which pigments are extracted during processing is also an indicator of the degree of maturity; pigments are more readily extracted from more mature fruit.

Red winegrape samples are best prepared by crushing, destemming, and macerating on the skins for one to two hours at room temperature before pressing. Red grapes at optimal maturity, on crushing and pressing, rapidly release the anthocyanin pigments from the vacuoles in the epidermal layers of the skin. Color and phenols can also be extracted from a berry sample using heat and an alcohol extraction buffer. Extraction methods can be used to monitor color and phenolic "maturation" during ripening. An extraction procedure for phenolic analysis was recently developed by ETS Laboratories (ETS Laboratories, 2001).

Analysis of Soluble Solids. Sugars accumulate rapidly in ripening berries from veraison to harvest. Sucrose is produced in the leaves by photosynthesis and is transported to the berries through the phloem. Within the berries, sucrose (a nonreducing sugar) is hydrolyzed (inverted) to glucose and fructose (reducing sugars), which are the primary sugars present at harvest. The glucose/fructose ratio changes during ripening from about 2.0 at veraison to 0.9–1.0 at full maturity. Glucose and fructose are the most abundant sugars, but several other sugars are present including sucrose, rhamnose, and two major pentose sugars: arabinose and xylose.

The percentage of sugar in the juice is estimated from the soluble solids content and is commonly expressed as degrees Brix (degrees Balling) or as grams of sucrose per 100 g of juice or must. Scales other than degrees Brix are also used to measure sugar content of juice and musts. Degrees Oeschle is used in Germany; it corresponds to the first three figures in the decimal fraction of the specific gravity of the juice/

must as measured by a specific gravity hydrometer. For example, a 20° Brix solution has a specific gravity of 1.082 and a degree Oeschle of 82 [(1.082-1) x 1000]. Degrees Baumé is used in France and Australia. Each degree Baumé equals approximately 1.8° Brix and corresponds approximately to the potential alcohol content. For example, a juice at 12° Baumé produces a wine with about 12% alcohol when fermented to dryness. Conversion tables can be used to simplify conversion of Brix to specific gravity, degrees Oeschle, and degrees Baumé (Ribereau-Gayon et al., 2000; Zoecklein et al., 1995).

The soluble solids content of juice or must is a good measure of grape maturity and is also the basis for calculating the approximate potential alcohol yield and the need for amelioration (chaptalization). Approximately 90–95% of the total soluble solids in ripe grapes are fermentable sugars, primarily glucose and fructose. The balance consists of organic acids and their weak acid salts, nonfermentable sugars, nitrogenous compounds, tannins, pectins, pigments, mineral salts, and other compounds. Remember that berry shriveling and dehydration lead to abnormal increases in sugar concentration (as well as acidity), and hydration gives a dilution effect with an apparent decrease in sugar content.

Procedures for determining soluble solids are found in several textbooks (Iland et al., 2000; Ough and Amerine, 1988; Zoecklein et al., 1995). The degrees Brix is most commonly measured with a hydrometer or refractometer. Hydrometers measure the soluble solids by comparing the density of juice to that of water. Accurate hydrometers are calibrated to narrow ranges of 5–10° and are subdivided to 0.1° units. Inexpensive hydrometers often cover a wide range such as 0–30° and have other scales such as "potential alcohol." These hydrometers are not very accurate and can be relied on only for rough estimates of sugar content.

Refractometers measure the refractive index, the degree to which light is refracted (bent) by the liquid, which is proportional to the dissolved solids content. The higher the dissolved solids content, the more the light is refracted. Both hydrometer and refractometer readings are usually calibrated at 20°C (68°F) and must be corrected for the actual temperature of the juice. If the sample temperature is colder than 20°C, a correction is subtracted from the reading to compensate for the increase in density. If the sample temperature is warmer, a correction is added to compensate for the decrease in density. Tables of correction factors are available in several wine textbooks (Iland et al., 2000; Ough and Amerine, 1988; Zoecklein et al., 1995).

Hydrometer readings are taken by lowering a hydrometer into settled juice contained in a hydrometer cylinder. Suspended solids can cause falsely high readings, so it is important to use well-settled or centrifuged juice. The equipment must be clean, the cylinder level, and the hydrometer floating freely. When the hydrometer has come to equilibrium in the juice, a reading is taken by placing the eye at the level of the liquid surface and reading the hydrometer stem at the bottom of the meniscus (the meniscus is the curved upper surface of the juice around the hydrometer stem).

Hydrometers are commonly used to monitor the rate of fermentation. Spinning the hydrometer gently removes entrained air bubbles that can cause the stem to rise, especially during fermentation from carbon dioxide evolution. Note that hydrometer readings during or after alcoholic fermentation do not reflect true sugar levels because of the effect of alcohol on must density. The rate of change in degrees Brix as measured by hydrometer is an accurate measure of the rate of fermentation (rate of decrease of sugar).

Hand-held refractometers can be used in the winery or in the field. Juice samples do not have to be settled as they do for hydrometer measurements. A drop of juice is placed on the glass surface, covering the surface entirely. The prism box is closed and the instrument held toward the light. A graduated scale in degrees Brix is read where the boundary of the dark and light field meet. Refractometers are calibrated by using a drop of distilled water at 20°C and adjusting the instrument to read 0° Brix. The zero position can be adjusted with the zero-adjust screw on most instruments. Common errors with refractometer measurements include failing to calibrate with distilled water and not making the necessary temperature corrections, especially when samples are taken in the field. Some refractometers come with thermometers attached, and others are self-correcting over a range of about ±10°C of the calibration temperature. Refractometers work only for juice or must samples and cannot be used for monitoring rates of fermentation, because the refractive index of water-alcohol mixtures is very different from that of water solutions.

Modifying Sugar and Potential Alcohol Content at Harvest

In Oregon, sugar additions of up to 2% (2 grams per 100 ml or 20 g/l) are normally allowed, and up to 4% may be used in very difficult vintages. To raise the alcohol level by 1% by volume, about 17 g/l of sugar needs to be added to white juice and about 20 g/l to red must. Greater additions are necessary for red wine to compensate for evaporation of alcohol during warmer red fermentations (Peynaud, 1984).

Sugar additions to red wine musts are best calculated on the anticipated yield of wine at pressing (approximately 160 gallons per ton) rather than on the must volume during fermentation. The potential alcohol content can be estimated by multiplying the degrees Brix of red musts by about 0.55 and white juice by about 0.60. For example, the estimated alcohol content of a white wine produced from 22° Brix juice is about 0.60 x 22 = 13.2% alcohol (if fermented to dryness), whereas the estimated alcohol content of a red wine produced from 22° Brix must is about 0.55 x 22 = 12.1% alcohol.

Analysis of Acidity and pH

Titratable acidity and pH are important harvest parameters because they affect acidity, taste balance, and the microbial and chemical stability of wines. Tartaric and malic acid represent about 90% of the acids in winegrapes. Both acids are secondary products related to sugar metabolism and are synthesized primarily in the grapes and also in leaves, although there is no evidence of transport of either acid from leaves to the fruit. Winegrapes are unusual in that few other fruits accumulate significant quantities of tartaric acid. Both tartaric acid and malic acid increase rapidly in the berries during the green, herbaceous growth period. From veraison to harvest the tartaric acid content (percentage) of the berries decreases due to the increase in berry size, then remains relatively constant during the later stages of ripening. Malic acid, on the other hand, decreases due to increasing berry size and also to utilization as an energy source by the berries (respiration). The decrease in malic acid during the later stages of ripening is more rapid during periods of warmer weather. The tartaric/malic acid ratio varies with grape variety, the degree of ripening, and environmental conditions during ripening. The overall net decrease in acidity during the later stages of ripening is primarily due to the decrease in malic acid content. Other acids present in juice at harvest include citric and ascorbic acids, amino acids, and weak phenolic acids. Additional acids found in wines include acetic, succinic, and lactic acids, which are fermentation by-products of yeast and bacteria.

To determine the total acidity of a juice or must, all of the acids present would need to be measured separately. Instead, the total acidity is normally estimated by measuring the titratable acidity (TA) by neutralizing a juice sample with a sodium hydroxide solution of a known normality. Whereas pH is a measure of the hydrogen ion content, TA is a measure of the hydrogen ions released (primarily from the organic acids tartrate and malate) when a juice/wine sample is titrated with a standardized base to a defined endpoint. Tartrate and malate are both dicarboxylic acids that are partially dissociated at pH values within the normal range of juice and wine. The pK values (the pH at which 50% of the hydrogen ions are dissociated from each of the two carboxyl groups) are 3.04 and 4.34 for tartaric acid and 3.46 and 5.1 for malic acid. For example, at pH 3.0 only about 50% of the first carboxyl group of tartaric acid has dissociated to release its hydrogen ions into solution. The TA of a juice/wine, therefore, depends not only on the organic acid content but also on the pH of the sample. The TA is always lower than the total acidity within the normal range of juice/wine pH, and the lower the pH the greater the number of hydrogen ions will be titrated from a given concentration of the organic acids present.

Tartaric acid and malic acid (anions) also combine with cations, primarily potassium and to a lesser extent calcium, to form weak acid salts. At harvest, grapes are often saturated with the weak acid potassium salt of tartaric acid known as potassium acid tartrate (also known as potassium bitartrate and cream of tartar). The concentration of the organic acids and their weak acid salts largely determines the pH and the buffering capacity of the juice. The buffering capacity is a measure of the resistance to change in pH by addition of either acids or bases. Juice and wine are highly buffered systems, primarily due to tartaric and malic acid and their equilibria with their weak acid salts. The buffering capacity of an individual juice/wine determines the change in pH that occurs with chemical acidification and deacidification, and with the completion of malolactic fermentation.

Titratable acidity is commonly expressed as grams per 100 ml (percent acidity) or as grams per liter expressed as tartaric acid. Procedures for TA analysis are found in several textbooks (Iland et al., 2000; Ough and Amerine, 1988; Zoecklein et al., 1995). A common method for determining the TA is to add a 5 ml juice sample to about 100 ml of distilled water in a beaker or flask, then titrate with a standardized base solution (0.1 N NaOH) using a calibrated burette until the weak organic acids are completely neutralized at a pH of about 8.2–8.4. A calibrated pH meter can be used, or a few drops of a 1% phenolphthalein indicator solution can be added to determine the neutralized end point. In either case it is important to neutralize the water to the desired end point prior to adding the sample. Add a few drops of strong base to the water until the pH reaches 8.2–8.4, or until the indicator turns a faint pink color. The juice sample is then added to the pH-adjusted water and titrated with the standardized base solution. The number of milliliters of standardized base used in the titration multiplied by 1.5 equals the TA in grams per liter (for 5 ml juice samples using 0.1

N NaOH). The percent acidity (grams per 100 ml) equals the grams per liter divided by 10.

Alternatively, a solution of 0.667 N NaOH can be used, which enables the TA to be read directly from the burette as grams per liter (with each ml equivalent to 1 g/l). In other countries, different standards are used for expressing TA. In France, the percent acidity (g/100 ml) is expressed as sulfuric acid. The percent acidity expressed as sulfuric acid multiplied by 1.5 approximately equals the percent acidity expressed as tartaric acid based on the ratio of the equivalent weights of the two acids. In Germany, the TA is often simply expressed as the volume 0.1N NaOH required to neutralize 100 ml of sample.

Common sources of error in titratable acidity measurements include careless pipetting of the sample (use a 5 ml volumetric pipette), failure to neutralize the acidity in the water before adding the juice sample, overtitration (past the first appearance of a pink color with the indicator, or to a pH greater than 8.4 using a pH meter), and failure to calibrate the pH meter properly.

Tartaric and malic acid content can also be measured directly by colorimetric, enzymatic, and chromatographic methods. Malic acid is commonly measured in an enzymatic analysis. A semiquantitative analysis for malic and tartaric acid can be done with a simple paper chromatography method for detecting organic acids. This method is also used to detect the conversion of malic to lactic acid in wines during and after completion of malolactic fermentation. The enzymatic assay for malic acid and the paper chromotagraphy method for organic acids are described in detail in several wine texts (Iland et al., 2000; Zoecklein et al., 1995). Samples are also commonly sent to wine analytical labs for determination of the organic acid content of fruit at harvest.

The pH is a measure of the intensity of the acidity rather than the quantity of acids present as measured by titration. The pH is a direct measure of the total hydrogen ion (H^+) content in solution and is expressed on a scale of 0–14. A solution with a pH of 7 is a neutral solution, having an equal number of acid and base ions (10^{-7} moles/l of H^+ and 10^{-7} moles/l of OH^-). As the pH decreases below 7, the acidity (hydrogen ion concentration) increases; as the pH increases above 7, the acidity decreases and the solution becomes more basic. The pH of a solution is defined as pH = - log (hydrogen ion concentration). It is important to understand that the pH scale is a logarithmic expression of an exponential function, so that a change of one pH unit represents a tenfold change in concentration of free hydrogen ions. Thus, a juice with a pH of 3.0 has ten times the acid intensity (H^+ concentration of 10^{-3} moles/l) of a juice with pH 4.0 (H^+ concentration of 10^{-4} moles/l). The pH of grape juice and wine usually ranges from 3.0 to 4.0, corresponding to hydrogen ion concentrations that are 10,000- and 1,000-fold greater than water at pH 7, respectively.

The pH is measured by a meter with a pH-sensitive electrode. A pH meter with an accuracy of at least ±0.05 pH units is required. The meter should be carefully calibrated prior to each use with standard buffer solutions (usually pH 4 and pH 7). An inexpensive standardizing buffer for juice and wine can be prepared by adding about 1 g of cream of tartar (potassium acid tartrate) to 100 ml of water. This makes a saturated solution with a pH of 3.56 at room temperature. The temperature control of the pH meter should be set at the temperature of the standard buffer solutions, and the juice samples should be at this same temperature when measured. Common errors in pH measurement include the failure to standardize the pH meter properly, disregarding temperature correction, and the use of worn or insensitive electrodes. Electrodes should be cleaned frequently and stored in an appropriate solution according to the manufacturers instructions.

Modifying Acidity and pH at Harvest

Titratable acidity of winegrapes in Oregon typically ranges from 5.0–10.0 g/l at harvest, with pH levels from 2.8 to as high as 3.8. Tartaric acid and malic acid ranges are 3–7 g/l and 1–5 g/l, respectively, and vary with maturity, variety, site, climate, and vintage. During ripening, the TA decreases and the pH increases because of dilution resulting from the increase in berry size, conversion of organic acids to their weak acid salts, and respiration of malic acid for energy. Juice and wines with lower pH levels tend to have better microbial stability, color stability, and oxidative stability. In addition, sulfur dioxide is more effective at lower pH levels. The pH also affects both heat (protein) and cold (potassium acid tartrate) stability of wines. Wines with low to moderate pH also tend to have more crisp, fruity flavors and tend to age better than wines with higher pH levels.

Desired pH levels at harvest range around 3.25–3.45 and finished pH levels of wines are about 3.40–3.65. Winegrapes from warmer sites or warmer seasons have lower TA and higher pH than wine grapes from cooler sites or cooler seasons. Grapes harvested from warmer sites or warmer vintages may require acid additions, and grapes from cooler sites or cooler vintages may require acid reductions in order to achieve the desired TA and pH balance in the juice and the wine.

Acidification is primarily done to juice or must prior to fermentation, or to new wines soon after fermentation, to increase the acidity and to lower the pH to the desired level. Food-grade organic acids including tartaric, malic, and citric acid can be added provided the final TA does not exceed 8.0 g/l (Bureau of Alcohol Tobacco and Firearms regulation). Tartaric acid, malic acid, or tartaric/malic acid blends are the preferred acidulants. Citric acid can be metabolized to acetic acid by bacteria and should be used only after completion of both yeast and malolactic fermentation. Malic acid can also be metabolized by lactic acid bacteria to produce lactic acid.

Tartaric acid is the most common acidulant and is added primarily to lower the pH to a desired level. Laboratory bench trials should be conducted to determine the effect of acid additions on TA and pH. The pH change resulting from acid additions is not predictable, because of the varying buffer capacities of different juice/must samples. Commonly, 0.5–2.0 g/l of tartaric acid are added to samples, which are then stored for about 24–48 hours at refrigerator temperature to reequilibrate and precipitate excess potassium acid tartrate. Approximately 50% of tartrate added precipitates as potassium acid tartrate. The TA, pH, and taste balance can then be evaluated after decanting from the tartrate precipitate and bringing the sample back to room temperature.

When making acid/pH adjustments to red musts, it is important to sample the must after at least 24 hours of maceration on the skins prior to the onset of fermentation. During maceration, the TA decreases and the pH increases because of the better dissolution of the acids and their weak acid salts from the berry juice/pulp. If a target finished wine pH of a red wine is about 3.65 (after malolactic fermentation is complete and the wine is cold stablized with respect to potassium acid tartrate), then the must pH should be adjusted to about 3.35–3.45 prior to fermentation. Modest adjustments can be made to the must, then fine-tuning adjustments are done later after completion of fermentation and before cold stabilization.

Deacidification may be necessary in cooler seasons when the acidity of the juice/must is higher than desired. Chemical deacidifications are commonly done with either calcium carbonate or potassium carbonate. An addition of 0.67 g/l of calcium carbonate reduces the acidity of a juice/wine sample by 1.0 g/l through precipitation of calcium tartrate. An addition of 0.90 g/l of potassium carbonate reduces the TA by 1.0g/l through precipitation of potassium acid tartrate. Bench trials should always be done to determine the amount to use and the effect on TA, pH, and taste balance.

Calcium carbonate deacidifications should be done to juice or young wines so that there is ample time for precipitation of calcium tartrate prior to bottling. This avoids later calcium tartrate instability problems. The natural calcium content of winegrape juice is about 80 mg/l or less. Deacidification procedures that add calcium carbonate may increase levels to as much as 300–500 mg/l. Wines are not considered stable with respect to calcium tartrate precipitation until the levels drop back to 100 mg/l or less. The residual calcium tartrate deposition may take up to three months to reach equilibrium and is not sped up significantly by lowering the temperature, as is the case with potassium acid tartrate. If wines are not stable with respect to calcium tartrate, precipitation occurs after bottling.

Deacidification by the addition of potassium carbonate increases potassium content and thereby affects the equilibria reactions between tartaric acid and its weak acid salt, potassium acid tartrate. A reequilibration is required before a decrease in acidity occurs. For this reason, potassium carbonate additions should be done to juice or new wines prior to cold stabilization. Addition to wines after cold stabilization increases the wine pH and potassium content and induces tartrate instability. Bench trial samples should be refrigerated for at least 24 hours after potassium carbonate additions in order to reach reequilibrium before evaluating.

Analysis of Fermentable Nitrogen Content

Nitrogen compounds are required by yeast for the production of cell biomass and the synthesis of proteins and enzymes necessary for the biochemical process of fermentation. The readily fermentable (easily assimilable) nitrogen compounds in juice and must consist primarily of ammonia and alpha-amino acids, particularly arginine, serine, glutamate, aspartate, threonine, and lysine. The amino acid proline is abundant in juice and must at harvest, but it is poorly utilized by yeast except under oxidative conditions. Arginine, usually the most abundant yeast-assimilable amino acid in grape juice and must, contains four nitrogen atoms and can potentially provide more nitrogen per molecule than other alpha-amino acids. Arginine is, however, also a precursor to the formation of ethyl carbamate, also known as urethane, which is a potential carcinogen. Ethyl carbamate is a naturally occurring compound found in all fermented foods and beverages. The wine industry is very interested in reducing ethyl carbamate in wines to levels below the current voluntary regulatory limit of 15 mg/l (Austin and Butzke, 2000).

From veraison to harvest, the ammonia concentration in the juice of the berry decreases at the same time that the amino acid content significantly increases. In unripe fruit the ammonia content may represent up to 50% of the nitrogen content of the juice/pulp, whereas at full maturity the amino acid content may represent up to 90% of the juice nitrogen. The juice/pulp of mature berries contains up to 20% of the total berry nitrogen, the remainder being distributed in the skins and seeds.

The total amount of yeast-fermentable nitrogen present in juice and must can be approximated by the sum of the nitrogen available from ammonia and nitrogen available from the alpha-amino acids present at harvest (Bisson, 1991; Jiranek et al., 1995). The ammonia content can be measured with a spectrophotometer at 340 nm using an enzymatic ammonia diagnostic kit (Sigma, Procedure No. 171-UV) or by using an ammonia ion selective electrode (Zoecklein et al., 1995). The enzymatic assay has the advantage of being fast, accurate, and inexpensive.

The alpha-amino acid content can be measured by spectrophotometric assay (Dukes and Butzke, 1998). The assay is based on the derivatization of primary amino nitrogen groups with an o-phthaldialdehyde/N-acetyl-L-cysteine (OPA/NAC) reagent (both proline and ammonia are insensitive to the assay). The resulting stable isoindole derivatives rapidly form at room temperature and are measured with a spectrophotometer at 335 nm. The procedure uses a juice "blank" to account for the absorbance of non-derivative forming compounds such as phenolic compounds. The assay is called NOPA for "nitrogen by OPA" and is reliable and fast. A recent modification of this procedure known as "arginine by ophthaldialdehyde," or ARGOPA, uses ion exchange to isolate arginine from the other amino acids and then measures arginine using the standard NOPA assay. This assay allows easy monitoring of arginine levels and provides additional information on the nitrogen status of juice (Austin and Butzke, 2000).

Modifying Yeast-assimilable Nitrogen Content

Low levels of yeast-assimilable nitrogen content (YANC) in grape juice and must at harvest have been associated with sluggish or stuck fermentations that often leave undesirable levels of residual sugar in wines. Problem fermentations are also sometimes accompanied by production of hydrogen sulfide and other reduced sulfur odors (Jiranek et al., 1995; Kunkee, 1991). Drought conditions commonly lead to low levels of yeast-assimilable nitrogen in fruit at harvest. Early loss of varietal character and development of other undesirable flavors have been observed in wines produced from fruit grown under drought and stressed conditions. This "untypical" aging syndrome is commonly referred to as UTA (Sponholz et al., 2000).

Slow and stuck fermentations have been commonly reported in Oregon. Winemakers often report fermentation problems with fruit from specific vineyard blocks over numerous vintages, suggesting a relationship with grapevine nitrogen status. Petiole analysis of Oregon vineyards over a three-year study indicated that a high percentage of the vines were consistently deficient in nitrogen, based on bloom-time and veraison standards (Watson et al., 2000). Analysis of more than two hundred juice samples from commercial wineries during the same three-year period also showed a high percentage of juice and must samples to be low in YANC at harvest. The nitrogen available from ammonia ranged from 0 to 145 mg N/l, and the nitrogen available from the alpha amino acids ranged from 35 to 380 mg N/l. The total estimated YANC ranged from 38 to 500 mg N/l. Differences were observed among varieties and vintages. The percentage of commercial Chardonnay juice samples with less than 140 mg N/l was 80% in 1997, 79% in 1998, and 17% in 1999. The percentage of commercial Pinot noir juice samples with less than 140 mg N/l was 37% in 1997, 34% in 1998, and 8.6% in 1999.

Variations in climate, soil, cultivation practices, soil moisture content, and fertilization practices may have a significant impact on juice and must nutrition (Butzke, 1998; Ingledew and Kunkee, 1985). In an experimental trial at a commercial vineyard in the Willamette Valley, cover crop management was shown to have a significant effect on YANC in Pinot noir grapes at harvest. Tilling of alternate rows in early spring to encourage nitrogen utilization and to reduce nutrient and water competition with the vines resulted in a 39% increase in YANC compared to untilled treatments. Wines from the tilled treatments also fermented more rapidly than wines from untilled treatments (Watson et al., 2002).

Winemakers often add supplements to juice and fermenting wines to balance perceived nutritional deficiencies (Bisson, 1991; Montiero and Bisson, 1992). Recommended levels of YANC needed for healthy fermentations are reported to vary from as low as 140 mg N/l to as high as 500 mg N/l or more. Suggested YANC concentrations are 200 mg N/l for 23° Brix juices, 250 mg N/l for 25° Brix juices, and 300 mg N/l for 27° Brix juices. Nutrient supplements often include combinations of diammonium phosphate (DAP), yeast extracts, and vitamins. Addition of 1 pound of DAP per 1,000 gallons of juice increases the YANC by about 25 mg/l. Addition of yeast extracts increases the YANC by half or less than the comparable weight of DAP, but yeast extract preparations are also a source of amino acids, vitamins, and minerals.

Table 1. Summary of Maturity Assessment

Factors	*Observations/Importance*
°Brix	Potential alcohol content, calculation of sugar additions
Titratable acidity	Acidity, flavor balance, and wine style
pH	Intensity of acidity, chemical and microbial stability of wine
Sensory evaluation	Development of characteristic varietal aromas and flavors
Fruit color	Color intensity and uniformity among clusters, ease of extraction during maceration/processing
Color of seeds, stems	Transition from green to brown during later stages of ripening
Fermentable nitrogen	Fermentation rate; deficiencies may affect production of sulfide odors and accelerated wine aging (UTA)
Condition of fruit	Berries soften at full maturity after reaching maximum size; pronounced shriveling/berry shatter indicates overripeness; fruit should be free of mold, rot, and insect and bird damage
Vine condition and weather	Assessment of further ripening potential; extremes of weather can delay or arrest maturation, excess heat/drought can cause premature berry shriveling (dehydration), excess rain can cause berry swelling (dilution)

Some researchers have reported increases in fruity aroma intensity and wine quality with moderate nitrogen supplementation of musts, particularly in white wines (Rapp and Versini, 1995). By contrast, excessive levels of nitrogen may be undesirable and could lead to increased formation of ethyl carbamate (Austin and Butzke, 2000; Butzke, 1998; Kunkee, 1991; Spayd, 1997).

Summary

Determining the best time to harvest requires experience and a careful assessment of winegrape maturity. Each variety at each vineyard site should be monitored carefully using proper sampling techniques, field observations, sensory evaluations, and accurate measurements of degrees Brix, titratable acidity, malic acid content, pH, and the nutritional status of the fruit (Table 1). A record should be kept each year of the maturity observations, sensory and chemical analysis, and condition of the fruit at harvest. Fruit maturity and quality are greatly influenced by the unique weather patterns of each harvest season, so it is also desirable to collect temperature and rainfall data for each site. A maturity and weather database will be invaluable in future years for determining the best time to harvest for optimal wine quality.

Note

1. Several excellent reference books were frequently used in the preparation of this chapter but are not cited specifically except when referring the reader to analytical procedures or reference tables. These texts should be consulted for thorough discussions of winemaking principles and practices and analytical procedures (Boulton et al., 1996; Iland et al., 2000; Ough and Amerine, 1988; Ribereau-Gayon, 2000; Zoecklein et al., 1995).

References

Austin, K. T., and C. E. Butzke. 2000. Spectrophotometric assay for arginine in grape juice and must. *American Journal of Enology and Viticulture* 51(3):227-232.

Bisson, L. F. 1991. Influence of nitrogen on yeast and fermentation of grapes. In J. M. Rantz (ed.), *Proceedings of the International Symposium on Nitrogen in Grapes and Wines.* American Society for Enology and Viticulture, Davis, Calif.

Boulton, R. B., V. L. Singleton, L. F. Bisson, and R. E. Kunkee. 1996. *Principles and Practices of Winemaking.* Chapman and Hall, London.

Butzke, C. E. 1998. Survey of yeast assimilable nitrogen status in musts from California, Oregon, and Washington. *American Journal of Enology and Viticulture* 49(1):220-224.

Dukes, Bruce C., and C. E. Butzke. 1998. Rapid determination of primary amino acids in grape juice using an o-phthaldialdehyde/N-acetyl-L-cysteine spectophotometric assay. *American Journal of Enology and Viticulture* 49(1):125-134.

ETS Laboratories. 2001. *Grape Extracts for Monitoring Phenolics.* ETS Laboratories. St. Helena, Calif..

Iland, P., A. Ewart, J. Sitters, A. Markides, and N. Bruer. 2000. *Techniques for Accurate Chemical Analysis and Quality Monitoring during Winemaking.* Wine Promotions, Campbell Town, Australia.

Ingledew, W. M., and R. E. Kunkee. 1985. Factors influencing sluggish fermentations of grape juice. *American Journal of Enology and Viticulture* 36(1):65-76.

Jiranek, V., P. Langridge, and P.A. Henschke. 1995. Amino acid and ammonium utilization by *Saccharomyces cerevisiae* wine yeasts from a chemically defined medium. *American Journal of Enology and Viticulture* 46(1):75-83.

Jiranek, V., P. Langridge, and P.A. Henschke.1995. Regulation of hydrogen sulfide liberation in wine producing *Saccharomyces cerevisiae* strains by assimilable nitrogen. *Journal of Applied Microbiology* 61:461-467.

Kennedy, J. A., Y. Hayasaka, S. Vidal, E. J. Waters, and G. P. Jones. 2001. Composition of grape skin proanthocyanidins at different stages of berry development. *Journal of Agricultural and Food Chemistry* 49:5348-5355.

Kennedy, J. A., M. A. Matthews, and A. L. Waterhouse. 2000. Changes in grape seed polyphenols during fruit ripening. *Phytochemistry* 55:77-85.

Kunkee, R. E. 1991. Relationship between nitrogen content of must and sluggish fermentation. In J. M. Rantz (ed.), *Proceedings of the International Symposium on Nitrogen in Grapes and Wines.* American Society for Enology and Viticulture, Davis, Calif.

Miranda-Lopez, R., L. M. Libbey, B. T. Watson, and M. R. McDaniel. 1992. Odor analysis of Pinot noir wines from grapes of different maturities by a gas-chromotography-olfactometry technique (Osme). *Journal of Food Science* 57(4):985-1019.

Montiero, F. F., and L. F. Bisson. 1992. Nitrogen supplementation of grape juice: I. effect on amino acid utilization during fermentation. *American Journal of Enology and Viticulture* 43(1):1-10.

Ough C. S., and M. A. Amerine.1988. *Methods for Analysis of Musts and Wines,* 2d ed. John Wiley and Sons, New York.

Peynaud, E. 1984. *Knowing and Making Wine.* John Wiley and Sons, New York.

Price, S. F., P. J. Breen, M. Valladao, and B. T. Watson. 1995. Cluster sun exposure and quercetin in Pinot noir grapes and wine. *American Journal of Enology and Viticulture* 46(1):187-194.

Rapp, A., and G. Versini. 1995. Influence of nitrogen compounds in grapes on aroma compounds in wines. In *Proceedings: The Composition of Musts and Influence on Stuck Fermentations.* Geisenheim Research Institute Department of Microbiology and Biochemistry, Geisenheim, Germany.

Ribereau-Gayon, P., et al. 2000. *Handbook of Enology,* Vol. 1: *The Microbiology of Wine and Vinifications;* Vol. 2: *The Chemistry of Wine Stabilization and Treatments.* John Wiley and Sons, New York.

Singleton, V. L. 1982. Grape and wine phenolics: background and prospects. In A. D. Webb (ed.), *University of California Davis Grape and Wine Centennial Symposium Proceedings.* Department of Viticulture and Enology, University of California, Davis.

Smart, R., and M. Robinson. 1991. *Sunlight into Wine. A Handbook for Winegrape Canopy Management.* Winetitles, Adelaide, Australia.

Spayd, S. E. 1997. Nitrogen components and their importance in grape juice fermentations: a Primer.WSU Wine and Grape Research Newsletter, Supplement No.1, Irrigated Agricultural Research and Extension Center, Washington State University, Pullman.

Sponholz, W. R., W. Grossman, and T. Huhn. 2000. The untypical aging (UTA) in white wines-its origins and prevention. Fifth International Proceedings on Cool Climate Viticulture and Oenology, January 16-20, 2000, Melbourne, Australia.

Watson, B. T., E. Hellman, A. Specht, and H. P. Chen. 2000. Evaluation of nitrogen deficiencies in Oregon grapevines and musts. *Oregon Wine Advisory Board Progress Reports 2000-2001.* Agricultural Experiment Station, Oregon State University, Corvallis.

Watson, B. T., M. Godard, and H. P. Chen. 2002. Manipulating soil moisture and nitrogen availability to improve fermentation behavior and wine quality. Part II: Fermentable nitrogen content, must and wine composition. *Oregon Wine Advisory Board Progress Reports 2001-2002.* Agricultural Experiment Station, Oregon State University, Corvallis.

Zoecklein, B. W., K. C. Fugelsang, B. H. Gump, and F. S. Nury. 1995. *Wine Analysis and Production.* Chapman and Hall, London.

31

Harvest

Edward W. Hellman and Mark L. Chien

Harvest is the culmination of the entire year's effort, and it is by far the most intense, hectic, and often worrisome part of the season. The basic goal of harvest is deceptively simple: to deliver the grape crop to the winery. Satisfactory achievement of this goal, however, requires a coordinated logistical operation that can become quite complex for large vineyards. Careful advanced planning and preparation with attention to detail are absolute requirements. Perhaps above all is the need for preparation and flexibility to adapt to unplanned circumstances, especially bad weather. A well-executed harvest will deliver the grapes to the winery at their peak condition. But if harvest is poorly managed, it can diminish an otherwise high-quality crop to a level of mediocrity.

This chapter discusses the major considerations and activities involved with planning and carrying out a successful harvest. Most winegrapes in Oregon are harvested by hand, so this discussion does not address the unique concerns of machine harvesting. Harvest practices are very similar regardless of whether the vineyard is large or small; the main differences are in the number of pickers, equipment and supplies, and delivery logistics.

Harvest Forecasting

It is impossible to predict harvest dates at the beginning of the season accurately; too many variables influence the rate of fruit ripening. However, with experience and good record keeping, it is possible to make reasonably accurate projections as the season progresses. At a minimum, the dates of major phenological (developmental) stages such as budburst or bloom, along with harvested yield and fruit quality parameters, should be recorded for each block. Accompanying local weather records and degree-day accumulations facilitate year-to-year comparisons.

Forecasting gets serious after veraison. Cluster sampling to monitor fruit ripening (see Chapter 30) should begin three to four weeks after veraison. The subsequent frequency of sampling depends on proximity to the desired ripeness and the rate of fruit ripening, which is greatly influenced by weather.

Coordination between Vineyard and Winery

The basis for a successful harvest is good communication and cooperation between vineyard and winery. To minimize misunderstandings and potential conflict, it is highly recommended that the vineyard and winery enter into a grape purchase contract that details the responsibilities and performance of both parties. A properly constructed contract should include several performance and quality items that relate to harvest activities and delivery of the fruit (see Chapter 15).

The logistics for fruit delivery should be discussed with the winery at an early stage in the business relationship. It is very helpful for the grower to visit the winery and see firsthand the receiving and processing areas and equipment. Some of the issues that should be addressed are the ability of the winery to handle the size of truck or trailer the vineyard intends to use. The winery's preference, if any, or ability to handle certain fruit containers (variously sized bins and boxes) is an important issue. It is also important to learn the winery's fruit-processing rate, the hours of operation for the crushing crew, and the turnaround time for emptying fruit containers. The combination of these factors determines the amount of fruit that can be harvested and delivered in a day.

As the season progresses, winemakers commonly visit the vineyard one or more times to monitor crop development. These can be important times to review the expectations of both parties so there are no big surprises at harvest. During the later stages of ripening, fruit maturity parameters should be monitored and the results communicated to the winery in a timely manner. The projected harvest time can be fine-tuned based on periodic fruit sampling, and the vineyard and winery can agree on a tentative harvest schedule. As harvest time gets closer, frequent communications with the winery become more important. Often the winery must coordinate fruit delivery from several different vineyards with their processing and fermentor capacities.

Harvest Preparations

Preparations for the current harvest should begin immediately after the previous year's harvest. A systematic review of the operation identifies the areas of planning and execution that need improvement, and these should be noted in a vineyard record book. Plans should be made to purchase or replace harvest-related supplies, materials, and equipment. Winery and labor contracts should be reviewed and modified where necessary.

Equipment

Equipment supply and service issues should be addressed well before harvest begins. Proper maintenance is critical to the smooth flow of harvest; an equipment failure on harvest day can bring the entire operation to a standstill. Perform a thorough check of all equipment and vehicles that will be involved with harvest including trucks, tractors, forklifts, trailers, pickups, and four-wheelers. Check all belts, fluids, lubrication, and air pressure. Properly licensed and trained drivers should be identified for all vehicles. Be sure that all vehicles are filled up with fuel at the start of each day, and maintain an adequate fuel supply at the vineyard.

Trucks and Trailers. The typical large-equipment needs for harvest include flatbed trucks or trailer-trucks to transport harvest bins to the winery. Since these trucks are typically used only at harvest time, rental trucks are usually the best option for nearby winery deliveries. Commercial truck haulers are often more appropriate for delivery to more distant wineries. Trucks of the proper size should be selected to fit your needs for delivery to the wineries. A 24-foot flatbed truck with a minimum of 33,000 gross vehicle weight can carry up to 24 large harvest bins (4 by 4 by 2 feet), double-stacked, which is 8–10 tons of fruit. Trucks should come equipped with a minimum of six 2-inch straps, preferably with ratchet straps, to secure the bins to the trailer. The proper use of straps is critical to ensure load safety and security. This applies to ropes as well. Learn to tie a trucker's knot to achieve maximum tautness of tie-down ropes.

Be sure to reserve trucks with the rental agencies or services with a commercial carrier at least six weeks prior to harvest, and follow up to confirm reservations as harvest time gets close. Rental trucks should be picked up a few days before the anticipated start of harvest to ensure availability and to give drivers time to familiarize themselves with the vehicle.

Each truck should be supplied daily with the materials the driver will need, including adequate cash to pay for weight scales and the driver's lunch, maps, rags for wiping windows, gloves and a list of weight scales. All truck drivers must have a Farm Endorsement on their driver's license, which allows them to operate CDL (commercial driver's license) equipment within a 250-mile radius of the farm without a CDL license. A tare weight on each truck should be obtained at a licensed scale, with the truck completely empty.

Additional trailers, preferably tilt-type, are pulled by tractors to shuttle bins back and forth from the picking area and the truck staging area. These trailers typically hold two to four 4- by 4-foot grape bins, which enables them to be pulled through vineyard headlands or even down wider rows.

Pickups. A pickup truck is commonly the vineyard manager's mobile office, so it contains the harvest record books, picking tickets, ticket/punch card pouches, and a scale to check picking bucket weights. Pickups also serve as portable supply depots and should be outfitted with a first aid kit, rubber gloves, drinking water and cups, soap and water for hand washing, paper towels, toolbox with necessary tools for equipment repair, sharpening stones, extra picking buckets and picking shears, and other miscellaneous items as described below.

Tractors. Tractors with forklift attachments are used to move and load bins onto trailers. They may also be used to pull smaller trailers for transportation of bins or boxes within the vineyard. Forks should be installed on tractors before harvest; be sure to lubricate and test for proper operation. It is also a good idea to supply each tractor with a small first aid kit, a jug of drinking water, towels, WD-40 or similar lubricant, shears, and a bucket for the driver to pick while waiting. Prior to harvest, tractor drivers should practice loading bins and other operating skills such as turning and backing up with trailers. Safe operating procedures should be reviewed, especially braking and turning on hills and moving with people in the rows.

Other Vehicles. A multi-passenger van or small bus may be needed to transport a picking crew within a large vineyard. Four-wheelers are very handy throughout the year and can be particularly useful for quick transportation of winemakers into the field, grape sampling, bird patrol, checking harvest crews, and getting to and from the vineyard office.

Supplies and Materials

Harvest requires many different materials and supplies that may not be used at other times of the year. The required number of each item depends on the size of the harvest and size of the crew.

Bins and Boxes. Harvest bins are constructed of wood or food-grade plastic and are available in several sizes. The most common sizes used in vineyards are (L:W:H dimensions in feet) 4:4:2 (capacity 900 pounds), 4:4:1.5 (675 pounds), and 4:4:1 (450 pounds).

Plastic liners are available for wooden bins, and they make cleaning bins much easier. One-piece molded plastic bins have lower maintenance costs than wooden bins and are also easy to clean. Some wineries prefer that grapes, particularly Pinot noir, be delivered in smaller containers to minimize fruit breakage during harvest and transportation. These boxes (also called totes) are plastic, stackable, and have ventilation and drainage holes. They come in different dimensions and capacity but commonly hold up to 40 pounds of grapes. These boxes require a different set of harvest logistics and are generally more complicated and time consuming to use. Often the vineyard imposes a surcharge to cover the additional expense. In some cases wineries supply the containers in which they want the grapes to be delivered. Arrangements should be made to pick them up before harvest.

All bins should be cleaned before and after harvest. Wooden bins often require maintenance to repair loose brackets or replace bad wood or plastic liners. It is a good idea to have extra rivets and nails on hand for quick repairs during harvest.

Picking Containers. Fruit is picked into small containers such as 5-gallon buckets (food grade plastic), then transferred into larger bins for transport to the winery. Buckets should have drainage holes drilled into the bottoms. Commonly, enough buckets are kept on hand to supply two buckets per picker, with some extras for replacement or in case the pickers have to wait to dump their fruit into bins. Alternatively, smaller boxes or totes can be used as both the picking and transporting container. Picking buckets or boxes should be cleaned at the end of each day. Wooden bins should be stored indoors or under tarps when not in use.

Picking Shears. Picking crews sometimes have a preference for a certain type of picking shear. This should be discussed with the crew leader or labor contractor prior to harvest. Shears should be oiled and sharpened before harvest and periodically thereafter. Sharpening stones and a good cleaner/lubricant such as WD-40 should be kept on hand throughout harvest. Picking knives can easily damage delicate Pinot noir clusters and their use is not recommended.

Picking Tickets. Every grower must have a method of compensating the grape harvesters. The easiest way is to pay by the hour, but many vineyards pay by the pound or bucket. This usually involves the use of tickets or punch cards to monitor the number of picking units each harvester has accumulated on each picking day. Typically, a monitor stands by the grape bins and hands a picker a ticket, or punches their card, for each bucket dumped in the bin. At the end of the day, pickers count and report their total tickets or holes punched. A random check of a picker's count assures accuracy and honesty. Ideally, the total weight of the recorded tickets matches the tare weight of the delivery for the corresponding load of grapes. Experience has shown that the tare is usually exceeded by ticket totals. This is why it is essential that quality control monitors in the field insist that buckets be completely full before they are dumped.

Computerized versions of the picking ticket method are available from some agricultural software companies. These typically involve a worker-worn badge containing a unique barcode. The picking monitor reads the barcode with a handheld scanner, and this information is later downloaded directly into the payroll accounting software. Additional reports that summarize vineyard yields and worker performance can usually be generated from the picking data.

Maturity Monitoring Tools. The vineyard should have all of its own equipment and supplies to monitor fruit maturity. This should include, at minimum, a hand-held refractometer, a pH meter with accompanying buffers for calibration, and a titration apparatus. All equipment should be calibrated before harvest and pH meters should be recalibrated daily.

Health Items. Several health-related items are required to be provided for all workers, including portable toilets, washing facilities, and drinking water. First aid kits should be readily accessible and well stocked with bandages of various sizes, antiseptic, aspirin, and medicines for bee-stings and poison oak.

Phones and Radios. Frequent communications among key personnel are critically important for a smooth harvest operation, but often these people are spread out in different locations. A cell phone is indispensable for the vineyard manager, who must stay in close contact with the wineries and truck drivers, weather forecasters, as well as vineyard personnel. Thus it is very helpful to have cell phones for all key personnel, including the vineyard crew supervisor and truck drivers. Radios can be used for shorter within-vineyard communications.

Miscellaneous. Numerous assorted items are also necessary or useful for harvest, including trash cans, rags or paper towels, tarps and rope for rainy days, and soap and brushes for cleaning buckets.

Establishing the Crew

The size of the picking crew depends on the amount of fruit to be picked in a given day. This, of course, is determined in conjunction with the winery and depends on its receiving and processing capacity for the day. The number of needed pickers can be estimated by dividing the day's planned harvest quantity by the average amount picked per worker in a day. Picking rates can vary considerably among individuals and are influenced by picker experience, the accessibility of fruit clusters, condition of fruit (amount of green or diseased), and cluster size. Regardless of the picking crew size, adequate supervisory personnel is required to keep harvest progressing at the desired rate and to ensure quality control of the picked fruit. A typical, relatively large picking crew might consist of a crew chief, two tractor drivers, four quality control monitors/ticket distributors, and twenty to twenty-five pickers.

Small vineyards may be able to find pickers through friends, neighboring vineyards, or a local temporary employment office. Requirements for larger crews are usually met by making arrangements with a licensed labor contractor. A current list of licensed contractors is available from the Oregon Bureau of Labor and Industries. It is also helpful to ask neighboring vineyards for contractor recommendations.

Labor contractors supply the picking crew and bear the burden of all government forms and other employer-required paperwork. It is, however, very important to understand that the grower has the ultimate responsibility for proper payment to the picking crew. Therefore, the grower should carefully inspect the contractor's paperwork, including the license, required bond, workers' compensation coverage and, if transportation is provided to the crew, a current certificate of vehicle inspection. A contract should always be executed between the grower and the labor contractor, and it is recommended that a lawyer be consulted to prepare an appropriate contract. The outline below itemizes some of the major responsibilities and provisions related to harvest that should be included in a contract between the vineyard and a labor contractor.

Contractor's Responsibilities and Provisions

1. Ensure that the pickers meet all state and federal requirements.
2. Supply the requested number of pickers for specific times and dates.
3. Provide each picker with grape harvest shears. (can be a grower responsibility)
4. Limit access to the vineyard to working personnel.
5. Be present with the pickers during harvest operations, or designate a representative to do so.
6. Pay the pickers and supply vineyard with a complete list of picker names prior to receiving payment.
7. Monitor quality and completeness of picking; if unsatisfactory, a penalty is deducted from the total payment.
8. Coordinate and control distribution of pickers among rows consistent with the harvest methods of the grower.
9. "Hold harmless" the vineyard from any act or accident while employees are on vineyard property.
10. Keep to an absolute minimum inclusion of leaves and material other than grapes.

Grower's Responsibilities and Provisions

1. Supply picking receptacles.
2. Operate trucks, tractors, and trailers in an efficient manner to minimize delays for the harvest crew.
3. Assist pickers in emptying containers into bins or other receptacles and sort the picked grapes for quality.
4. Supply the crew with drinking water, toilet, and washing facilities as required by Oregon Agricultural Code.
5. Pay contractor at specified rate.
6. Make available to the contractor copies of weight receipts or other weight records.
7. Pay contractor within the specified time after completion of harvest.
8. Base payment on the weight receipts from grapes delivered to wineries.

If you are supplying your own labor, be sure to obtain proper documentation from your workers (see Chapter 28). Do not plan to do this on the first day of harvest, because the crew will be too anxious to begin picking. On the day before harvest begins, assemble the crew and fill out I-9 forms for all new workers. Collect and photocopy all the necessary documents from your workers, including I-551 cards and Social Security cards. Explain exactly how, how much, and when they will be paid. Explain all harvest and safety procedures in detail. Be sure they know who their supervisors are. Have all of the required documents posted, and be aware of possible spot checks by the Immigration and Naturalization Service, Department of Labor, or the Oregon Bureau of Labor and Industry.

The importance of safety cannot be overemphasized. Safety must be stressed to all workers, from equipment operators to pickers, every day. Anticipate dangerous situations and develop procedures to mitigate the danger. Remain in compliance with all Worker Protection Standard regulations. The safety of your crew should be a paramount concern.

Field Operations

The day before harvest, all equipment should be moved into the field and staged in the proper locations. Bins should be placed in the blocks to be picked. All supplies and materials should be available in adequate numbers and in proper operating condition.

Harvest usually begins early, at first light. Each day of harvest, the crew should receive picking instructions, including reminders to pick each vine thoroughly and to pick up any dropped fruit. Picking buckets or boxes must be completely full if the contract calls for piecework payment. Emphasize that green second-crop, unripe, and diseased fruit should not be picked and that leaves and other foreign material should be kept out of the grapes. Safety precautions and reporting procedures should be emphasized and reviewed daily.

The design and size of a vineyard influences the harvest procedure, particularly the handling of bins or boxes. One common method is to load a small tilt-trailer with two bins and move it through the vineyard, staying in close proximity to the picking crew. Pickers dump their full buckets into the bins, collect a receipt for each bucket, and resume picking. When the bins are full, they are transported to a central truck staging area and deposited close to the truck. Another tractor with front forks loads the bins onto the delivery truck. A second tractor/trailer, operating behind the first, keeps harvest progressing by providing empty bins for the pickers. Meanwhile, the first tractor/trailer reloads with empty bins and returns to the picking site. Many modifications of this method are possible, but the objectives are to position the bins as close as possible to the crew, to always have empty bins in position for the pickers (they dislike waiting), and to transport and load full bins onto the delivery truck quickly.

Quality Control. A key aspect influencing the quality of grapes delivered to the winery is the sorting practices used during harvest. Pickers are routinely instructed to exclude green fruit and to minimize leaves and other plant debris in the picking bucket. Still, additional sorting of fruit is commonly done during harvest of premium winegrapes. In addition to removing green or underripe fruit, clusters are closely examined for diseases. Most wineries have very low tolerance for powdery mildew–infected fruit. Clusters with botrytis bunch rot may be trimmed to remove the damaged fruit, or discarded whole. There is often some level of tolerance for botrytis in white winegrapes, but essentially zero tolerance for red grapes. Any foreign material included with the harvested grapes is generically referred to as MOG (material other than grapes). Commonly this includes grape leaves and canes, but paper, cans, and other trash can make their way into picking bins without diligent quality control procedures and supervisory personnel. Grape harvest contracts commonly emphasize that MOG should be minimized and may provide for a penalty if a specified percentage is exceeded.

Record Keeping. The vineyard manager should record all pertinent harvest data, including the vineyard block number or name, the row numbers that were picked, the quantity of fruit picked, number of bins, fruit quality parameters, and name of the purchasing winery. Also record the number of workers and the number of hours worked. Additional observations on the vineyard, the fruit, or the harvesting process should be recorded for future reference. Yield per acre is important data, and harvest cluster weights are useful for crop prediction procedures.

Delivery. Picking should end early enough to leave ample time for delivery of grapes. Typically, a harvest day may end by noon or early afternoon. The driver should be with the truck when it is loaded and do a safety walk around it, paying special attention to the security of the load and the estimated weight. It is important to avoid overloading a truck. Prior to departure, call the winery to let them know the truck is leaving and provide an estimated time of arrival. Find out if there is any problem or delay at the crush pad. Be sure the driver has all the proper directions and maps to the winery.

Most important, the driver must weigh the load and keep the weight ticket. Do not give the ticket to the winery. The weight ticket is the same as cash and must be kept for the vineyard's records. When the truck returns, it should be hosed off and staged at the point of loading for the next picking session.

Clean-up and Preparations for Next Harvest. After each day of harvest, buckets should be rinsed and picking shears should be cleaned, oiled, and sharpened. Tractors and trailers should be hosed off. Preparations for the next picking should be made by refueling vehicles, loading empty bins onto trailers, and staging all equipment in its proper location. Move portable toilets to the new field to avoid long walks and down time. All consumable supplies should be replenished. Keep track of the location of bins and retrieve them from the wineries as soon as possible so they are available again for harvest. There must always be enough bins to accommodate the next harvest day. Grape sampling should continue to monitor ripeness and forecast harvest dates.

At the completion of harvest for the year, all equipment and tools should be cleaned and stored. Grape trailers should be placed on blocks. Rented trucks and other equipment should be cleaned and returned. And, of course, invoices should be sent to the wineries.

Finally, take a well-deserved, lengthy vacation.

EPILOGUE: LOOKING TO THE FUTURE

Oregon's Winegrowing Challenges

Ted Casteel

It has now been a decade since the predecessor of this book, the fourth edition of the *Oregon Winegrape Grower's Guide,* rolled off the presses and into the hands of grape growers worldwide. In the course of that decade, Oregon has secured its place as one of the world's fine wine regions—no small accomplishment. Planted acreage, tons harvested, and price per ton have all doubled or better, while our scores in major wine publications and bottle prices have been climbing steadily. In many ways our industry has emerged as a model industry in Oregon agriculture—a case study in how to do it right. Certainly the *Grower's Guide* has played an important, albeit small, role in helping to attain our highly valued place in Oregon agriculture.

In looking forward to the next decade, I would like to focus on some of the challenges we must face if we are to consolidate our gains and continue to grow and mature as an industry.

• We need to continue to get better at what we do. The best wines of five years ago would be considered only average wines today. The bar is being raised every year. If we get comfortable with what we have achieved, our new-found status will be fleeting. We need to continue to invest in research, both collective (Oregon Wine Advisory Board) and private. And we must also invest time and energy in our newcomers. Every diseased grape that finds it way into a bottle, every bad Oregon wine that disappoints a consumer, diminishes us all. The new Chemeketa Community College Vineyard Management program and Northwest Viticulture Center, as well as this book are examples of the commitment that needs to continue.

• We need to focus on quality. We are too close to the margins of viticulture to be a high-volume industry. The warmer regions of both the New World and the Old have too much of an advantage in the big production game. Besides, quality sells, and it is much more satisfying.

• Related to the quality issue is the challenge we face when Mother Nature lets us down—not every year is the vintage of the decade. We need to invest energy and resources to improve our ability to elevate quality in more challenging years, to become more consistent. Robert Drouhin once said, "There are no bad vintages, only lighter vintages." We need to find out what it takes to say that with equal confidence.

• We need to remember what got us here. There is common impression among participants in the International Pinot Noir Celebration and Pinot Camp: what they find most appealing about us and most unusual is not just our great Pinot noir, or our passion for viticulture; rather, it is the strong sense of community and collegiality that has characterized our industry from the beginning. And for most Oregon winegrowers, this spirit of cooperation and common cause is what makes our industry such a satisfying and rewarding place to work.

• If we want the next decade to be as successful as the last, we need to continue on our path toward sustainable farming. Ten years ago concerns about sustainability were barely on our radar screen. Calculated neglect was the operating principle when it came to our soils and fertilization practices. Ten years ago the prevailing attitude was that organic wines were inferior, short-lived, and often spoiled—a beverage reserved for hippies and vegans. Thanks to the voluntary effort of a great many people, since that time great strides have been made toward sustainability—whether it be the LIVE program, organic or biodynamic programs, or others. Today most of us believe that healthy soils and healthy vines grow the best grapes, and that sustainability is an essential element of wine quality. Regardless of which branch of the movement growers choose to affiliate with, the direction is good and true. In the end, sustainability is not about orthodoxy or zealotry. It is about attitudes and habits that will allow us to continue to live the life we came here to find, and to pass it along to those who come after us.

Southern Oregon's Untapped Potential

H. Earl Jones

Although Oregon is still in the nascence of its viticultural development and understanding of the correct varieties to grow in its varied landscape and diverse climates, it has made great strides over the past forty years. This is nowhere more evident than in southern Oregon, a region that will likely play an increasingly larger role in the future of the state's young wine industry. The principal reason has to do with its regional climate, which is broadly conducive to production of a range of cool- to warm-climate grape varieties. Of Oregon's viticultural areas, only the Umpqua, Rogue, and Applegate Valleys of southern Oregon have such a wide range of viticultural potential.

Climate is the major factor that determines both where grapes can be grown and the quality of the wine produced from those grapes. The finest wines are typically produced from grapes grown at or near the coolest climate that will just ripen a given grape variety. Southern Oregon has a considerable amount of potential viticultural land that is blessed with a warm mesoclimate. Within the region, there are smaller areas, typically found at higher elevations or under a maritime influence, which enjoy a cool mesoclimate. There are obviously transitional zones that also have an intermediate mesoclimate, but at present such areas are ill defined and not well developed viticulturally. There are no significant blocks of land with viticultural potential that have a truly hot mesoclimate in southern Oregon. Exploration of the viticultural potential of these lands is underway, and the process will likely take decades for completion. Those who ferret out the optimal mesoclimate for a particular winegrape variety will be rewarded with very high quality fruit from which fine wines can be produced.

Those exploring the viticultural diversity of southern Oregon must adhere to sound fundamental rules for marginal-climate viticulture to find an optimum match of variety and climate. They must (1) plant on south-facing slopes; (2) plant in well-drained sites; (3) plant in different soil types (remembering point 2); (4) control springtime vine vigor; (5) utilize drip irrigation later in the season for control of fruit quality and vine health; and (6) Limit yield to three or fewer tons per acre.

Viticulturists have already identified cool or intermediate mesoclimates in southern Oregon that have produced excellent Müller-Thurgau, Pinot gris, Gewürztraminer, Pinot noir, Chardonnay, and Riesling wines. I predict that even better Riesling growing sites will be identified, assuming that the wine regains popularity and becomes commercially viable once again.

As an example of finding an ideal match between a variety and its climate, the world benchmark for high-quality Sauvignon blanc has recently shifted to Marlborough, New Zealand, where an intensely perfumed, fruity refreshing wine is now made from that grape. The climate of Marlborough, which is cooler than the world's other Sauvignon blanc producing areas, is arguably the factor most relevant to quality. Obviously, that climate existed there long before alert viticulturists matched the mesoclimate to that variety and enjoyed their success.

Although Sauvignon blanc (and Sémillon) are grown in southern Oregon, neither has been planted widely, but I predict that the number of good sites available for these varieties will change that in the future. In the complex topography of southern Oregon, what other hidden cool and intermediate mesoclimates await discovery and matching to one of these interesting grape varieties?

The early exploration of the warm mesoclimate lands of southern Oregon has been exciting. Not only are there commercially significant quantities of these lands available, but the lands also have a variety of very different soils situated around a variety of slopes, aspects, and mesoclimates. This is southern Oregon's great resource, and the potential is huge. Already, Cabernet Franc, Cabernet Sauvignon, Merlot, Syrah, Viognier, Dolcetto, and Tempranillo wines have been produced successfully. The quality level of some of these early wines is indeed world class. These achievements are the accomplishments of many, all of whom have successfully matched the variety to the climate. Southern Oregon viticulturists are honing their skills and actively seeking the optimal mesoclimate (or should we say terroir?) for these and other warm-climate varieties.

Another example of matching variety with climate can be found with Malbec, a French variety of former grandeur in the area of Cahors, which may have reached a new plateau of quality thanks to viticultural trials in the unique climate of Mendoza, Argentina. This variety has also done well in preliminary trials in at least three warm sites in southern Oregon. Success will likely be realized more slowly and with great difficulty for those varieties that require a warm to hot mesoclimate. Yet, let us not forget the novel Sauvignon blanc or Malbec experiences and the possibility that some of these warm- to hot-climate varieties (e.g., Grenache, Sangiovese, Nebbiolo, Zinfandel) may find their optimal mesoclimates in the southern Oregon landscape.

What are the greatest perils along this risky road to vinous success? Many would say Pierce's disease or phylloxera. Although these are real and serious problems, I do not think they are the greatest threats to successful commercial production of world-class wines in southern Oregon. The greater risks appear to be these:

- Severe winter cold that can kill cold-sensitive varieties or damage the graft union of plants, resulting in crown gall. Over the past seventy years such severe winter weather has occurred in southern Oregon approximately one year in ten. Vineyard sites situated in broad valleys with good air drainage suffer the least from this unavoidable peril.
- Late spring frosts that reduce crop yield or shorten the growing season.
- Early fall frosts that can prematurely terminate the ripening process.
- Mismatching the site climate and the planted variety such that the mesoclimate of the ripening month is either excessively warm or excessively cool. Both extremes are detrimental to wine quality. If a variety is to show its best, it needs favorable environmental circumstances during the ripening period.
- Fall rains, thankfully, are not a major problem for southern Oregonians.

Growers and winemakers who adhere to the fundamental rules of marginal climate viticulture should also be able to produce exceptional wines consistently. This means avoiding the perils mentioned above on an annual basis. If these seemingly mundane problems become serious enough to affect wine quality, what does one do? Remedial measures, aside from grafting over to a variety better suited to the site, are at the least expensive, often impractical, and sometimes impossible. All of these potential problems must be addressed during site selection.

Producing high-quality wines in fewer than seven vintages out of ten will likely undermine the consumer's opinion of the variety, vineyard, or winery. This would not have been true in the marketplace twenty years ago. Today the wine world communicates extensively, and wines from anywhere and everywhere are readily available at the market or delivered to your door. This places great emphasis on each wine's quality factor. High-quality vintages must be produced consistently, or one can expect price resistance, market erosion, and eventually a tarnishing of the viticultural area's reputation. Given the myriad of rare mesoclimates and landscapes available for viticulture in southern Oregon, few if any areas have greater potential to create finer wines.

Enhancing Economic Success and Professionalism

Allen Holstein and Rollin Soles

We believe that the Oregon wine industry faces an important crossroads. Many growers and winemakers will soon make decisions, or not make decisions and live with the consequences, about the future direction of our industry. We are offering the following "white paper" to generate discussion and, we hope, influence the Oregon winegrape industry to strive for a higher level of professionalism and economic success.

The oldest vineyards in Oregon are now about thirty years of age, and all are own-rooted. The discovery of phylloxera in Oregon in 1990 was a watershed event, triggering the realization that all these vineyards must be replanted. In effect, we are in the process of completing the first generation of wine growing in Oregon in the modern era. The second generation requires the use of vines grafted to phylloxera-resistant rootstocks and, we believe, significant other changes to remain globally competitive.

Unlike California's industry, smaller vineyards operating with relatively modest capital outlays have characterized our wine industry. Now, with replacement at hand, the cost of vineyard establishment has increased dramatically. During the past ten years, significant improvements in viticultural techniques have been introduced in Oregon. These include high-density vine spacing, rootstocks, smaller-clustered clones, and drip irrigation. These developments have improved grape quality, but they have also increased the cost of replanting or entry into the wine business. New vineyards now face significantly higher land price values. When you also consider increased equipment costs and the popularly supported minimum wage increase, it becomes clear that the risks of entering and thriving in the Oregon wine industry are greater than just ten years ago.

Because of the increased cost of grape growing, the Oregon wine industry must shift our focal point from a local, regional one to a more national and global context with an emphasis on high-priced wines. It should be recognized that the Oregon industry is greatly influenced by the larger wine-producing companies and regions of the world. In the past ten years, the world wine stage has shifted from a focus on the quality of Old World wines to recognizing the potential quality of New World wines. For the first time, consumers have accepted the concept of "icon status" wines from New World producers—wines with very high prices and limited production that tend to be uninfluenced by economic factors. In addition, Australia will soon outsell France in the United Kingdom, and at a higher average price per liter. This raises the possibility of other New World wine producers succeeding in the export market. Chile and Australia are now among the top five of countries importing wine into Oregon and the United States. Additionally, California has dramatically increased production of Pinot noir and other Oregon-grown varieties. Below, we outline some quality-increasing techniques for Oregon vineyards and present a multipoint proposal that encourages specialization, outside investment by high profile, capitalized wine companies, and expansion of the market for Oregon grapes.

In addition to increased development and production costs, Oregon vineyards will always produce yields below that of other New World producers. We have to admit that one of our main selling points, cool climate, is also one of our disadvantages because it limits our yield potential. We suggest that Oregon's niche on the world stage will be one of high-quality wines and high prices. We recognize that in certain vintages and certain situations we will have to live with lower-priced wines, but that should not be the goal.

If high price is the best strategy for "Brand Oregon," producers must make the extra effort in the vineyard to ensure that their product has a high quality to price ratio. These additional efforts will put additional economic strain on independent growers, potentially to the point of challenging their existence. We can already see a trend whereby growers are having their grapes custom-crushed at a winery and selling their own wine in an effort to receive a sustainable return on their vineyard investment. This trend runs counter to the global trend of winery consolidation. We believe wineries will have to invest more heavily in their own vineyards to ensure adequate supplies of high-quality fruit. As the economics of vineyards and wineries become more integrated, more resources will become available to develop a new vineyard or replant properly.

Phylloxera-resistant rootstocks have demonstrated the additional benefit of reducing vigor and increasing quality. This is a good argument for replacing own-rooted vineyards sooner rather than later. There has been some debate over the impact of phylloxera on wine quality, but the reality is that infested vineyards are in decline and will soon become unprofitable to farm. There has been a wider controversy about what kind of balance between vine growth and fruit yield is ideal for high wine quality. Vine stress is often mentioned in popular wine literature as a quality-enhancing factor. But stress associated with phylloxera, aggressive cover crops, and Oregon's summer drought conditions exceeds the limits romanticized by popular literature. Drip irrigation can often lead to higher-quality wines, despite the contrary reputation. Drip irrigation decreases the mortality of young vines and increases productivity over the course of the vineyard's life. It can be considered a kind of insurance policy for the high-dollar vineyard of the present. Despite increased regulatory pressure, water availability will be a major issue with respect to site selection.

Another strategy to create high-priced wines with good quality to price ratios involves the use of high plant densities. On first consideration, you may ask, Why should we add to our already high establishment and production costs by increasing plant density? The reason is that properly executed, high-density plantings create more leaf surface and create it earlier in the season than low-density plantings. This results in maximizing the heat and light the season has to offer. It makes less difference in warm years than it does in cooler vintages, but we have found that high-density plantings generally produce a more stable yield year to year. High density also gives more even ripening within the vine row. Further work is required to fine tune the cost/benefit ratios associated with vine density. We have found high-density plantings to be better viticulture. Whether this is better business is dependent on downstream factors.

The small size of Oregon's industry and its unique climatic conditions suggest that our research programs should have a narrow focus on one area of discovery. Many grape-growing regions of Oregon lack sunlight and heat compared to competitors in California, Australia, South America, and even France. Therefore, we believe research should be concentrated on understanding the plant physiological mechanisms affecting the efficiency of sunlight, heat, and water utilization. These discoveries could lead to management practices that best utilize water and the limited light and heat our vines receive.

Some current research could be construed as leaning toward enhancing the marketing image of our industry while neglecting to show how this research improves actual fruit quality and Oregon's stature in the academic arena worldwide. The Oregon Wine Advisory Board, whose funds are limited, supports most of our research. It is our contention that the limited funds would be best used by researching how, why, and when a grapevine most efficiently utilizes inputs of light, heat, and water to improve quality significantly. We could develop our expertise in vine physiology in a cool climate while accessing the efforts of other viticulture research institutions for other concerns such as disease and pest management.

In most regions of Oregon, the grower is faced with selling to a small market and producing low yields to ensure ripeness. This situation favors vineyards that are directly connected to wine producers and disadvantages independent growers. This is not, however, the trend in major wine-producing regions like California and Australia. Let us present three points of discussion to both enhance Oregon's wine grape value and support demand for our fruit.

- Each Oregon wine region must focus on finding the one or two varieties that have a discernible potential for high quality compared with the best examples in the world. The Willamette Valley has fortunately found that Pinot noir, Pinot gris, and soon Chardonnay are varieties that most distinguish it internationally. What will be the exciting varieties and clones for Oregon's other viticulture regions?
- It may be likely that established, high-quality, well-financed wine companies can be encouraged to develop wineries in Oregon growing regions, especially if they believe the region produces wine among the highest quality of which a variety is capable. This could be encouraged by all of our individual actions and openness to the visits of interested parties, for example, inviting and welcoming outside wineries to our International Pinot Noir Celebration. Oregon's wine industry has already benefited greatly from investment in the Willamette Valley by Maison Joseph Drouhin in 1988 and by Australia's Petaluma Winery in 1987. The future could bring a California sparkling wine company to the Illinois Valley, for example, or a Pomerol wine company to an Oregon region known for Merlot. This would greatly validate the Oregon region chosen by the outside wine company and also provide a new, expanded market for Oregon grapes and increased security to our growers.
- A new Pacific appellation should be considered by Oregon growers and promoted by the Oregon Wine Advisory Board. A multistate appellation that encompasses California, Oregon, and Washington could greatly increase the market and price stability for Oregon-grown grapes. We have already seen complex, exciting wines blended from Oregon and California fruit (from Au Bon Climate and Bonny Doon, for example). Our grape prices are linked economically to those of California and Washington and, like it or

not, in today's wine marketing world demand for wine is often driven by brand image over grape origin. This runs counter to Oregon's intensive focus on the importance of vineyard designation. There is a current trend for global consolidation of wine producers, and a strict reliance on vineyard-designated marketing puts Oregon in danger of being overlooked.

A Pacific appellation would enable Oregon growers to take advantage of both world trends and Oregon's focus on vineyard designation. Growers would have viable marketing options whether they have a known high-quality vineyard site or a site with a reputation for average quality. Similarly, growers would have the option to keep yield down for high-quality wine designation or up for volume blending. For example, a Willamette Valley Pinot noir or Chardonnay may be just the component to make a Sonoma wine's flavor something special. A southern Oregon Syrah could be used by a premium California winery to blend North America's first Grange Hermitage or Chateauneuf du Pape. Alternatively, one could imagine a less-ripe, higher-acid, 4-ton per acre Oregon crop being attractive to a California producer who could use it to balance a wine with high alcohol and less fruit acid. Additionally, at 4 tons per acre the grower may come out with a per-acre dollar return to encourage more planting.

We believe that the Oregon wine industry should look to other parts of the world to forecast some of our impending challenges. New and exotic pests will continue to present themselves. Labor availability will be cyclical but overall will continue to trend toward less available and more expensive. As the farming community becomes a smaller part of the electorate, regulatory pressure will grow. Easing of development restrictions and growing population will increase land prices to growers, thus limiting expansion of our industry. The current diverse makeup of our wine industry will also make effective industry actions and communication with the market and electorate more difficult.

The Oregon wine industry has come a long way in the past fifteen years. Today more than ever, we must present a profitable, quality-oriented, environmentally sound, and progressive image to Oregon citizens, and, more important, to the increasingly small world of wine. We hope this discussion encourages us all to strive continually to reach a higher level of professionalism in our industry. As one famous American puts it, "If we don't succeed, we run the risk of great failure."

Index